Brakes

Wiley Automotive Series

John Remling **Brakes**
John Remling **Steering and Suspension**

Forthcoming titles

John Remling **Basics**

Brakes

John Remling
Board of Cooperative Educational Services
Valhalla, New York

John Wiley & Sons
New York Chichester
Brisbane Toronto

It was set in Helvetica Light by York Graphic Services.
Joan Tobin supervised production.

Cover and text design by Edward A. Butler.
Drawings on chapter opening pages by Mario Stasolla.
Cover painting by Edward A. Butler.

ASBESTOS DUST HAZARD WARNING
Brake linings contain 33% to 73% asbestos. Breathing asbestos dust may cause asbestosis and cancer. During brake servicing, an air purifying respirator should be worn during all procedures starting with the removal of the wheels and including reassembly. Prior to disassembly, dust should first be cleaned from all brake parts using an industrial vacuum cleaner equipped with a high efficiency filter system. *Under no circumstances should compressed air or dry brushing be used for cleaning.*

This text was developed with
Continuing Education Systems, Inc., Hinsdale, IL

Library of Congress Cataloging in Publication Data:

Remling, John, 1928–
Brakes.

(Wiley automotive)
Includes indexes.
1. Automobiles—Brakes—Maintenance and repair. I. Title.

TL269.R45 629.2′46′028 77-7262
ISBN 0-471-71645-6

Printed in the United States of America

10 9 8 7 6 5 4

To the Student

The field of automotive repair service offers many opportunities for interesting, gainful employment. If you have the desire and the ability to be an auto mechanic, education and practical training will help you become a good one.

You may think that all it takes to be a good mechanic is the ability to remove and install a few parts—that all the skills needed are in your hands. It is true that a mechanic must have some highly developed hand skills. But hand skills are useless unless you know when and where to apply them. How will you know which part to replace or adjust? How will you know what to do to keep that part from failing again? As a mechanic you must check automotive systems to find the causes of many kinds of problems. But that is not enough. You must check related systems and parts to be sure that the problems will not occur again. Such checking requires a skill called *diagnosis.* Diagnosis is the basis of all repair, and it requires knowledge.

As a mechanic you must have a working knowledge of all the systems that make up an automobile. You must know the ways in which those systems work and relate to one another. Without this knowledge you cannot make an accurate diagnosis. Without an accurate diagnosis you may do unnecessary work, and you may replace parts that do not need replacing. As an auto mechanic you will use your knowledge to diagnose problems and to determine needed repairs. You will then use your hand skills to make the repairs.

As an auto mechanic you will need knowledge, diagnostic skills, and repair skills. Specifically, you will need—

Knowledge of:

1. The function and operation of automotive systems and their parts
2. The names of parts
3. Automotive theory
4. Measurement and related mathematics and science

5. Hand tools and other equipment
6. Shop practices and safety

Diagnostic skills such as:

1. Recognizing malfunctions
2. Isolating sources of trouble
3. Using test equipment
4. Interpreting test results
5. Analyzing failure
6. Evaluating completed repairs

Repair skills such as:

1. Lubricating and adjusting parts
2. Repairing, overhauling, or replacing parts

This book can help you acquire some of this knowledge and some of these skills.

John Remling

Contents

*For the remaining chapters we list in a developmental sequence, jobs whose learning objective is either knowledge (K), diagnostic skills (D), or repair skills (R).

Brakes

Introduction

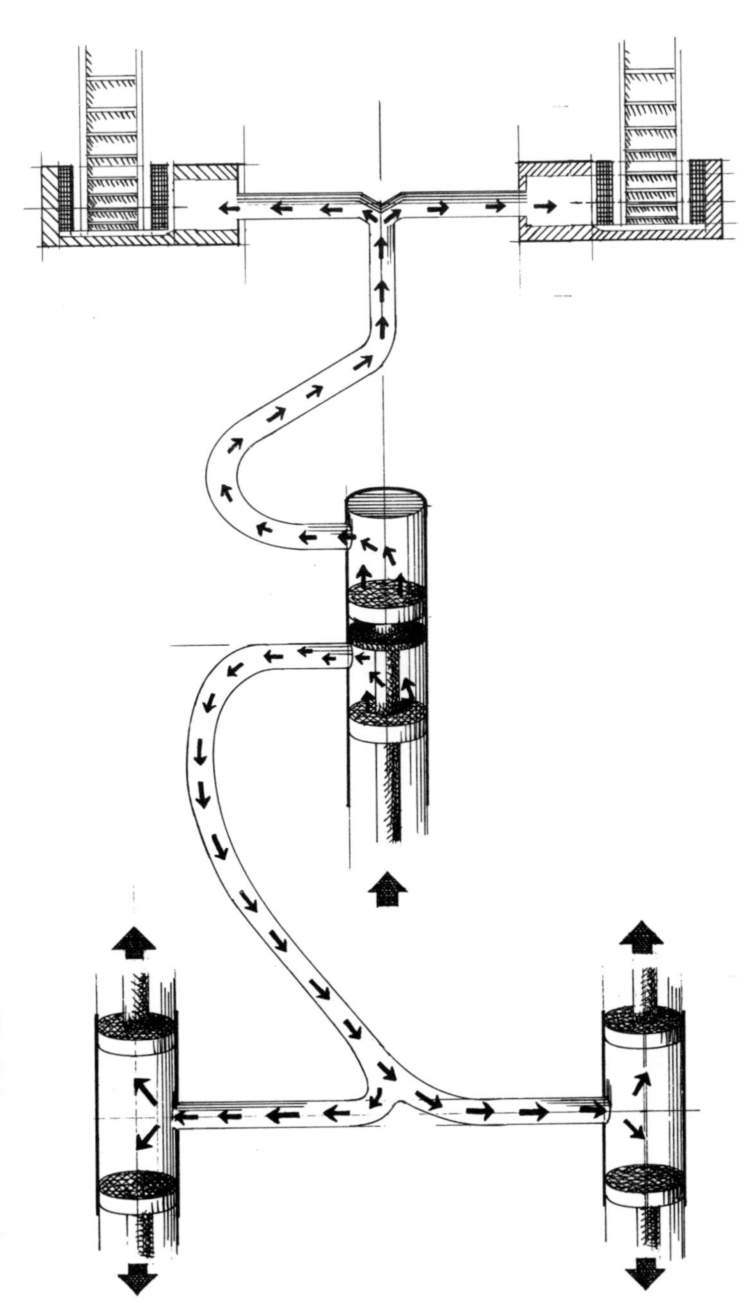

If all the systems in a car were listed in their order of importance, the brake system would most likely be placed at the top of the list. Most drivers have experienced the inconvenience caused by an engine that would not start. They remember a winter night when the heating system in a car gave them nothing but cold air, and they have experienced system failures and realize the cost in comfort, convenience, time, and money. A failure in the brake system, however, frequently can cause property damage, serious injury, and even death. A brake mechanic must always remember that the quality of any brake service is measured only in terms of safety.

Because of the importance of the brake system, there are many federal and state regulations concerning brakes, and many of them apply to both mechanics and car manufacturers. The makers of cars and quality brake parts are well aware of the necessity of providing safe brake systems. They have adopted many federal and state regulations as standards.

These regulations and standards are included here, and this book is one of several in a series covering a broad range of automotive service areas. Upon completion of this text, you should have achieved:

1 Sufficient knowledge, diagnostic skills, and repair skills to enable you to service automotive brake systems to meet the standards of the industry.

2 The basic knowledge, diagnostic skills, and repair skills to enable you to successfully pass the Certification Examination for Brake Mechanics given by the National Institute for Automotive Service Excellence (NIASE).

In studying the brake system you will find many words with which you are familiar, but some of them will have meanings that are different from those that you know. For example, brake shoes in no way resemble the shoes you wear on your feet. You will also find some words with which you are not familiar. Although many of those words are descriptive, what they describe is not always clear. A metering valve could be defined as a valve that meters or measures something. But what does it measure? Where does it measure it? Why does the thing it measures have to be measured? As in any other technical field, automotive service has its own language. A mechanic may use the language of a layperson, especially when explaining to a car owner the need for a certain repair, but that same mechanic must understand and use the language of the trade when talking with other mechanics or with parts dealers and when reading repair manuals.

You will find that the language of the trade is easy to learn. When a new word or phrase is first used in the text, it is printed in *italic* type, and a definition or explanation is provided. To aid you in adding the new word or phrase to your vocabulary, I recommend that you study the pictures related to the text. This will help you to:

1 Better understand the definition or explanation.
2 Recognize various automotive parts when you see them on a car.
3 Understand the function or operation of the parts and their relationship to one another.

Throughout the text you will be advised to consult the service manual for the car on which you are working. Although all car makers use similar brake systems, each maker has adopted certain variations. Sometimes the variations are necessary in order for the system to meet design requirements. At times a special tool is required, and its use is shown and explained. Usually a special sequence is given that will save you time. Precautions are given so that you may perform the operations without causing damage to parts or injury to yourself.

Even if you knew the correct procedure for any given job, you would still need the car maker's service manual because it provides specifications. Specifications are measurements that must be taken before, during, and after assembling components of automotive systems. Only by working to specifications can you be sure that the completed job is correct.

In most chapters there are pages that present jobs for you to perform. Some of the jobs require you to identify parts or the function of parts. These jobs are tests to help you to determine whether you have learned the information presented in the text material that precedes them. They require you to furnish information within a certain amount of time. Do not attempt to perform a job until you are sure you understand the material that precedes it.

Other jobs require that you adjust, replace, or overhaul parts in a brake system. Those jobs are also tests. They examine your skills in performing service procedures to the specifications of the car manufacturer and require you to complete the procedure within a specified time. These jobs should be begun only after you have practiced them and gained the necessary skills.

Each job is a checkpoint that you must pass to achieve your goals. It is placed at a point where you can measure the knowledge, diagnostic skills, or repair skills you have gained. Since the material that follows each job is built on the preceding material, you should use the job to determine whether you are ready to advance. If you do not achieve a satisfactory performance on a particular job, you will know that you have not gained sufficient knowledge or skills to move on. You should review the material preceding the job and, when necessary, you should practice the procedures that revealed your weakness.

At the end of each chapter you will find a self-test. It is provided so that you can see how well you have learned the material presented in the chapter. All the self-test questions and incomplete statements are of the multiple-choice type. They are all similar in form and content to the questions used in the certification examinations given by NIASE. When taking the self-test, read each statement or question very carefully. Read each of the answers or completion choices so that you can be sure to chose the one that best answers the question or best completes the statement. After responding to all of the test items, check your responses against the answer key at the back of the book. If you have chosen an incorrect response, review the appropriate material. The answer key indicates where that material can be found.

Chapter 1 Overview

Different types of brake systems are used in different makes of cars. This chapter will acquaint you with those systems. It will also present an overview of the knowledge and of the diagnostic and repair skills that mechanics must have and that you will acquire from the proper use of this text.

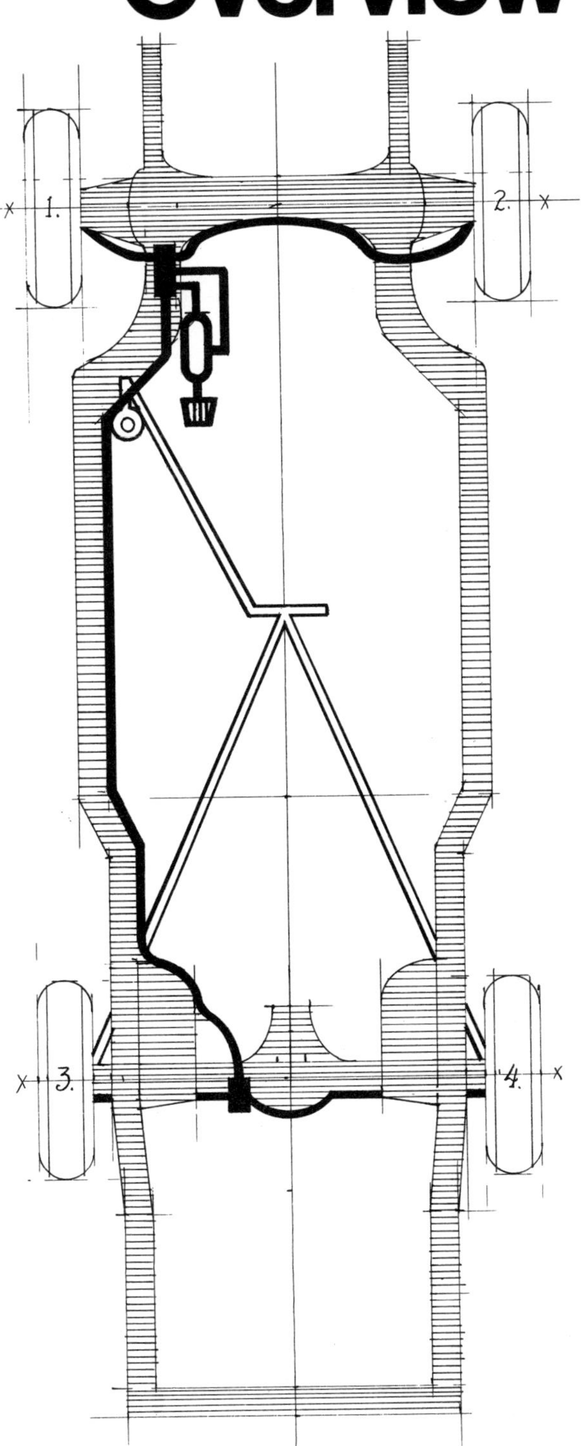

AUTOMOTIVE BRAKE SYSTEMS Every car has two brake systems. One system is used to slow or stop the car when it is moving. This system is called the *service brake system.* The other system is used to hold a stopped car in place. This system, which is not intended to stop a moving vehicle, is called the *parking brake system.*

The Service Brake System You may have heard the term "hydraulic brakes" used to describe service brake systems. Though this term is commonly used, it is technically incorrect. All service brake systems are mechanical. They use mechanical parts to bring the wheels to a stop. The driver applies the service brakes by pushing down on the service brake pedal. This action causes the brake shoes at all four wheels to be forced against drums or rotors behind the wheels. The contact of the shoes against the drums or rotors creates friction, which stops the wheels from rolling.

The force that operates this mechanical system, however, is carried to each wheel by a hydraulic system. Therefore, the service brake system actually consists of two systems, or subsystems. One is mechanical, the other hydraulic. Each has its own job to do, and each depends on the other. Diagnosing defects in the service brake system and correcting them requires knowing how both the mechanical and the hydraulic subsystems operate.

In this book you will study the function and operation of the mechanical system, and the service procedures related to that system. You will be introduced to all the service operations involved in replacing brake shoes and other components of the system. You will inspect parts for damage and wear, and will perform the measuring and machining operations necessary to restore a service brake system to the specifications of the car manufacturer. You will also study the function, operation, and service procedures related to the hydraulic system. You will perform the service operations required to repair, overhaul, and replace various components. You will test and inspect the system to ensure that your completed jobs will meet specifications.

The Parking Brake System The parking brake system is entirely mechanical. The driver applies the parking brakes by pulling a lever or stepping on a pedal. The parking brakes generally operate on the rear wheels only. Moreover, they use cables pulled by levers to carry the driver-applied force to the wheels and not a hydraulic system.

Though the parking brake system does not depend on the hydraulic part of the service brake system, it does depend on the mechanical part of that system. Parking brakes normally use the same brake shoes that the service brakes use on the rear wheels.

In this book you will study the relationship between the parking brake system and the service brake system. You will perform the operations required to adjust the parking brakes. You will also inspect, repair, and replace the various parts in the system.

TYPES OF BRAKES There are two types of service brakes in common use: the *drum brake* and the *disc brake.*

Drum Brakes In a drum brake, the brake shoes are forced against the inside surface of a drum. A drum is a ring that is attached to the inside of a wheel and rotates with it. The friction between the brake shoes and the drum causes the drum to slow down and stop turning (Figure 1.1). Since the wheel is bolted to the drum, the wheel stops rolling.

Many different designs of drum brakes have been used, but one design is now used by almost all manufacturers. Later in this book you will learn more about that design and perform the operations necessary to test, service, and repair such brakes.

Disc Brakes In a disc brake, two brake shoes are forced against opposite sides of a *rotor.* A

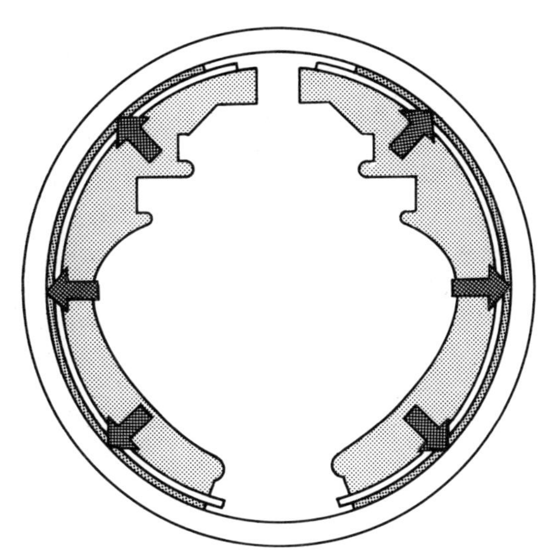

Figure 1.1 Braking forces in a drum brake system (Management and Marketing Institute).

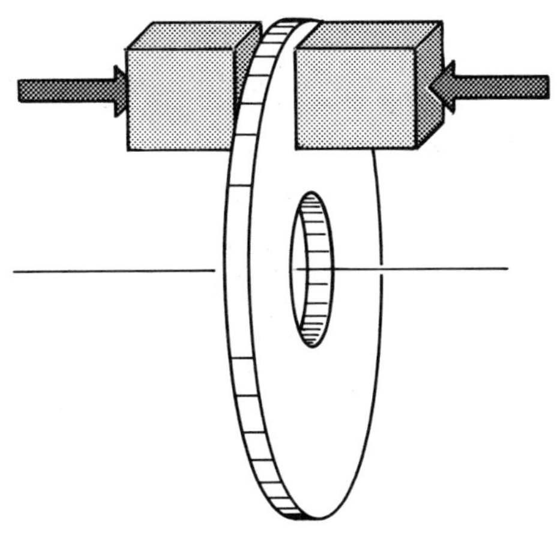

Figure 1.2 Braking forces in a disc brake system (Management and Marketing Institute).

rotor is a disc that is attached to the inside of a wheel and rotates with it (Figure 1.2). The brake shoes squeeze the rotor between them. The friction between the shoes and the rotor causes the rotor to slow down and stop turning. Since the wheel is bolted to the rotor, the wheel stops rolling.

There are many different types of disc brakes. But as in the case of drum brakes, one type is most popular. You will learn how this type of disc brake operates, and perform the operations required to service disc brakes. You will also perform the tests and measurements necessary to diagnose disc brake problems accurately.

Combination Brake Systems Most cars have combination brake systems. A combination system uses disc brakes on the front wheels and drum brakes on the rear wheels. Each type of brake has certain braking characteristics, including some advantages, that the other does not have. Combining the two types of brakes enables us to take advantage of both types. In servicing a combination system you must be competent in dealing with both disc and drum brakes. But that is not enough. Since both types of brakes have different braking characteristics, combination systems must be balanced. This means that both the disc brakes and the drum brakes must provide just the right amount of braking. They must do this over a wide range of speed and load conditions.

Combination systems are balanced by the use of various valves in the hydraulic system. This book will help you learn to identify those valves. You will learn their function, and how to diagnose problems caused by valve malfunction or failure. You will learn the service techniques for working on those valves, and how to replace the valves in combination brake systems.

Power-Assist Units As you have probably found out for yourself already, it is harder to stop a car when it is moving fast than when it is moving slowly. You have to press harder on the brake pedal, and the car travels a greater distance before it stops. It is clear that the speed of a car affects braking. The weight of a car also affects braking. A heavy car requires more braking effort than a light car.

The brake system on any car must be able to stop the car quickly in a short distance. It must do this regardless of the speed and weight of the car. The hydraulic system that carries the driver-applied force from the brake pedal to each wheel multiplies the force applied to the pedal. Even so, there are limits to its ability to multiply force. This could mean that only drivers with a great deal of physical strength could apply enough force to stop heavy cars driven at high speeds. Even if everyone had enough physical strength, driving could still be very hard work. Even a short trip would be tiring. Therefore, a means of boosting the force applied by the driver must be used.

One method of boosting the driver-applied force is to use a power-assist unit, or "booster." You have probably driven a car equipped with power brakes. If so, you have found that you could stop the car with very little effort. A power-assist unit reduces the amount of force you have to apply to the brake pedal. The mechanical and hydraulic systems you will study in this book are the same on cars with or without power brakes. The presence or absence of a power-assist unit is the only real difference.

In this book you will study the operation of the most commonly used power-assist units. You will learn how to test them and to diagnose their faults. You will perform needed adjustments, and you will replace power-assist units in power brake systems.

SUMMARY

This chapter has outlined the material you will study in this book. You now have a general idea of what you must learn if you want to diagnose defects in brake systems and to repair them. You must be skilled at diagnosis, regardless of whether you wish to become a brake specialist or a general auto mechanic. Now is the time to set your goals.

SELF-TEST

Each incomplete statement or question in this test is followed by four suggested completions or answers. In each case select the *one* that best completes the statement or answers the question.

1 Two mechanics are discussing brake systems.
Mechanic A says that service brake systems are hydraulically operated.
Mechanic B says that parking brake systems are mechanically operated.
Who is right?
A. A only
B. B only
C. Both A and B
D. Neither A nor B

2 A brake system stops a moving car by creating friction between the brake shoes and a moving
I. drum
II. rotor
A. I only
B. II only
C. Both I and II
D. Neither I nor II

3 Which of the following statements are true?
I. Most parking brake systems use the same brake shoes as the service brake system.
II. Brake systems use friction to stop a moving wheel.
A. I only
B. II only
C. Both I and II
D. Neither I nor II

4 When the parking brake is applied, the driver's force is carried to the brake shoes by
I. cables
II. levers
A. I only
B. II only
C. Both I and II
D. Neither I nor II

5 Two mechanics are discussing brake systems.
Mechanic A says that combination brake systems use drum brakes at the front wheels and disc brakes at the rear wheels.
Mechanic B says that combination brake systems are balanced by the use of valves.
Who is right?
A. A only
B. B only
C. Both A and B
D. Neither A nor B

Chapter 2 Basic Principles of Braking

How do brakes work? Why does your car come to a stop when you step on the brake pedal? Since repairs are determined by diagnosis, and diagnosis is based on knowledge, you will need to know the answers to these questions. After you complete this chapter, you will have a working knowledge of the basic principles of braking and you will have started building a firm foundation for diagnostic and repair skills related to brakes. You will be on your way toward becoming a brake mechanic.

In this chapter, your objectives are to identify the definition of terms that refer to:

1
The basic principles of braking.
2
The operation of brake systems.

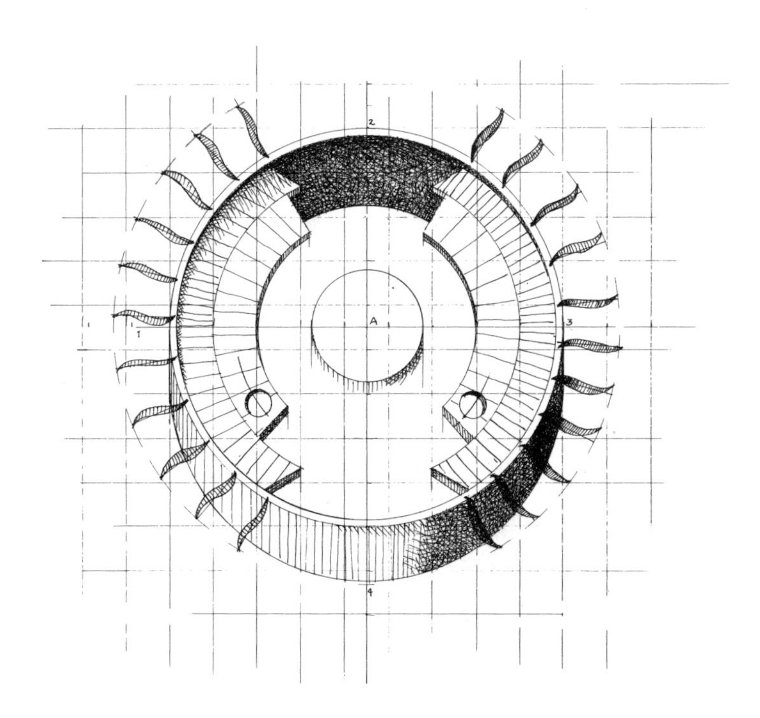

ENERGY It takes energy to operate your car. Energy can be defined as the ability to do work. People cannot produce energy, but they can change it from one form to another. It can also be used in various ways.

When you buy gasoline, the energy in it is converted to power by the car's engine. The engine produces this conversion by burning the gasoline and using the released heat energy to turn a crankshaft. This motion is passed on to the wheels, which turn and move the car.

A car is kept in motion by two forces. One of those forces is the power that started the car moving, and the second force is the weight and speed of the car. The combination of these two forces is called *kinetic energy.* It is the energy of motion.

The heavier a vehicle is and the faster it is moving, the more kinetic energy it has and, therefore, the harder it is to stop. In fact, a car in motion would never stop unless it were made to do so by another force. That force must eliminate the kinetic energy. The brake system provides that other force.

Just as energy cannot be produced, it also cannot be destroyed. The brake system, however, changes the kinetic energy of a moving car to another form of energy. The kinetic energy was obtained by releasing heat energy in the engine. The best way to stop a car is to change the kinetic energy back to heat energy. The braking system does this through the use of friction.

FRICTION Friction is the resistance to motion that exists between two objects in contact. There are many types of friction, but the dry, sliding type provides the greatest resistance to motion.

Friction creates heat. You have probably found that you can warm your hands by rubbing them together. The rubbing changes the kinetic energy of your moving hands to heat energy. This change is caused by friction.

As shown in Figure 2.1, braking systems force brake shoes against drums or discs attached to the wheels. This contact creates dry, sliding friction between the revolving drums or discs and the brake shoes, which do not revolve. The friction, in turn, converts the kinetic energy of the car to heat energy and brings the wheels and, therefore, the car to a stop. The greater the pressure of the brake shoes against the drum or disc, the greater the friction.

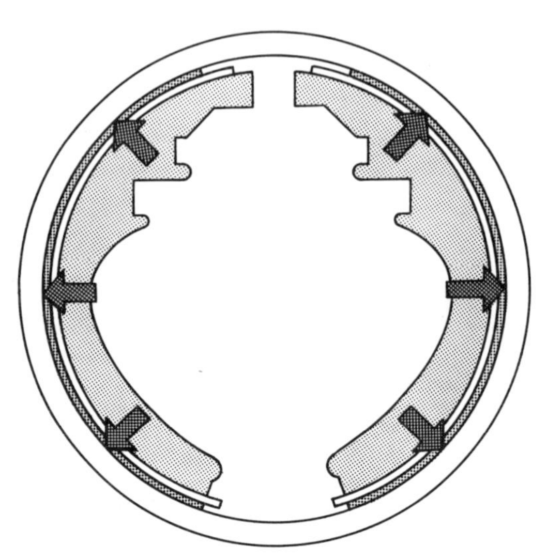

Figure 2.1 Braking force of brake shoes on a brake drum.

Friction also depends on the materials in contact. A basketball player wearing leather-soled shoes would have a difficult time on the court. The friction between the player's shoes and the floor would not be great enough to prevent slipping. By wearing rubber-soled sneakers, the player can move about with less chance of sliding because the friction between the sneakers and the floor is much greater than the friction between the shoes and the floor.

Differences in the amount of friction can be measured and are called *coefficients of friction.* The coefficient of friction between rubber-soled sneakers and the floor is greater than the coefficient of friction between leather-soled shoes and the floor.

The two objects in contact in a brake system are the brake shoes and the drum or disc. Therefore, the coefficient of friction in a brake

Job 2a

IDENTIFY THE DEFINITION OF TERMS THAT REFER TO THE BASIC PRINCIPLES OF BRAKING

SATISFACTORY PERFORMANCE

A satisfactory performance on this job requires that you do the following:

1 Place the number of the four brake terms in front of the phrases that best define these terms.
2 Complete part 1 in 10 minutes.

PERFORMANCE SITUATION

1 Energy
2 Kinetic energy
3 Friction
4 Coefficient of friction

______ The resistance to motion between two objects in contact.

______ A measurement of the amount of friction.

______ The ability to do work.

______ The transfer of heat.

______ The energy of motion.

system depends not only on the driver's force in applying the brakes but also on the materials used to make the shoes and the drums or discs.

Car manufacturers have no way of knowing how much force a driver will apply to the brake pedal. However, they can select the materials used to make shoes, drums, and discs. Brake shoes are made of steel, and most brake drums are made of cast iron; both materials provide the strength needed to withstand the pressure required for braking. On the other hand, they have a poor coefficient of friction when used together. If the metal shoes are forced against the revolving metal drums or discs, the metal-to-metal contact provides poor braking. It also causes excessive wear to both parts. Therefore, the steel shoes are covered with a *brake lining,* which provides the proper coefficient of friction between the shoes and the drums or discs, but does not cause excessive wear.

BRAKE LINING The material used in making brake lining must be carefully chosen. If the coefficient of friction it provides is too low, the brakes will not stop the car fast enough. If it is too high, the wheels will lock and the car will skid. Another factor to consider is the resistance of the lining to heat. Since the brakes convert the kinetic energy of the car to heat energy, the brake lining should be made of materials that will withstand very high temperatures. For this reason, the material most commonly used in brake lining is asbestos. Asbestos provides a relatively high and stable coefficient of friction when used with cast iron drums and discs.

As you can see in Figure 2.2, many types of brake lining are available. Each type is designed for a specific use, and the coefficient of friction provided by the lining can be changed by adding other materials to the asbestos. Powdered asbestos is blended with small amounts of other materials, such as powdered rubber or metal. These materials are then mixed with a liquid cement called a *binder,* and the mixture is formed to the desired shape and is dried by baking.

There are other factors that must be considered in selecting materials used in brake lining. Brake lining must resist moisture; it must not be affected by changes in humidity; it must not soak up water when a car is driven through puddles or flooded streets; it must wear well; it must not cause noise; it must not produce odors when hot; and it must resist *brake fade.*

BRAKE FADE Brake fade is one of the problems you will diagnose as an automotive mechanic. It takes place in drum brakes when they overheat. If brakes are working efficiently, they can stop a car in a short distance, but when they do, they become extremely hot. As drum brakes overheat, two changes may take place. The first change may take place in the lining. For some types of lining the coefficient of friction between the lining and the drum decreases as the temperature increases. Thus repeated brake applications result in less and less efficient braking, and the driver must apply more and more force to the brake pedal in order to stop the car. The second change may take place in the brake drum; the drum receives most of the heat developed during braking. As you probably know, most metals expand when heated and, because a brake drum is metal, it expands as the temperature of the drum goes up. Since the drum is round in shape, its expansion causes it to increase in diameter, and the drum actually moves away from the brake shoes. As a result, the shoes must be pushed farther and farther out to contact the drum. For this reason, the driver must push farther and farther down on the brake pedal to stop the car. The expansion of the drum plus the decrease in the coefficient of friction between the lining and the drum can progress until the driver is unable to stop the car.

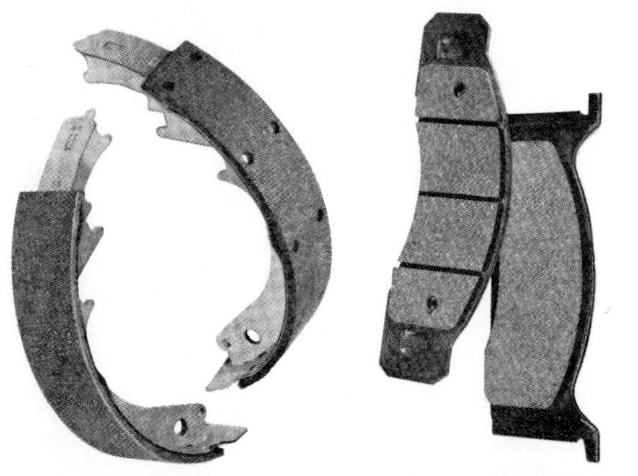

Figure 2.2 Various types of brake lining attached to brake shoes (Raybestos Division, Raybestos Manhattan, Inc.).

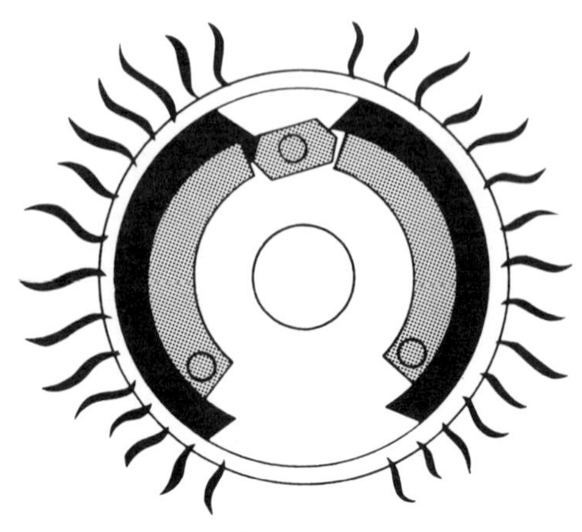

Figure 2.3 Heat dissipation from a brake drum. The heat energy developed in the brake is transferred to the atmosphere largely through the exposed surface of the drum (Ford Motor Company).

HEAT DISSIPATION In order to prevent brake fade, brakes must eliminate heat as fast as possible. This removal is accomplished by *heat dissipation,* which is the transfer of heat. In braking, heat is transfered from the brakes to the atmosphere, as you can see in Figure 2.3. The rate of dissipation depends largely on two

Figure 2.4 A brake drum with cooling fins (General Motors).

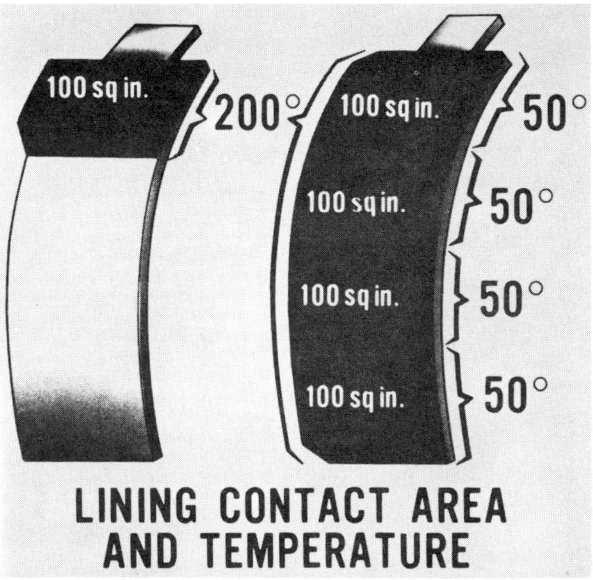

100 sq in. 200°
100 sq in. 50°
100 sq in. 50°
100 sq in. 50°
100 sq in. 50°
LINING CONTACT AREA
AND TEMPERATURE

Figure 2.5 More lining area means cooler brakes and longer lining life (Ford Motor Company).

things—the amount of air flow and the size of the surface over which the heat is distributed.

The next time you change a wheel on a car, notice how the wheel practically encloses the drum. The small-diameter, wide-rimmed wheel in common use today almost forms a cover over the drum. It restricts the amount of air that can flow over the drum surface. There are also some fenders that are designed in such a way that they do the same thing. To help make up for these problems, many drums are cast with cooling fins or rings so that a greater drum surface is exposed to the air (Figure 2.4). The greater surface promotes faster heat dissipation.

When you start working on various brake systems, you will notice that large, heavy cars have larger brake shoes than do small, light cars. Early in this chapter you learned that a heavy car has more kinetic energy than a light car when both are moving at the same speed. This means that stopping a heavy car produces more heat than stopping a light car. The larger braking surfaces provided by larger shoes promote better heat transfer and dissipation (see Figure 2.5). The result is a more stable coefficient of friction, less drum expansion, and therefore less brake fade.

Job 2b

IDENTIFY THE DEFINITION OF TERMS THAT REFER TO THE OPERATION OF BRAKE SYSTEMS

SATISFACTORY PERFORMANCE

A satisfactory performance on this job requires that you do the following:

1 Place the number of the three brake terms in front of the phrases that best define these terms.
2 Complete in part 1 in 10 minutes.

PERFORMANCE SITUATION

1 Brake lining
2 Brake fade
3 Heat dissipation

______ The transfer of heat to the atmosphere.

______ A material made of asbestos.

______ A metal shell placed inside the drum.

______ A loss of braking caused by heat.

SUMMARY

You have now learned the basic principles of braking. You know how friction between brake shoes and brake drums or discs converts kinetic energy to heat energy, and why the heat energy in drum brakes must be dissipated.

SELF-TEST

Each incomplete statement below is followed by four suggested completions. In each case select the *one* that best completes the statement.

1 Energy can be defined as
   A. the ability to do work
   B. a conversion of power
   C. a combination of forces
   D. the release of heat

2 The braking system converts kinetic energy to heat through the use of
      I. friction
      II. hydraulics
   A. I only
   B. II only
   C. Both I and II
   D. Neither I nor II

3 The coefficient of friction between any two objects is determined by the
      I. materials in contact
      II. force holding the objects together
   A. I only
   B. II only
   C. Both I and II
   D. Neither I nor II

4 Brake lining is made of
   A. cast iron
   B. asbestos
   C. steel
   D. rubber

5 The temporary loss of braking caused by overheating is called
   A. sliding friction
   B. wet friction

C. brake fade
D. coefficient loss

6 Heat dissipation is largely dependent on
I. the flow of air over the brake parts
II. the amount of braking area
A. I only
B. II only
C. Both I and II
D. Neither I nor II

7 Larger brake shoes and drums are used on heavy cars because they
I. have a higher coefficient of friction
II. absorb and transfer more heat
A. I only
B. II only
C. Both I and II
D. Neither I nor II

Chapter 3 Drum-Brake Assembly and Operation

Automobiles in the past have used many different types of drum brakes. But the type most commonly used today is the *self-energizing duo-servo brake.* To understand the full meaning of this term, you must first understand the assembly and operation of the brake.

Your objectives in studying this chapter will be to identify:

1
The parts of a typical duo-servo brake.
2
The function of the parts in a duo-servo brake.
3
The terms that refer to the operation of drum brakes.

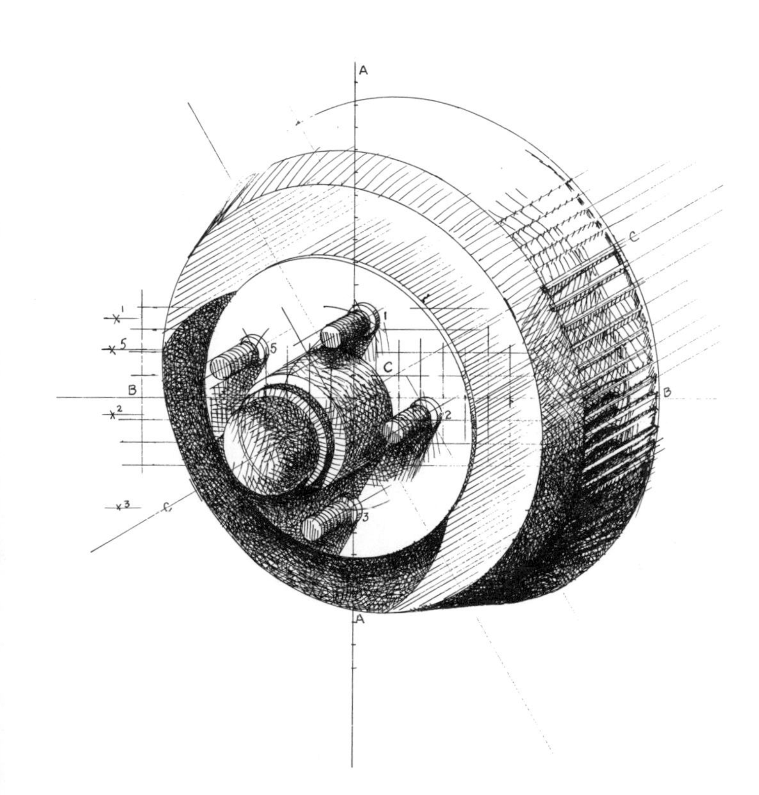

THE SELF-ENERGIZING DUO-SERVO BRAKE Except for slight changes to enable different self-adjuster parts to be fitted, all manufacturers use a brake of the type shown in Figure 3.1. This brake is typical of those used on front wheels. All the parts are mounted on the *backing plate,* which is a round piece of pressed steel that is attached to the car. At the front wheel the backing plate is bolted to the *spindle.* The spindle is the small axle, or shaft, on which the front wheel spins. At the rear of the car, backing plates are bolted to the rear axle housing.

The backing plate holds all of the brake parts except the drum, which is bolted to the wheel. An *anchor pin* is located at the top of the backing plate. The anchor pin takes the full load of the braking forces and must be securely attached to the backing plate. Many anchor pins are welded or riveted in place. The backing plate has six little raised platforms called *bosses.* These bosses, shown in Figure 3.2, contact the edges of the brake shoes and support the shoes when they are installed.

Hold-down assemblies hold the brake shoes to the backing plate. The *retaining spring* holds the bottom of the brake shoes in contact with the *star wheel adjuster.* The star wheel adjuster is a threaded device that can be expanded to push the shoes farther apart as the lining wears.

The *wheel cylinder,* mounted near the top of the backing plate, is a hydraulic device. It contains two pistons that move outward when the driver steps on the brake pedal. This outward movement pushes the brake shoes against the brake drum. The *retracting springs* pull the shoes away from the drum and force the pistons back in the cylinder when the driver releases the brake pedal.

In studying Figure 3.1, you probably noticed that the brake shoes were labeled *primary shoe* and *secondary shoe.* Both shoes are designed in the same way. Both serve as a mounting for the brake lining. But the linings attached to them are quite different because the shoes are subject to different forces in braking. Therefore, each lining must provide a different coefficient of friction with the brake drum.

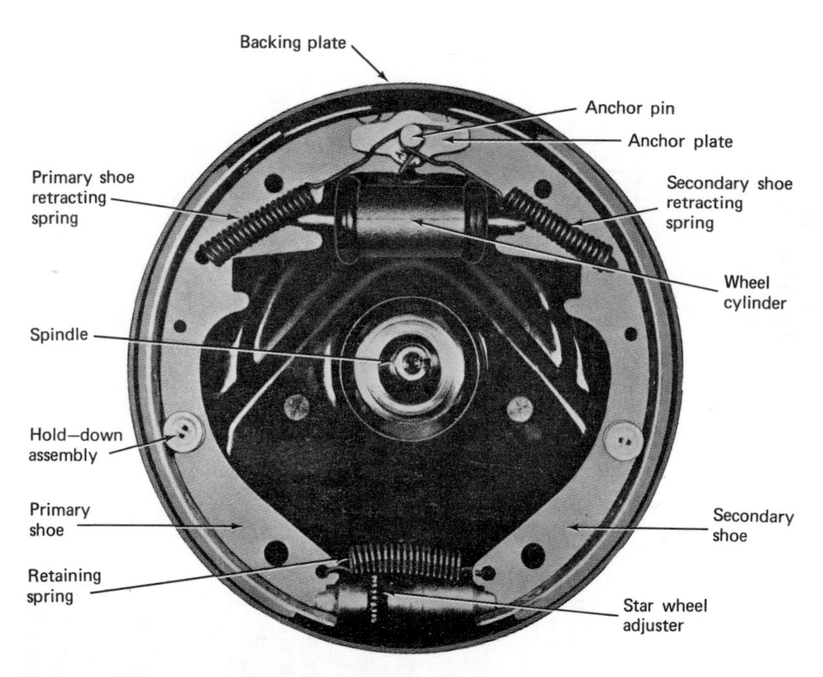

Figure 3.1 The self-energizing duo-servo brake. This photograph shows only the parts that are common to all brake systems (Raybestos Division, Raybestos Manhattan, Inc.)

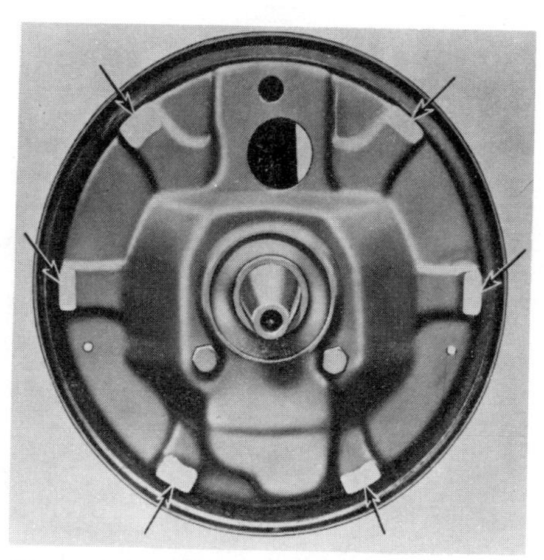

Figure 3.2 A typical backing plate. Note the bosses indicated by the arrows (courtesy Chevrolet Service Manual, Chevrolet Motor Division).

Job 3a

IDENTIFY THE PARTS OF A TYPICAL DUO-SERVO BRAKE ASSEMBLY

SATISFACTORY PERFORMANCE

A satisfactory performance on this job requires that you do the following:

1 Identify the numbered parts on the drawing by placing the number for each part in front of the correct part name below.
2 Correctly identify all of the parts within 15 minutes.

PERFORMANCE SITUATION

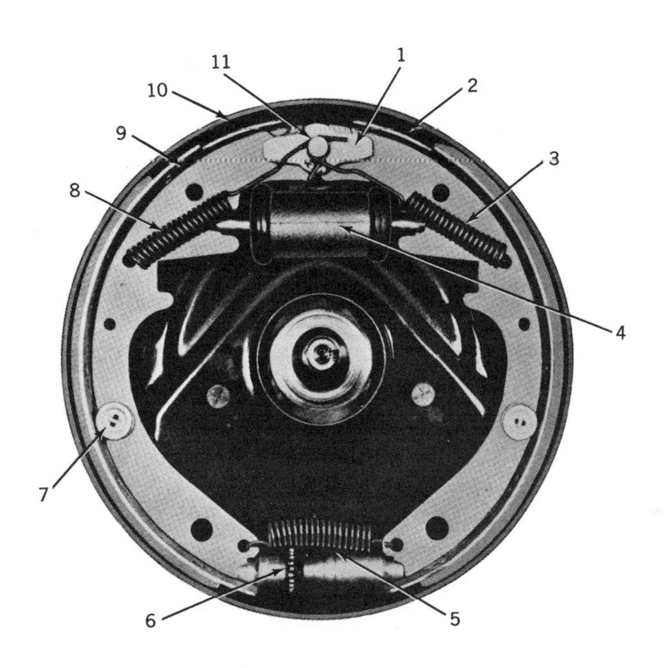

_______ Backing plate

_______ Secondary shoe Retracting spring

_______ Secondary shoe

_______ Star wheel adjuster

_______ Wheel cylinder

_______ Cam plate

_______ Primary shoe

_______ Primary shoe Retracting spring

_______ Anchor pin

_______ Retaining spring

_______ Hold-down assembly

_______ Anchor plate

Job 3b

IDENTIFY THE FUNCTION OF PARTS IN A DUO-SERVO BRAKE ASSEMBLY

SATISFACTORY PERFORMANCE

A satisfactory performance on this job requires that you do the following:

1 Identify the function of the following listed brake parts by placing the number of each part in front of the term that best describes its function.

2 Correctly identify the function of all the parts within 15 minutes.

PERFORMANCE SITUATION

1 Hold-down assembly	_______	provides a means of expanding the assembled brake shoes.
2 Brake shoe	_______	holds the shoes together at the star wheel adjuster.
3 Retracting spring	_______	provides the proper coefficient of friction with the drum.
4 Star wheel adjuster	_______	pulls the shoes away from the drum when the brakes are released.
5 Anchor pin	_______	keeps the shoes from turning with the brake drum.
6 Brake lining	_______	holds the shoes flat against the backing plate.
7 Retaining spring	_______	provides a mounting for the brake lining.
8 Wheel cylinder	_______	holds the wheel cylinder against the anchor pin.
	_______	pushes the brake shoes out into contact with the drum.

DRUM-BRAKE OPERATION When brakes are applied to a wheel in motion, the friction between the rotating drum and the brake lining tries to pull the shoes around with the drum. The anchor pin prevents the shoes from rotating and causes the pulling effort to tug the shoes out into tighter contact with the drum. The positions of the anchor pin and of the star wheel adjuster determine the ease with which the shoes yield to this pulling action. By carefully positioning these parts, the manufacturer can obtain high braking forces without requiring the driver to exert much effort on the brake pedal. This is the *self-energizing* feature of this type brake. It uses friction together with the placement of parts to increase the force holding the shoes against the drum.

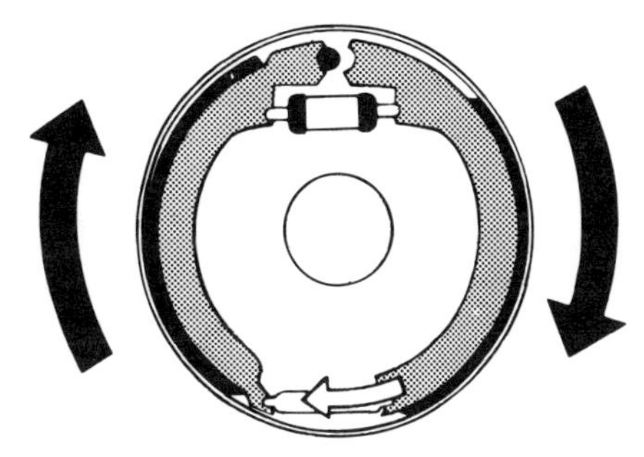

Figure 3.3 Servo action when brakes are applied to a drum moving clockwise (Ford Motor Company).

The motion of the rotating drum jams the secondary shoe against the anchor pin. The same motion of the drum pulls the primary shoe away from the anchor pin and forces it against the star wheel adjuster (Figure 3.3). The star wheel adjuster passes this push along to the bottom of the secondary shoe. This extra push adds to the force holding the secondary shoe against the drum. In this manner the primary shoe "serves" the secondary shoe. This is called *servo action.*

When a car is moving backward, the drum rotation is reversed; the function of the shoes is also reversed. The secondary shoe is pulled away from the anchor pin and is forced against the star wheel adjuster. The push is then passed along to the bottom of the primary shoe. This action adds to the force holding the primary shoe against the drum.

As you can see, the drum brake you are studying provides servo action regardless of the direction of wheel rotation. This means that the brake is as effective in stopping a vehicle moving backward as it is in stopping one moving forward. Since servo action is provided in both directions, the brake is termed *duo-servo.*

Job 3c

DEFINE TERMS THAT REFER TO DRUM BRAKE OPERATION

SATISFACTORY PERFORMANCE

A satisfactory performance on this job requires that you do the following:

1 Define the terms that refer to drum brake operation by placing the number of the term in front of the correct meaning of the term.

2 Correctly define all the terms within 10 minutes.

PERFORMANCE SITUATION

1 Self-energizing	______	The primary shoe exerts a force on the secondary shoe.
2 Servo action	______	The secondary shoe acts as a primary shoe when the brakes are applied while the car is moving backward.
3 Duo-servo action	______	Friction causes the brake shoes to attempt to turn with the brake drum.
	______	Friction between the drum and the shoes produces heat that causes the drum to expand.

IDENTIFYING BRAKE SHOES AND LINING The position of the primary and secondary shoes in a drum brake is very important. Because the lining on each shoe is different, shoes that are incorrectly installed will have to be installed again.

You will recall that the secondary shoe is energized not only by its own contact with the drum but also by the servo action of the primary shoe. This means that the secondary shoe performs about 70 percent of the braking and receives about 70 percent of the wear. If both primary and secondary shoes were lined with the same kind of lining, the lining on the secondary shoe would wear out twice as fast as the lining on the primary shoe.

The proper position of the brake shoes is determined by the direction of wheel rotation. Most manufacturers mount wheel cylinders at the top of the backing plate, with the forward shoe serving as the primary shoe.

Most brake linings are identified with a rubber stamp. You may have seen the *PRI* and *SEC* markings on the surface of some new linings. As the linings are fitted for installation, these markings are sometimes ground off. Even so, there are several other ways to identify unmarked linings.

As stated before, the primary shoe does much less work than the secondary shoe. To balance the wear on the linings of the two shoes, most brake-shoe manufacturers make the primary lining shorter than the secondary lining. Often the primary lining is softer. Sometimes it is also thinner. Besides helping to give the two types of lining equal life, the differences between them help you to identify each type.

SUMMARY

The self-energizing duo-servo brake is the most commonly used drum brake. It uses friction to increase its own effectiveness. It uses two shoes that expand to contact the inside of a rotating drum. They are called primary and secondary shoes. They have different functions and use different linings. You know why different linings are used and ways of identifying them.

SELF-TEST

Each incomplete statement or question below is followed by four suggested completions or answers. In each case select the *one* that best completes the statement or best answers the question.

1 In a self-energizing duo-servo brake the primary shoe exerts a force on the secondary shoe through the
   I. hold-down assembly
   II. retracting springs
A. I only
B. II only
C. Both I and II
D. Neither I nor II

2 In a drum brake, the retracting springs
   I. pull the shoes back from the drum
   II. push the wheel cylinder pistons back in
A. I only
B. II only
C. Both I and II
D. Neither I nor II

3 Which brake part keeps the brake shoes from turning with the drum when the brakes are applied?
A. Hold-down assembly
B. Star wheel adjuster
C. Wheel cylinder
D. Anchor pin

4 Two mechanics are discussing drum brakes. Mechanic A says that anchor pins are usually permanently attached to backing plates. Mechanic B says that rear backing plates are usually bolted to the rear axle housings. Who is right?
A. A only
B. B only
C. Both A and B
D. Neither A nor B

5 Two mechanics are discussing drum brakes. Mechanic A says that the secondary lining is usually longer than the primary lining. Mechanic B says that the primary lining is usually harder than the secondary lining. Who is right?
A. A only
B. B only
C. Both A and B
D. Neither A nor B

6 The drum brakes in most common use are called "duo-servo" because they
A. use two brake shoes actuated by one wheel cylinder
B. are as effective when the car is moving backward as when it is moving forward
C. are applied by two wheel cylinder pistons
D. have a retracting spring for each shoe

7 The platforms, or bosses, on the backing plates support the
A. brake shoes
B. star wheel adjuster
C. retracting springs
D. hold-down springs

Chapter 4 Wheel-Bearing Service

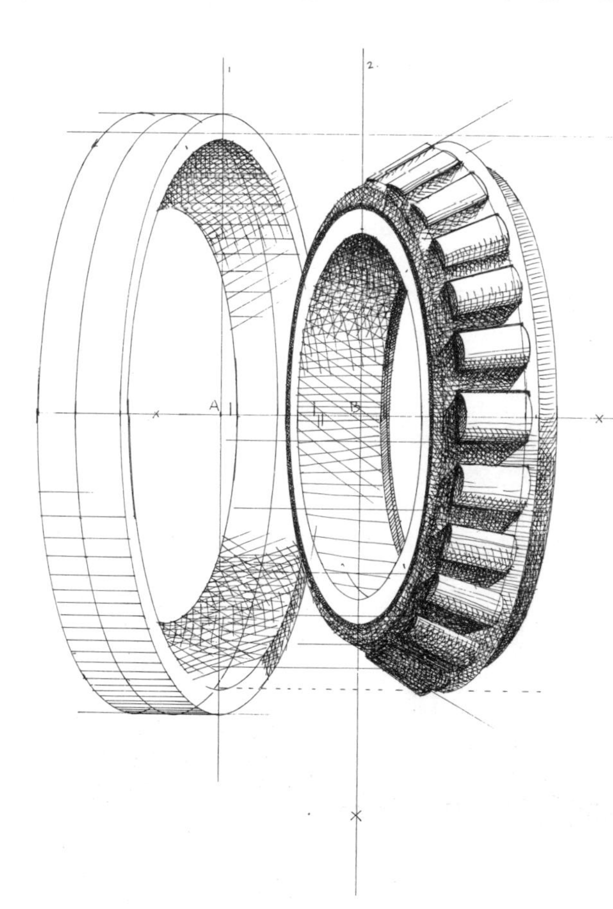

You know that the brake shoes expand to contact the inside of the rotating brake drum. Suppose the drum were wobbling as it turned? What kind of braking action would result? Worn or improperly adjusted wheel bearings allow a drum to shift and wobble and thus affect brake operation.

This chapter will help you realize how important the wheel bearings are to the brake system. It will also help you learn many things that are necessary for you to become a brake mechanic. You will learn to:

1
Raise a car and support it by the frame.
2
Raise a car and support it by the suspension system.
3
Remove and install wheels.
4
Identify the bearings and related parts of a front hub assembly.
5
Adjust front wheel bearings.
6
Repack front wheel bearings.
7
Replace front wheel-bearing cups.

Of the seven objectives listed, all but one are repair skills. Most of the operations you will perform in this chapter are part of every brake job.

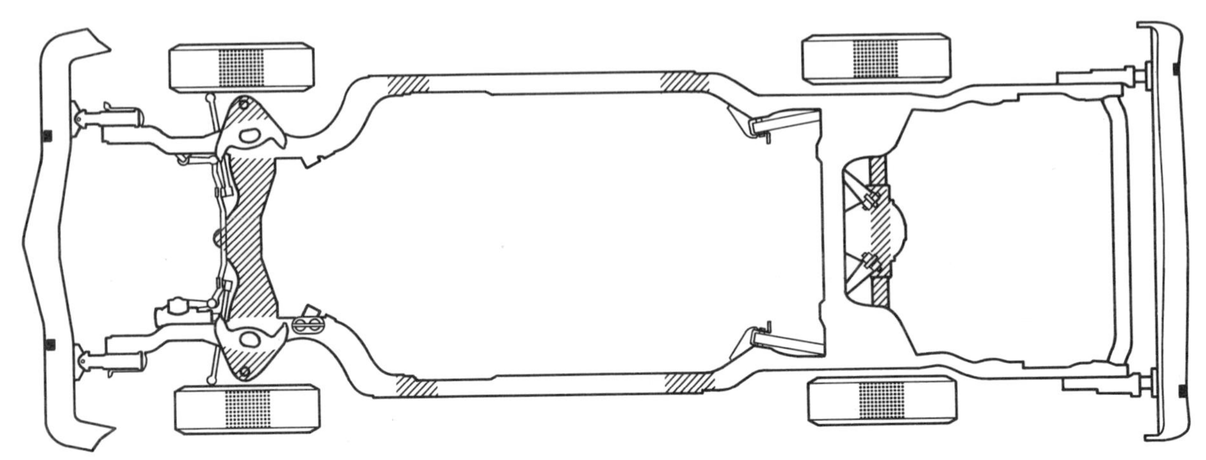

Figure 4.1 Drawing of a typical chassis, showing locations where it is safe to place a jack or car stands (courtesy Chevrolet Service Manual, Chevrolet Motor Division).

RAISING AND SUPPORTING A CAR

Since most brake work requires that the wheels and drums be removed, you must raise the car off the floor. This task is usually done with a floor jack. Whenever you raise a car with a floor jack, you must place the jack in the right spot. You would not want the car to slip off the jack, nor would you want to place the jack where it could damage any parts under the car. Many different kinds of frames and suspension systems are in use. Therefore, you should be sure to check the manufacturer's service manual to find the areas under the car that provide a safe, solid place for you to position a jack.

Jacks are made to raise cars, not to support them. Working on a car that is held off the floor by a jack is a sure way of getting hurt, since the car can slide off the jack. Anytime you raise a car from the floor, even for a few minutes, place car stands under the car to support its weight. For this, too, you should check the manufacturer's service manual. Figure 4.1 is a typical drawing found in service manuals showing areas where it is safe to place a jack or car stands.

Though there are many different areas under a car where stands can be positioned, there are only two ways to support a car. One is by the frame, and the other is by the suspension system. When a car is supported by the frame, the wheels are free to drop down to the limits of their suspension systems. This means you have more room to work on the brakes since the wheels drop down below the fenders. Supporting a car by the frame gives you another advantage. The car stands are not right behind the backing plates and can't get in your way.

RAISING A CAR AND SUPPORTING IT BY THE FRAME

The following steps outline the procedure for raising a car with a jack and supporting it with car stands placed under the frame. You should check the manual for the correct lifting and support points for the car on which you are working.

Front 1 Roll a jack under the front of the car. Raise the jack slightly, and adjust its position so that it is centered under the front crossmember. (See position A in Figure 4.2.)

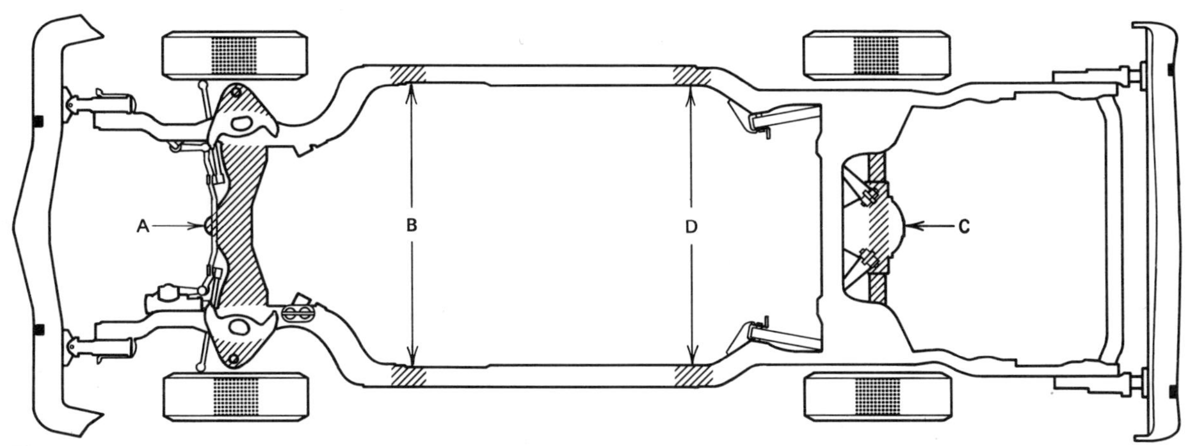

Figure 4.2 Drawing of a typical frame, showing locations for raising a car with a jack (A and C) and positions where car stands should be placed (B and D) for supporting the car by the frame (courtesy Chevrolet Service Manual, Chevrolet Motor Division).

2 Operate the jack to raise the car until the wheels clear the floor by about six inches (18 cm).

3 Place the car stands under the frame side rails just behind the bend in the frame. (See positions B in Figure 4.2.) Raise the car stands as close as possible to the frame.

4 Lower the jack so the car is supported by the stands.

5 Remove the jack.

Rear 1 Roll a jack under the rear of the car. Raise the jack slightly and adjust its position so that it is centered under the rear axle housing. (See position C in Figure 4.2.)

2 Operate the jack to raise the car until the wheels are clear of the floor by at least 10 inches (25 cm).

3 Place car stands under the frame side rails just ahead of the bend in the frame. (See positions D in Figure 4.2.) Raise the car stands as close as possible to the frame.

4 Lower the jack so the car is supported by the stands.

5 Remove the jack.

Job 4a

RAISE A CAR AND SUPPORT IT BY THE FRAME

SATISFACTORY PERFORMANCE

A satisfactory performance on this job requires that you do the following:

1 Using a hydraulic floor jack, raise a car from the floor and support it with car stands placed under the frame.
2 Following the recommendations of the car manufacturer regarding jack and car stand positioning, complete the job within 15 minutes.
3 Fill in the blanks under "Information."

INFORMATION

Vehicle identification ______________________________

Reference used ____________ Page(s) ____________

RAISING A CAR AND SUPPORTING IT BY THE SUSPENSION SYSTEM The following steps outline the procedure for raising a car with a jack and supporting it with car stands placed under the suspension system. Be sure to check the manual for the correct lifting and support points for the car on which you are working.

Front 1 Roll a jack under the front of the car. Raise the jack slightly and adjust its position so that it is centered under the front crossmember. (See position A on Figure 4.3.)

2 Operate the jack to raise the car until the wheels are clear of the floor by about six inches (18 cm).

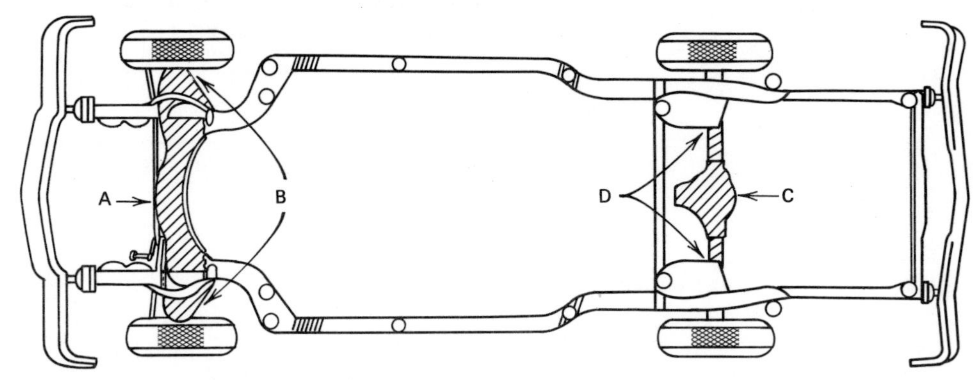

Figure 4.3 A typical frame, showing locations for raising a car with a jack (A and C) and positions where car stands should be placed (B and D) for supporting the car by the suspension system (courtesy Chevrolet Service Manual, Chevrolet Motor Division).

3 Place car stands under the lower control arms as close as possible to the wheels. (See positions B in Figure 4.3.) Raise the car stands as close as possible to the lower control arms.

4 Lower the jack so the car is supported by the stands.

5 Remove the jack.

Rear 1 Roll a jack under the rear of the car. Raise the jack slightly and adjust its position so that it is centered under the rear axle housing. (See position C in Figure 4.3.)

2 Operate the jack to raise the car until the wheels are clear of the floor by at least six inches (18 cm).

3 Place car stands under the axle housing as close as possible to the wheels. (See positions D in Figure 4.3.) Raise the car stands as close as possible to the housing.

4 Lower the jack so that the car is supported by the stands.

5 Remove the jack.

Job 4b

RAISE A CAR AND SUPPORT IT BY THE SUSPENSION SYSTEM

SATISFACTORY PERFORMANCE

A satisfactory performance on this job requires that you do the following:

1 Using a hydraulic floor jack, raise a car from the floor and support it with car stands placed under the suspension.

2 Following the recommendations of the car manufacturer regarding jack and car stand positioning, complete the job within 15 minutes.

3 Fill in the blanks under "Information."

INFORMATION

Vehicle identification ______________________________

Reference used ____________________ Page(s) ______________

Figure 4.4 A typical impact wrench (courtesy Snap-on® Tools Corporation).

REMOVING AND INSTALLING WHEELS

Removing Wheels Most brake services or repair jobs will require you to remove the wheels. Many shops use *impact wrenches* to remove the lug nuts that hold the wheels on the car (Figure 4.4). Impact wrenches are like drill motors, but instead of spinning a drill bit they spin a socket wrench with a rotating, hammering action. Impact wrenches are great time-savers. However, they are not always available, so in many instances you will have to use a muscle-powered lug wrench (Figure 4.5).

Have you ever tried to loosen a lug nut when the wheel is off the ground? As you turn the wrench the wheel turns. You can apply the parking brakes, but they hold only the rear wheels. Maybe you can get a friend to sit in the car and press down on the

Figure 4.5 A typical lug wrench (courtesy Snap-on® Tools Corporation).

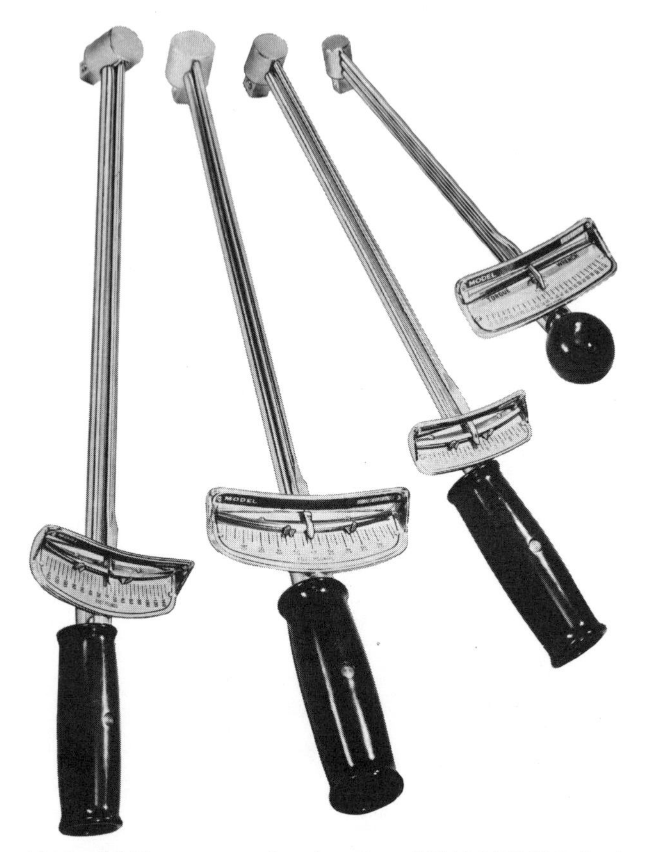

Figure 4.6 Torque wrenches (courtesy AMMCO TOOLS, Inc.)

brake pedal until you can loosen the nuts. But what if you are working alone?

If you are going to remove the wheels from a car, remove the wheel covers and loosen the lug nuts **before** you raise the wheels off the floor. This procedure is easy, and it requires far less energy than trying to loosen the lug nuts after the car is off the floor. Once the car is up in the air, you can spin the nuts off with little effort.

Installing Wheels As a brake mechanic, you will find that some brake problems are caused by improper wheel installation. If the lug nuts are left too loose, the wheel may come off. If they are tightened too much, the hub may be distorted. Distorting the hub will actually twist or bend the brake drum. This may cause erratic braking. Lug nuts, then, should be tightened just enough to keep them from loosening, but not so much that they will distort the hub. For this reason, lug nuts should be tightened to a torque specification. *Torque* can be defined as a turning or twisting

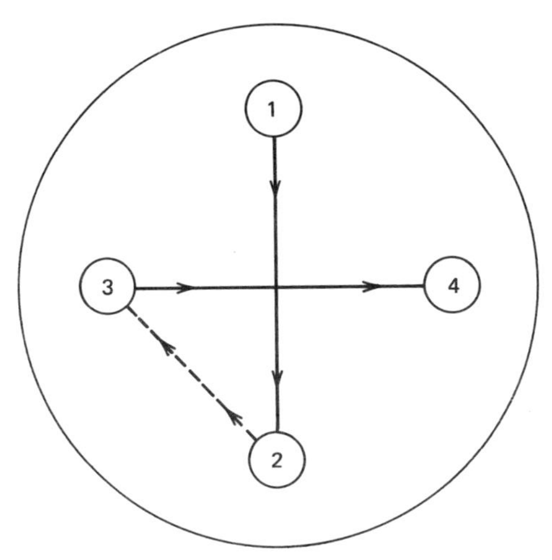

Figure 4.7 "Cross" pattern, recommended as the sequence for tightening lug nuts on a four-hole wheel.

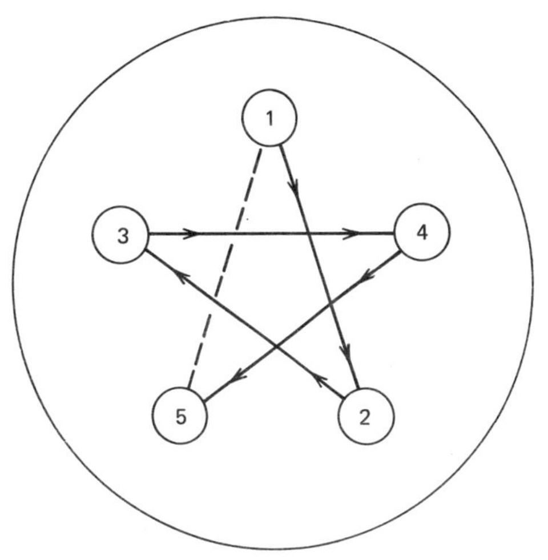

Figure 4.8 "Star" pattern, recommended as the sequence for tightening lug nuts on a five-hole wheel.

effort. When you turn a wrench, you are applying torque to a nut or a bolt. By measuring this torque you can tell how much you are tightening it.

To measure torque, you must use a *torque wrench.* A torque wrench is a wrench whose handle indicates how much torque, or twisting effort, is being applied to a nut or a bolt. There are many types of torque wrenches. Some have a wand, or pointer, that moves across a scale as you pull on the wrench. Others have a dial. And still others give off a loud click when a pre-set torque is reached. There are even some that have a lamp that lights when the nut is tight enough. Some of the many types of torque wrenches are shown in Figure 4.6.

Torque is measured in foot-pounds. Suppose you had a wrench that was 12 inches long and you applied a force of 20 pounds to the handle. Then the nut or bolt you were tightening would be tightened to 20 foot-pounds of torque. There are hundreds of nuts and bolts on a car that must be tightened to a specified torque. This is one of the reasons that the manufacturer's service manuals are so important. They contain the torque specifications that you need to turn out a quality job.

Tightening the lug nuts to the right amount of torque is only half the job. The nuts must also be tightened in a sequence. The proper sequence for four-hole and five-hole wheels is shown by the numerals and arrows in Figures 4.7 and 4.8. The proper sequence for four-hole wheels is represented by a "cross" pattern. For five-hole wheels it is represented by a "star" pattern.

An impact wrench can be used for installing wheels as well as removing them. Just remember that it should be used only to run the nuts up snug. After you lower the car to the floor, you should use a torque wrench to tighten the nuts to specifications.

Job 4c

REMOVE AND INSTALL WHEELS

SATISFACTORY PERFORMANCE

A satisfactory performance on this job requires that you do the following:

1 Remove and install all four wheels on the car assigned.
2 Following the steps in the "Performance Outline," complete the job within 30 minutes.
3 Fill in the blanks under "Information."

PERFORMANCE OUTLINE

1 Remove the wheel covers and loosen the lug nuts.
2 Raise and support the car.
3 Remove the lug nuts and wheels.
4 Install the wheels and the lug nuts.
5 Lower the car to floor.
6 Tighten the lug nuts to specifications in sequence.
7 Install the wheel covers.

INFORMATION

Vehicle identification ______________________________

Reference used ____________________ Page(s) ____________________

Lug nut torque specifications ______________________________

FRONT WHEEL BEARINGS Though most parts of a brake are mounted on a backing plate, a front brake drum is mounted on a *hub.* The hub provides the means for mounting the drum and the wheel so they will rotate on the spindle. The hub is made of cast iron and contains two bearings. These bearings allow the hub to rotate freely, even with the weight of the car pressing down on the wheels.

A typical hub and drum assembly is shown in Figure 4.9. Study this drawing carefully. As a brake mechanic you must be familiar with all the parts and their names.

Bearings The bearings used in front wheels are sometimes called antifriction bearings. These bearings use balls or rollers to provide rolling friction. Rolling friction allows the wheel to rotate on the spindle very easily. The bearings in Figure 4.9 are *tapered roller* bearings. The rollers themselves are tapered, and they roll on a tapered *race,* or ring, which is also called a *cone.* The rollers are held in place on the cone by a *cage,* which is a metal band with a slot for each roller. The cage also keeps the rollers from contacting each other. The cone, rollers, and cage are assembled as a unit (Figure 4.10). This allows you to remove the bearing as an assembly without having to work with separate pieces.

There are two cone-and-roller assemblies in each hub. The inner assembly is installed from the rear of the hub. It is the larger of the two.

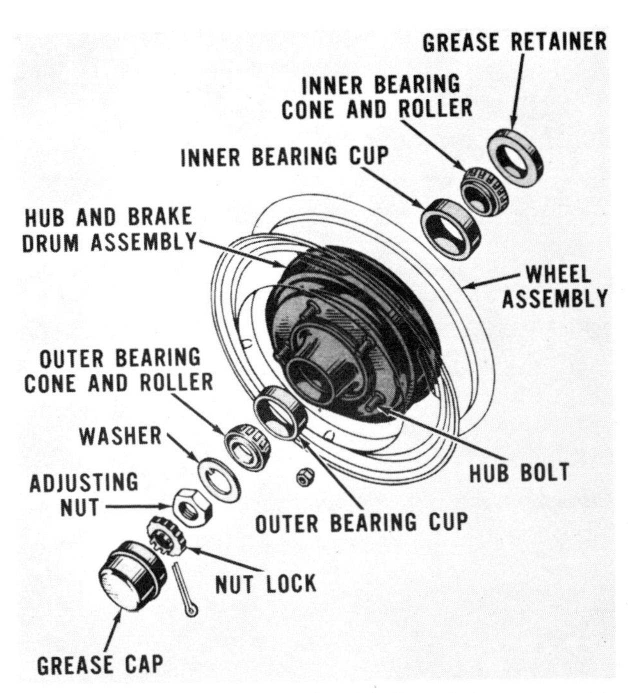

Figure 4.9 A typical front wheel hub and drum assembly, showing all component parts in the order of their assembly (Ford Motor Company).

The outer cone-and-roller assembly is the smaller. It is accessible without removing the hub from the spindle. The cone-and-roller assemblies are easily removed.

Figure 4.10 Cone-and-roller assembly.

Bearing Cups The bearing *cups* provide a hard, smooth surface on which the rollers can roll. They form the outer race and are tapered to match the taper of the rollers. You can see one of these races in Figure 4.11. Even though these races are separate from the cone-and-roller assembly, they are actually a part of the bearing. They should be replaced whenever the cone-and-roller assembly is replaced.

Figure 4.11 Bearing cup.

Grease Retainer The grease retainer, shown in Figure 4.12, is a seal that serves to keep the wheel-bearing grease from leaking out inside the brake drum. Such leakage would contaminate the brake lining and cause braking problems. The grease retainer fits into the rear of the hub, behind the inner bearing assembly. Grease retainers should be replaced each time the wheel bearings are repacked.

Figure 4.12 Grease retainer.

Washer The washer, which is flat, fits near the end of the spindle and is very important. In Figure 4.13 you will notice that the washer has a little *key,* or tab, on the inside. This key fits into a keyway, or groove, on the spindle and keeps the washer from turning. The washer separates the outer bearing assembly from the adjusting nut. If the bearing assembly were in contact with the nut, it could act to turn the nut.

Figure 4.13 The washer.

Adjusting Nut The adjusting nut, pictured in Figure 4.14, holds the parts of the hub assembly on the spindle and in the proper position. Turning the nut right or left tightens or loosens the bearings in their cups.

Figure 4.14 Adjusting nut.

Nut Lock The nut lock is a pressed steel cover that fits over the adjusting nut. This nut has *castellations,* or notches (Figure 4.15). These castellations allow the nut to be locked in any desired position. You will find some cars that do not have nut locks. Some adjusting nuts are castellated and thus do not require additional locks. However, the nut lock is used by most manufacturers, and it allows you to obtain a more accurate wheel-bearing adjustment than you can obtain with a castellated adjusting nut.

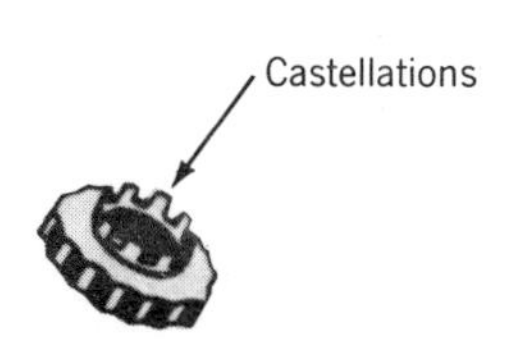

Figure 4.15 Nut lock.

Cotter Pin The cotter pin, shown in Figure 4.16, is used to secure the nut lock. It passes through a hole drilled in the end of the spindle, and holds the nut lock by its castellations.

Figure 4.16 Cotter pin.

Grease Cap The grease cap, or hub cap, seals the open end of the hub. Although it does serve to keep the grease in the hub, its primary job is to keep dirt and water out. A car should never be placed in service without grease caps. Figure 4.17 shows a typical grease cap.

Figure 4.17 Grease cap.

Job 4d

IDENTIFY WHEEL BEARINGS AND RELATED PARTS

SATISFACTORY PERFORMANCE

A satisfactory performance on this job requires that you do the following:

1 Identify the numbered parts on the drawing by placing each number in front of the correct part name.
2 Correctly identify all the parts within 15 minutes.

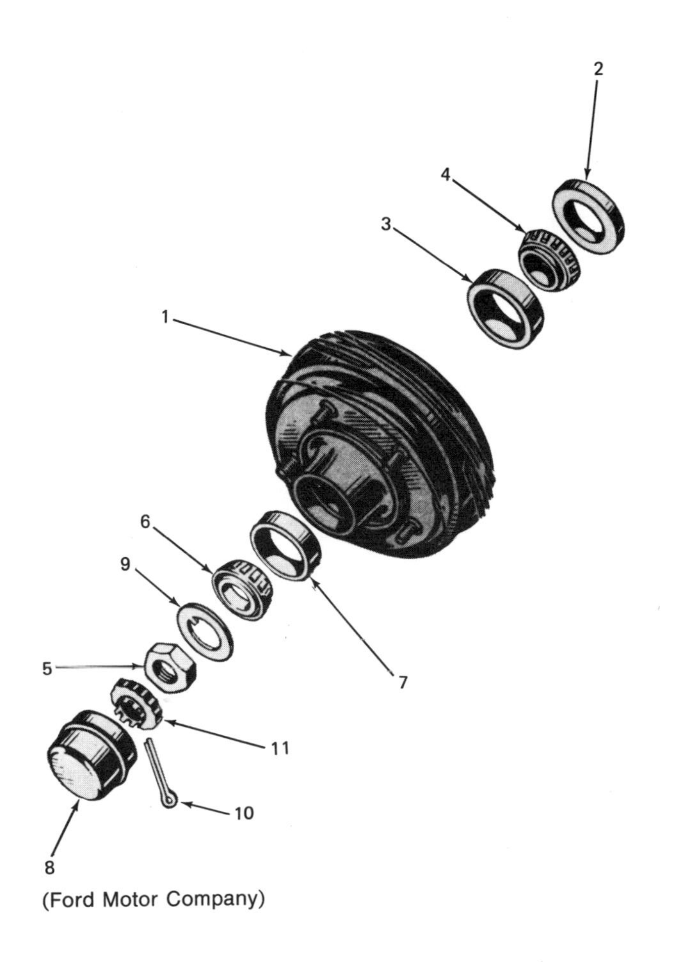

(Ford Motor Company)

PERFORMANCE SITUATION

_______ Inner bearing cup

_______ Cotter pin

_______ Spindle shim

_______ Washer

_______ Outer bearing assembly

_______ Inner bearing assembly

_______ Nut lock

_______ Adjusting nut

_______ Hub assembly

_______ Grease retainer

_______ Grease cap

_______ Outer bearing cup

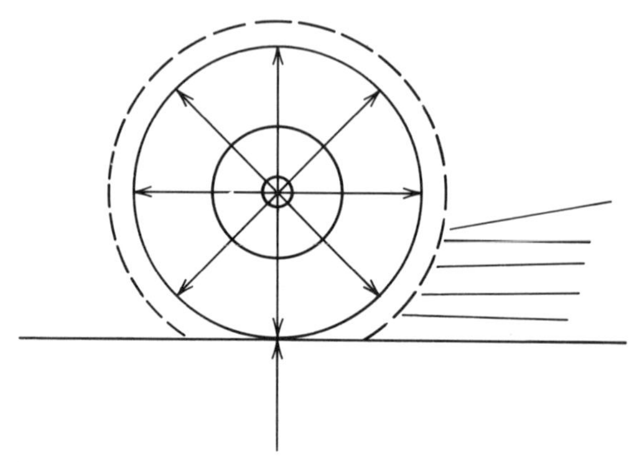

Figure 4.18 Radial load on a wheel.

SERVICING WHEEL BEARINGS Wheel bearings should be serviced about every 20,000 miles and as a part of every brake job. Wheel-bearing service is considered routine maintenance, but it provides a bonus for both the car owner and the mechanic. To get to the inner bearing, you must remove the brake drum. This gives you an opportunity to inspect the brake parts. If any worn or damaged parts are found, you should notify the car owner of the need for repairs. This is the car owner's bonus: a report on the condition of the brake system. Your bonus is that the owner will most likely have you perform the needed repairs.

The condition of the wheel bearings and their proper adjustment are very important to the operation of the brake system. Worn or loose bearings often cause a car to pull to one side when the brakes are applied. The wheel bearings should always be checked when you are diagnosing brake problems. When we discuss disc brake systems later in the book, you will find that worn or loose wheel bearings can cause other problems.

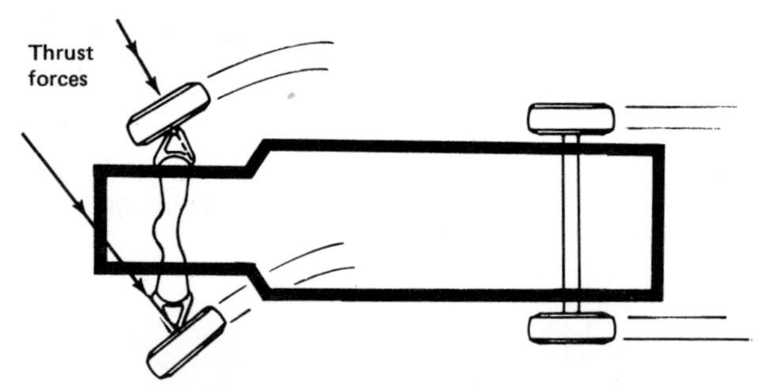

Figure 4.19 Side thrust forces on front wheels during cornering (courtesy American Motors Corporation).

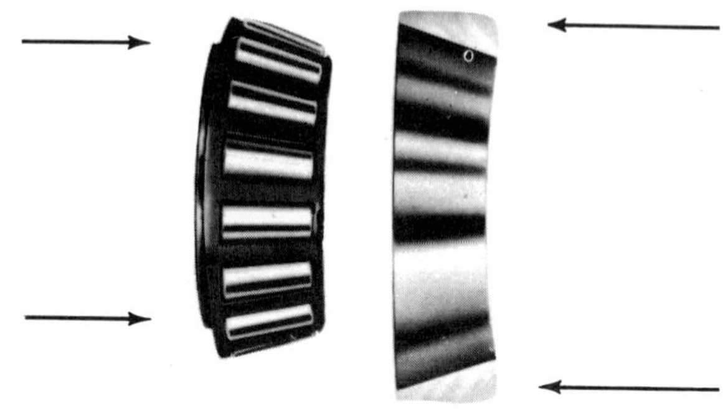

Figure 4.20 View of a cone and roller assembly and a sectioned cup showing where thrust loads are applied (General Motors).

Before you pull the bearings out of a hub, you will need to know certain things about them.

Radial and Thrust Loads Wheel bearings are subjected to two major forces. These are radial loads and thrust loads. A *radial* load acts perpendicular to the axis of the wheel. The weight of the car pushes down on the road in a straight line from the spindle to the road. When the wheel rotates, this load still pushes straight down, but the force is carried by another part of the wheel. If you could see these lines of force, they would look like spokes in a wheel. They would radiate from the spindle to the tread of the tire (Figure 4.18).

A *thrust* load acts parallel to the axis of the wheel. It tends to push the wheel off or farther onto the spindle. The wheels of a car are subjected to considerable thrust loads, especially when cornering (Figure 4.19). Because of these thrust loads, the wheel bearings must do much more than spin freely and support the weight of the car. They must keep the wheel from sliding in or out on its spindle.

Wheel bearings are designed to handle both radial and thrust loads. You may find ball bearings in the front hubs on some cars, but tapered roller bearings are generally used. Tapered roller bearings have long life and are capable of handling extreme loads. Figure 4.20 shows a view of a cone-and-roller assembly and a sectioned bearing cup. The arrows indicate where the thrust loads are applied. You can see that the tapered shape of the bearing assembly and its cup handle these loads.

WHEEL-BEARING ADJUSTMENT

Adjusting the wheel bearings properly is very important. Tightening the adjusting nut forces the outer bearing in against its cup. The cup then pushes against the hub, which pushes the inner cup against its bearing. If the nut is overtightened, both bearings will be jammed against their cups. Jamming increases the friction in the bearings and causes them to wear out very quickly. If the nut is not tightened enough, the bearings will not be in proper contact with their cups. Bearings that are too loose also wear out fast. Moreover, they allow the hub and, thus, the drum, to wobble. As this wobble constantly changes the position of the drum, braking may become erratic or may pull the car to one side.

The following adjustment procedure and specifications are typical. However, such procedures and specifications vary with different manufacturers. Therefore, you should always consult the car manufacturer's service manual for the procedure and specifications for the car on which you are working.

1 Raise the front of the car with a jack, and support the car with jack stands.

2 Remove the wheel cover.

3 Remove the grease cap. Lacking a special tool, you may use a pair of water-pump pliers for this task.

4 Using a pair of diagonal cutters, remove the cotter pin. First, straighten the legs of the pin. Then grasp the head of the pin, and with a prying action, pull the pin through the hole in the spindle.

5 Remove the nut lock.

6 Torque the adjusting nut to between 17 and 25 foot-pounds to seat the bearings (Figure 4.21). Rotate the wheel while tightening the nut.

Figure 4.21 Direction for torquing an adjusting nut (Ford Motor Company).

7 Place the nut lock over the adjusting nut so that the castellations on the lock are aligned with the cotter pin hole in the spindle.

8 Using water-pump pliers or the proper size wrench, back off on the nut and the nut lock together so that the next castellation is aligned with the cotter pin hole.

9 Install a new cotter pin to lock the nut in this position (Figure 4.22).

Figure 4.22 Installed cotter pin (Ford Motor Company).

10 Install the hub cap.

11 Install the wheel cover.

12 Repeat steps 2 to 11 at the remaining wheel.

13 Lower the car to the floor.

Job 4e

ADJUST WHEEL BEARINGS

SATISFACTORY PERFORMANCE

A satisfactory performance on this job requires that you do the following:

1 Adjust the front wheel bearings on the car assigned.
2 Following the manufacturer's procedure and specifications, complete the job within 30 minutes.
3 Fill in the blanks under "Information."

INFORMATION

Vehicle identification ____________________

Reference used ____________ Page(s) ____________

REPACKING WHEEL BEARINGS

Wheel bearings that are properly adjusted, properly lubricated, and kept free of dirt will usually outlast the car on which they are installed. A good mechanic understands and appreciates the importance of the so-called simple jobs. Adjusting wheel bearings is a simple job but, as you can see, it must be performed carefully.

You are now going to repack wheel bearings. This, too, is a simple job, but you must follow a step-by-step procedure to perform it properly. Be sure to check the adjustment procedure and specifications in the proper service manual. Here is a general procedure.

1 Remove the wheel covers.

2 Using a lug wrench, loosen all the lug nuts holding the wheels to the hubs.

3 Raise the front of the car with a jack and support the car with car stands.

4 Remove the lug nuts.

5 Remove the wheel.

6 Remove the grease cap.

7 Remove the cotter pin.

8 Remove the nut lock (if fitted), the adjusting nut, the washer, and the outer bearing. The outer bearing can be easily removed by hitting the edge of the drum with the heel of your hand. This action usually causes the bearing to slide out on the spindle where it can be easily grasped.

9 Grasp the brake drum with both hands, and with a twisting motion, slide it off the spindle.

Note. Never force a drum off if it will not slide off easily. Refer to the section on adjusting brakes for the procedure for loosening the brake adjustment to provide more clearance.

Note. The brake assembly and the inside of the brake drum may contain asbestos dust. Breathing asbestos dust may cause asbestosis and cancer. You should wear an air purifying respirator during all procedures where you may be exposed to asbestos dust.

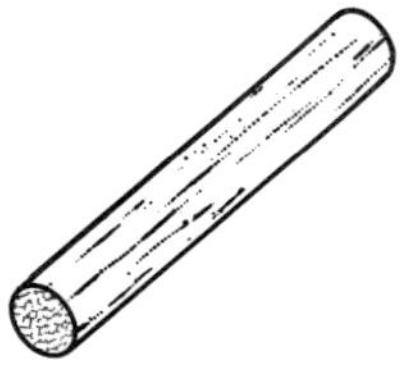

Figure 4.23 A wooden punch for removing inner wheel bearings.

10 Lay the drum on the floor with the open end down. Using a soft blunt punch, such as a piece of broom handle, drive out the inner bearing and the grease retainer (Figure 4.23).

Figure 4.24 Front wheel-bearing diagnosis (courtesy Chevrolet Service Manual, Chevrolet Motor Division).

11 Thoroughly wash all parts in a suitable solvent such as Kleer-Flo or Agitine.

12 Dry all parts.

Note. **If compressed air is used to dry the parts, be careful not to spin the bearings. Hold the bearings firmly, and direct the air between the rollers at the small end of the bearing. This will dry the bearing and blow out all the old grease. Spinning bearings with an air gun will cause damage to the bearings and may cause personal injury if a roller flies out of the cage.**

13 Inspect the bearings. (See Figures 4.24 and 4.25 for diagnosing bearing faults.) Turn the cone and rollers so all the roller surfaces can be seen. Replace any bearing that shows signs of pitting, discoloration, or scoring. Also check for the presence of any metallic powder or flakes. They usually indicate a defective cone.

Note. **Anytime a bearing is replaced, the cup into which it fits must also be replaced, even if it exhibits no apparent damage. Refer to the section on replacing bearing cups for the correct procedure.**

14 Carefully clean all traces of grease from inside the hub. Inspect the bearing cups. Also check for the presence of any metallic powder or flakes. Replace any cups that exhibit these faults.

Note. **Anytime a bearing cup is replaced, its mating bearing must also be replaced, even if it exhibits no apparent damage. Refer to the section on replacing bearing cups for the correct procedure.**

15 Repack the bearings. Many shops use a bearing packer similar to the one shown in Figure 4.26. Follow the instructions of the manufacturer if you use one of these devices. If a packing device is not available, the grease should be worked up between the rollers by drawing the bearing, large end down, across grease held in the palm of your hand. Push grease into the bearing in this manner until it oozes out at the small end.

Note. **Be sure to use only grease that meets the specifications stated in the manufacturer's service manual.**

16 Place the inner (large) bearing in its cup and install a new grease retainer. The new retainer should be installed so that the sharp edge of the seal is facing inward as shown in Figure 4.27. An installing tool should be used if available. This tool enables you to drive the retainer in place without danger of damaging it. Lacking an installing tool, you can carefully drive the retainer in place with a soft-face hammer. Most retainers should be driven in until their outer surface is flush with the surface of the hub.

17 Check to see that there is no dirt, grease, or foreign matter on the inner surfaces of the drum. Carefully slide the hub in place over the spindle.

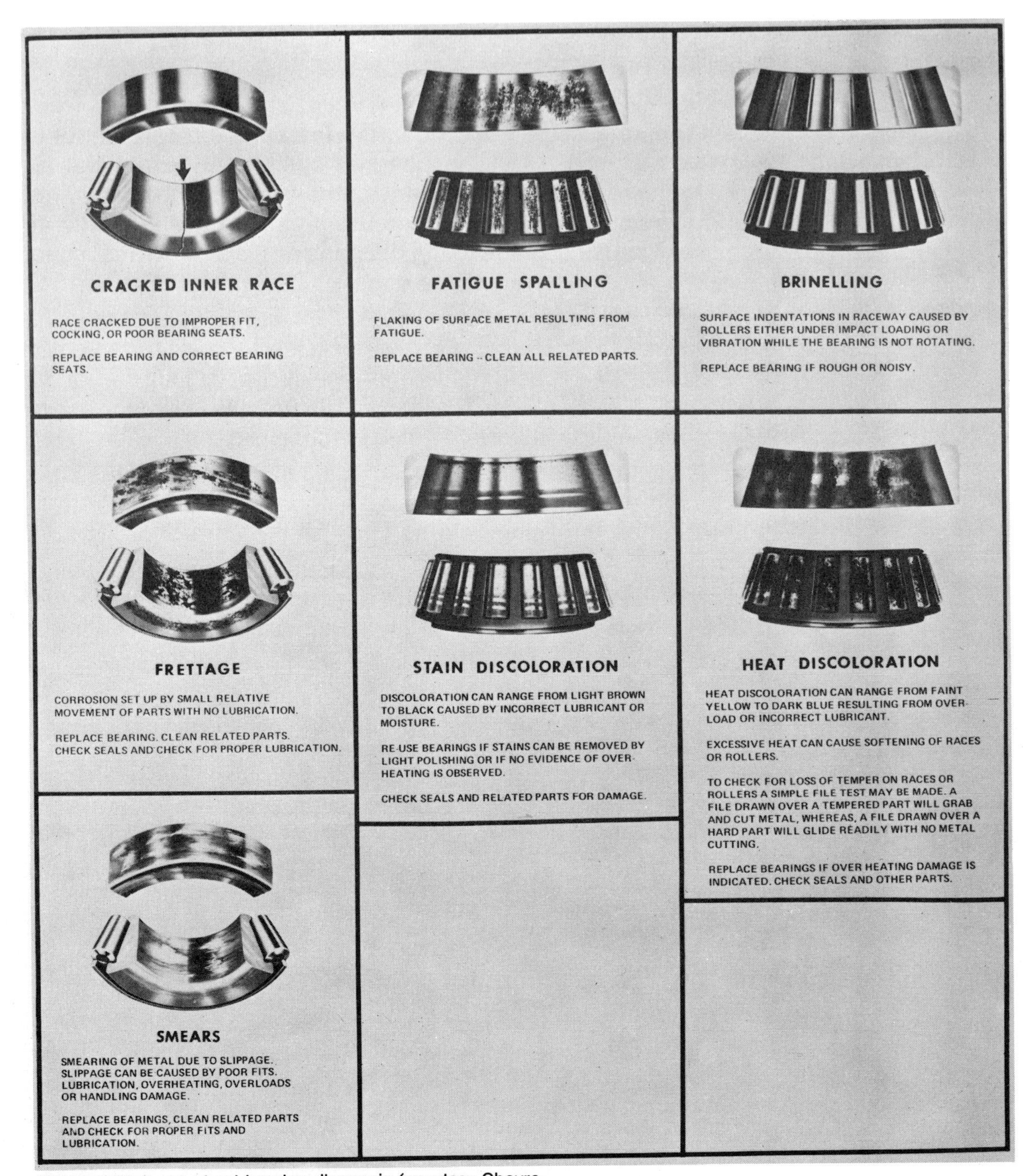

Figure 4.25 Front wheel-bearing diagnosis (courtesy Chevrolet Service Manual, Chevrolet Motor Division).

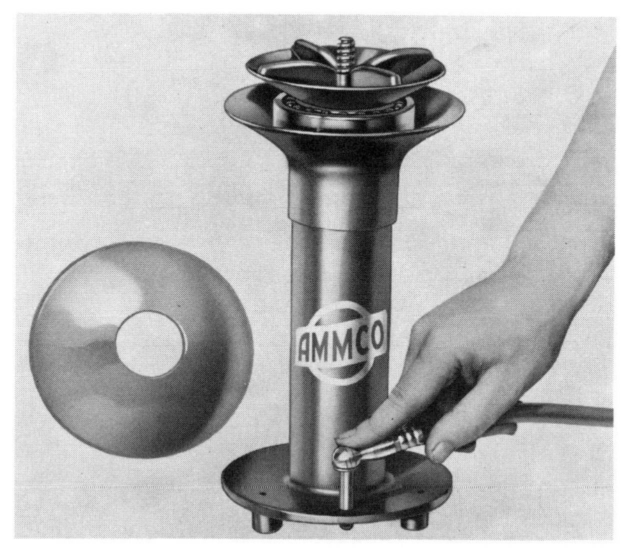

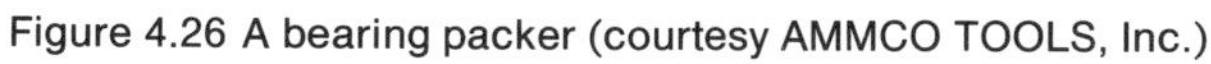
Figure 4.26 A bearing packer (courtesy AMMCO TOOLS, Inc.)

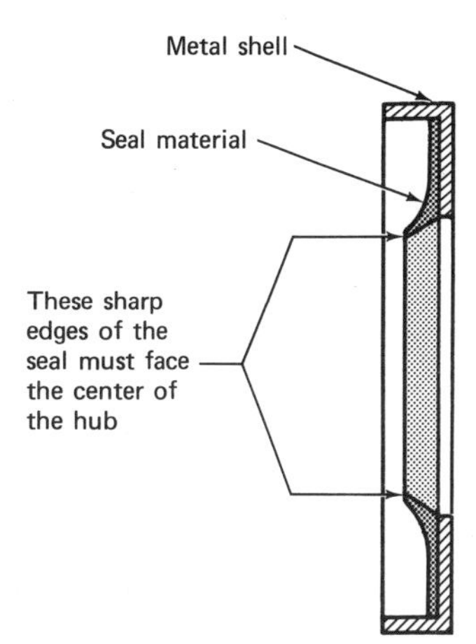

Figure 4.27 A cross section view of a typical front hub grease retainer.

Note. Use care to center the hub on the spindle. Othewise, the threads on the spindle could damage that new retainer you just installed.

18 Install the outer bearing on the spindle, sliding it into its cup.

19 Install the washer.

20 Install the nut.

21 Adjust the bearings and install the cotter pin according to the procedure and specifications of the manufacturer.

22 Install the grease cap.

23 Adjust the brakes if it was necessary to loosen the adjustment during disassembly. (Refer to the section on adjusting brakes in Chapter 5 for the correct procedure.)

24 Install the wheel and run the lug nuts up snug.

25 Repeat steps 4 to 24 on the remaining wheel.

26 Raise the car with a jack, remove the car stands, and lower the car to the floor. Tighten the lug nuts to the proper torque specification and in the sequence given by the manufacturer.

27 Install the wheel covers.

Job 4f

REPACK WHEEL BEARINGS

SATISFACTORY PERFORMANCE

A satisfactory performance on this job requires that you do the following:

1 Repack the front wheel bearings in one wheel of a designated car.
2 Following the steps in the "Performance Outline," complete the job within 60 minutes.
3 Fill in the blanks under "Information."

PERFORMANCE OUTLINE

1 Raise and support the car.
2 Remove the wheel. Remove the hub assembly.
3 Remove the retainer and the inner bearing.
4 Clean and inspect all the parts.
5 Repack the bearings.
6 Install the inner bearing and the retainer.
7 Install the hub assembly.
8 Install the outer bearing, washer, and nut.
9 Adjust the bearings to the manufacturer's specification.
10 Install the cotter pin and grease cap.
11 Install the wheel.
12 Lower the car to the floor.
13 Torque the lug nuts.

INFORMATION

Vehicle identification ______________________________

Reference used ____________________ Page(s) ____________________

Wheel lug nut torque ______________________________

Replacing Bearing Cups Since the rollers of a wheel bearing roll inside their cup, any wear on one part affects the other. If a worn bearing is replaced without replacing its cup, the new bearing will have a short life. Bearing cups, or races, are fitted very tightly in the hub. This prevents them from turning with the bearing. If you find a cup that is loose in its hub, the hub must be replaced. This is because the cup is made of very hard steel and the hub is made of cast iron. If the cup turns in the hub, the relatively soft cast iron will wear so that it will not provide a tight fit for a new cup.

Some manufacturers recommend the use of special pullers to remove bearing cups. These pullers have hook-like fingers, or jaws, that grab the inner edge of the cup. The jaws are expanded until they have a firm grip on the cup. Then the tool and cup are pulled out together (Figure 4.28).

Lacking special pullers, you can remove a bearing cup by carefully driving it out with a soft punch. The inner edge of the cup is raised a little above

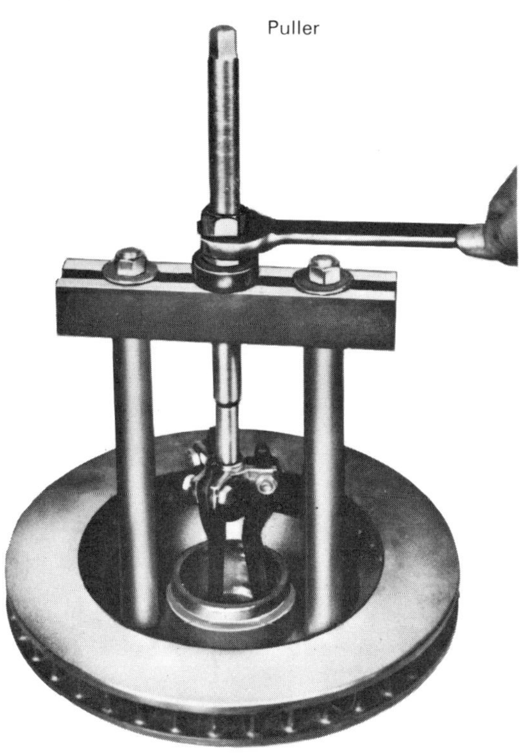

Figure 4.28 Removing an inner-bearing cup with a puller (Ford Motor Company).

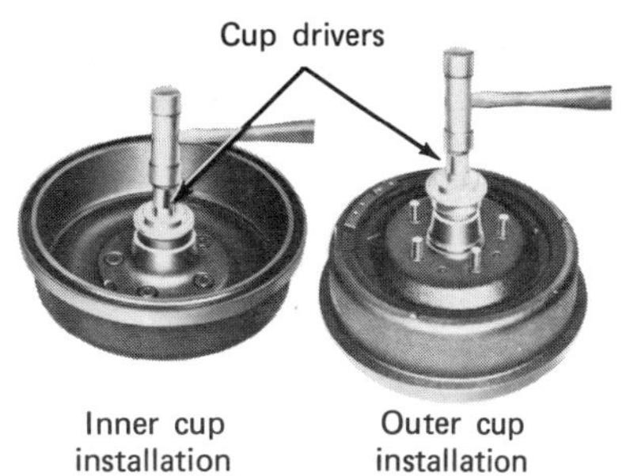

Figure 4.29 Installing front wheel-bearing cups with cup drivers (Ford Motor Company).

the inner surface of the hub. Some hubs have notches that expose portions of the edge of the cup. A brass punch with a flat end can be used to drive the cup out. The use of a hardened steel punch here could damage the hub. The edge of the punch is placed against the inner edge of the cup and struck with a hammer. The punch is then shifted to the other side of the cup and hit again. This eases the cup out of the hub. Do not try to drive the cup out too fast. You must move it slowly and keep it as straight as possible. By shifting the punch back and forth, you will not cock the cup in its bore. If the cup is cocked, it may damage the hub.

After the cup has been removed, the hub should be carefully cleaned. Special cup drivers are available to drive the new cup in place (Figure 4.29). The use of cup drivers is recommended. They make the job easier, and they guard the cup against damage.

Since all shops do not have cup drivers, you can use a brass punch to install a new cup too. When you install a new cup, be sure you have it facing the correct way. The thickest edge of the cup always goes into the hub first. Since the outer edge of the cup is much thinner than the inner edge, the punch is sometimes difficult to keep from slipping. If you keep the corners of your punch ground or filed square, the punch will be less likely to slip.

Another method of installing a new cup is to use the old cup as a driver. The old cup is held, thick side down, against the new cup. The new cup must be driven all the way down into its bore. You will know when it hits bottom by the different sound you will hear when your hammer hits the punch or old cup. If you use the old cup as a driver, you may have to drive it out after the new cup is all the way in. To do this, turn the drum around and drive the old cup out again. It will come out easily if you remembered to place the thick side down.

Job 4g

REPLACE A WHEEL BEARING CUP

SATISFACTORY PERFORMANCE

A satisfactory performance on this job requires that you do the following:

1 Replace a designated wheel-bearing cup in a front hub of the car assigned.
2 Following the steps in the "Performance Outline," complete the job within 60 minutes. The cup should be bottomed in its bore, and the cup and hub should exhibit no damage.
3 Fill in the blanks under "Information."

PERFORMANCE OUTLINE

1 Raise and support the car.
2 Remove the wheel and the hub and drum assembly.
3 Remove the retainer and the inner bearing.
4 Remove the assigned bearing cup.
5 Clean and inspect all the parts.
6 Install the replacement bearing cup.
7 Present hub for inspection.
8 Repack the bearings.
9 Install the inner bearing and the retainer.
10 Install the hub and drum assembly.
11 Install the outer bearing, washer, and nut.
12 Adjust the bearings.
13 Install the cotter pin and grease cap.
14 Install the wheel.
15 Lower the car to the floor.
16 Torque the lug nuts.

INFORMATION

Vehicle identification ______________________

Location of cup replaced ______________________

Reference used ______________ Page(s) ______________

SUMMARY

In this chapter you have learned some of the basic skills of a brake mechanic. You can now raise and support a car safely. You know the proper way to remove and install a wheel. You are familiar with all the parts in a front hub, and you can perform all the services required in maintaining and replacing front wheel bearings.

SELF-TEST

Each incomplete statement or question in this test is followed by four suggested completions or answers. In each case select the *one* that best completes the statement.

1 When supporting a car with car stands, the stands should be placed under the
   I. frame
   II. suspension system
   A. I only
   B. II only
   C. Either I or II
   D. Neither I nor II

2 After the wheels have been installed, the lug nuts should be tightened
   I. in the proper sequence
   II. to the proper torque specifications
   A. I only
   B. II only
   C. Either I or II
   D. Both I and II

3 Tapered roller bearings are used in most front hubs because they can handle
   I. radial loads
   II. thrust loads
   A. I only
   B. II only
   C. Both I and II
   D. Neither I nor II

4 Two mechanics are discussing the replacement of a wheel bearing cup.
   Mechanic A says the cup is threaded into the hub.
   Mechanic B says it is held in with a retainer.
   Who is right?
   A. A only
   B. B only
   C. Both A and B
   D. Neither A nor B

5 Wheel lug nuts should be tightened with
   A. a lug wrench
   B. an impact wrench
   C. a torque wrench
   D. a ratchet wrench

6 Wheel bearing cups can be removed with a
   I. puller
   II. flat punch
   A. I only
   B. II only
   C. Either I or II
   D. Neither I nor II

7 Anytime wheel bearings are repacked, you should replace the
   I. bearing cups
   II. grease retainers
   A. I only
   B. II only
   C. Both I and II
   D. Neither I nor II

8 You are attempting to repack wheel bearings. You have removed the outer wheel bearing but the drum will not come off the spindle. What should you do?
   I. Pull the drum off with a puller
   II. Loosen up the brake adjustment
   A. I only
   B. II only
   C. Either I or II
   D. Neither I nor II

9 Two mechanics are discussing wheel bearing service.
   Mechanic A says grease retainers should be driven in until they are flush with the hub.
   Mechanic B says bearing cups should be driven in until they bottom in the hub.
   Who is right?
   A. A only
   B. B only
   C. Both A and B
   D. Neither A nor B

10 While repacking wheel bearings, you discover that the outer wheel bearing is chipped. You should
   I. replace the outer wheel bearing
   II. replace the outer wheel bearing cup
   A. I only
   B. II only
   C. Either I or II
   D. Both I and II

Chapter 5 Brake Adjustment and Shoe Replacement

In this chapter you will work with parts of the mechanical system of drum brakes. Although all car manufacturers use self-energizing duo-servo brakes, some of the parts that make up those brakes are designed differently. A brake mechanic must understand these differences and the techniques necessary to service them.

When you have studied this chapter and have performed the tasks presented in it, you will be able to:

1
Adjust brakes.
2
Replace front brake shoes.
3
Replace rear brake shoes.

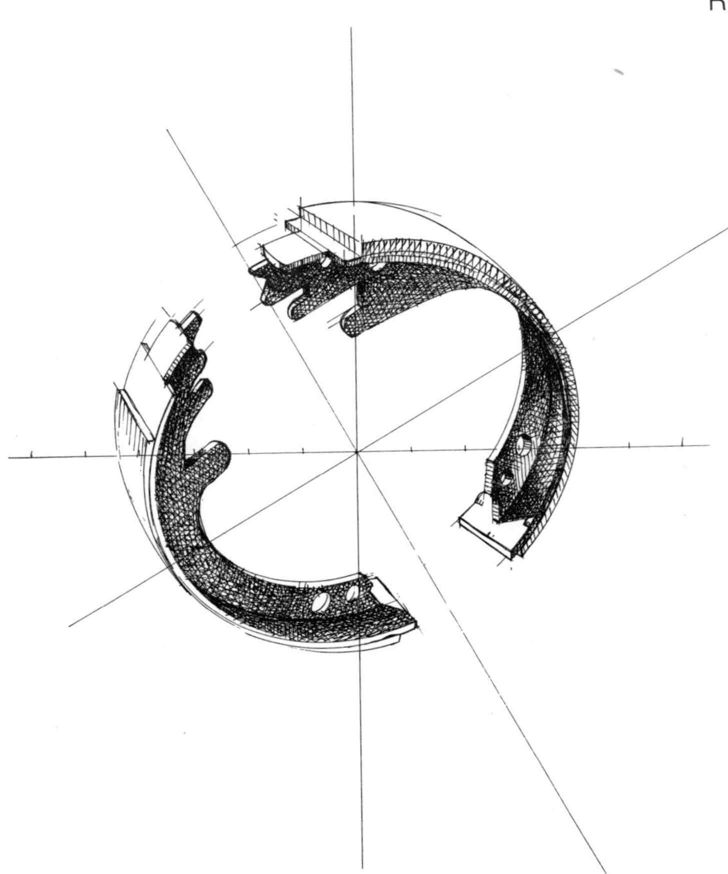

BRAKE ADJUSTMENT A well-adjusted brake system allows a driver to stop a car without pushing the brake pedal too far toward the floor. This brake condition is called *pedal reserve,* but most mechanics refer to it as "a high pedal." A high pedal is good to have for two reasons.

1 It allows for the possibility of brake fade or a system malfuncton.
2 It provides a fast braking action, cutting down the distance a car travels after the driver reacts to the need to stop.

To have a high pedal, a car must have brake linings that are very close to the drums when the brakes are released. Then, when the brake pedal is pressed down, the linings have only a short distance to travel before they contact the drums. As you know from Chapter 3, the brake shoes are held in position by the anchor pin and the star wheel adjuster. Since the retracting springs always pull the shoes back to contact the anchor pin, any wear on the linings increases the distance between the linings and the drums. However, as the linings wear, the star wheel adjuster pushes the shoes farther apart at the bottom of the drum assembly. This brings them closer to the drums and makes up for the lining that is worn away.

Star Wheel Adjusters Star wheel adjusters get their name from the star-like, or gear-shaped, appearance of the head of the adjusting screw. An example of a star wheel adjuster is pictured in Figure 5.1.

Most cars are fitted with self-adjusting brakes. When the self-adjusters are working properly, they advance the star wheel adjuster as the brake lining wears away. However, the use of self-adjusters has not eliminated the need for adjusting brakes by hand.

Figure 5.1 A typical star wheel adjuster.

Figure 5.2 A typical access slot in a backing plate (General Motors).

Access Slots Because brakes must sometimes be adjusted by hand, all manufacturers have provided a means by which you can gain access to the star wheel and turn it. *Access slots* are provided either in the backing plate or in the drum, as shown in Figure 5.2.

Access slots are covered by metal or rubber plugs. These plugs are sometimes called *dust covers.* They prevent dirt and water from entering the brake assembly (Figure 5.3). They should always be re-inserted after a brake adjustment.

Some manufacturers do not completely punch out the access slots in the drum or backing plate. They provide *lanced,* or semi-punched, slots. Such slots still have a piece of metal in them. They must be opened by punching out the piece of metal with a hammer and punch. Whenever you open a lanced slot, you must remove the drum and discard the piece of metal you have punched out of the slot. If the piece of metal were left in the brake assembly, it could get caught between the lining and the drum. This would prevent the brake from working properly, and it would damage the drum.

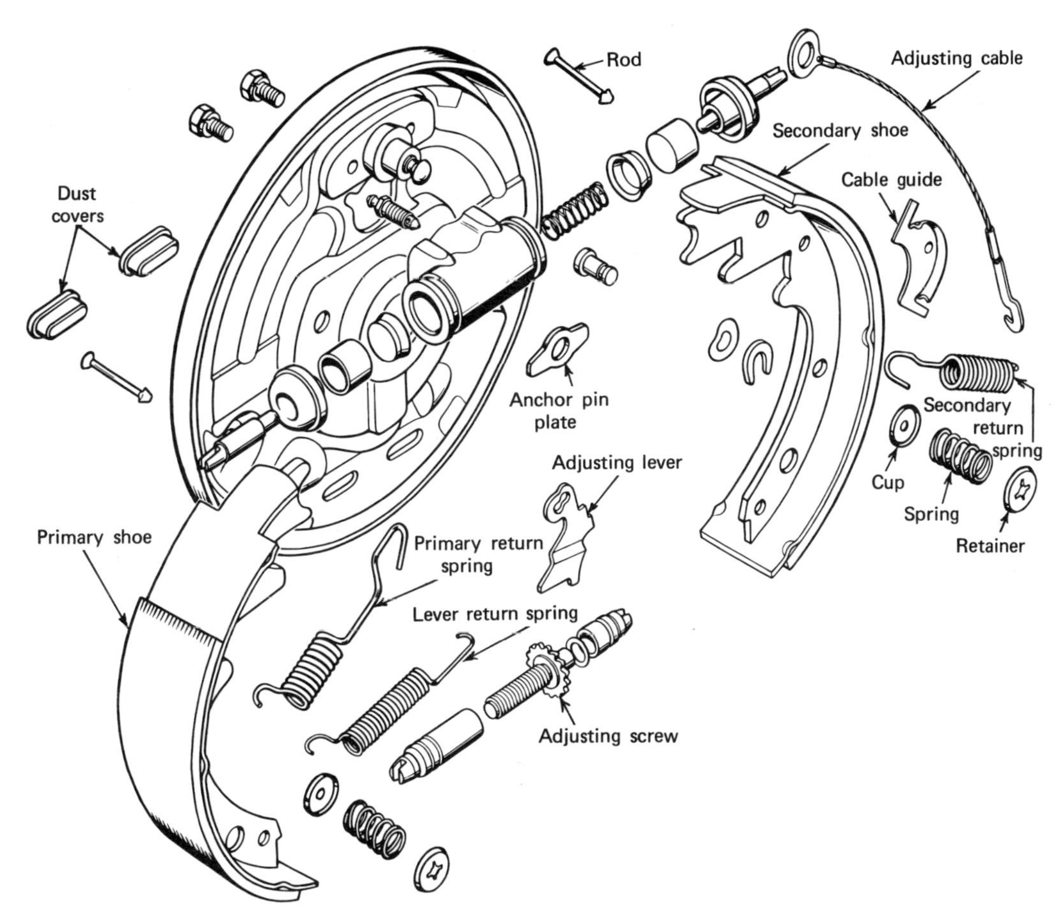

Figure 5.3 An exploded view of a brake assembly. Note the dust covers used to seal the access slots in the backing plate (Chrysler Corporation).

If the access slot is in the backing plate, a rubber or metal plug may be used to cover it. If the slot is in the drum, only a metal plug should be used.

Turning the Star Wheel If the access slot is in the drum, a screwdriver can often be used to turn the star wheel. If the slot is in the backing plate, a special tool called a *brake spoon* probably has to be used. A brake spoon is a curved metal lever. On many cars, parts of the suspension system obstruct the access slot. On those cars you have to use a brake spoon that is curved so that you can work around the obstruction. You will need several different types of spoons in your tool kit. One of the most widely used brake spoons is shown in Figure 5.4.

When you try to adjust brakes on a car equipped with self-adjusters, you will find that the star wheel can be moved in only one direction. The lever in the self-adjusting mechanism acts as a ratchet, and it allows you to tighten the adjustment but not to loosen it. Any attempt to loosen the adjustment will cause the lever to lock the star wheel. To loosen the adjustment, you must hold the lever away from the star wheel.

If the access slot is in the drum, a small hook can be inserted through the slot and used to

Figure 5.4 A brake spoon (Chrysler Corporation).

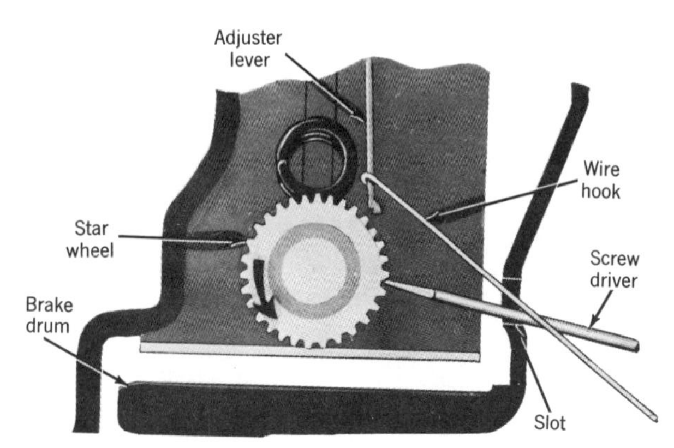

Figure 5.5 Loosening a brake adjustment through an access slot in the drum (courtesy Chevrolet Service Manual, Chevrolet Motor Division).

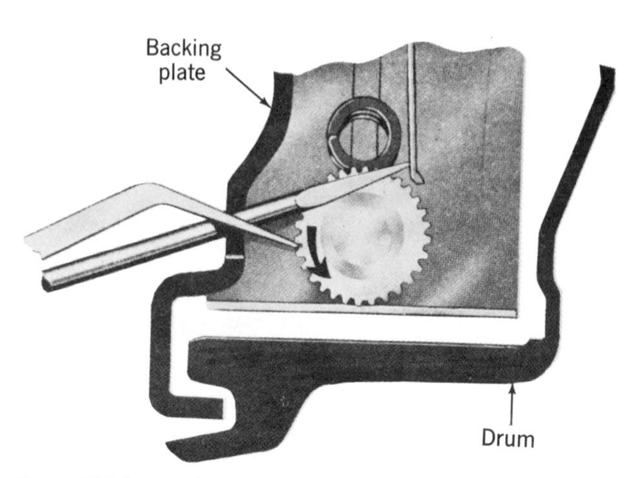

Figure 5.6 Loosening a brake adjustment through an access slot in the backing plate (courtesy Chevrolet Service Manual, Chevrolet Motor Division).

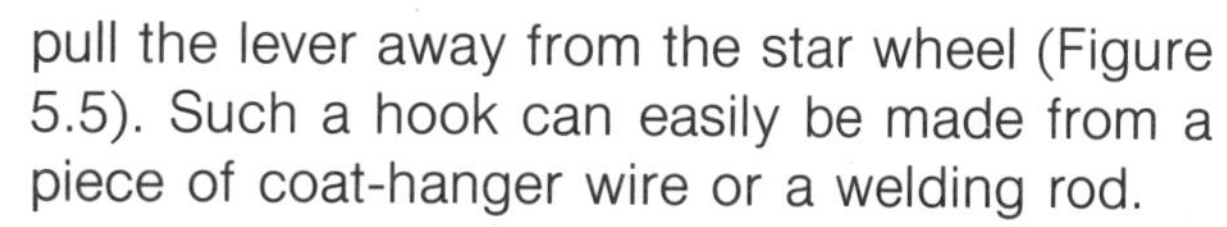

pull the lever away from the star wheel (Figure 5.5). Such a hook can easily be made from a piece of coat-hanger wire or a welding rod.

If the access slot is in the backing plate, the lever must be pushed away from the star wheel (Figure 5.6). A screwdriver can be used, but a piece of wire will give you more room for your brake spoon.

Adjusting brakes to obtain a high pedal is not the only reason for releasing the lever from the star wheel. Often you will try to remove a drum only to find that it will not come off, even though it appears to be loose. When this happens, it is usually because the drum has been badly scored or worn and the shoes have been adjusted outward so far that the edge of the drum will not slide over the lining. When this condition is found, you must *back off,* or loosen, the adjuster. This procedure will increase the clearance between the lining and the drum. Never use force to pull a drum over the lining.

ADJUSTING BRAKES

Below is a typical procedure for adjusting the service brakes on a car equipped with self-adjusters. You should consult the manufacturer's service manual for the specific procedure.

1 Raise the car with a jack, and support the car with car stands.

2 Remove the dust covers, or plugs, from the access slots in the backing plates or drums.

3 Insert a thin screwdriver or a stiff piece of wire into the access slot, and hold the adjuster lever away from the star wheel. By using a brake spoon, tighten the adjustment until you feel a heavy drag when you try to turn the wheel.

4 Continue to hold the adjuster lever away from the star wheel. Loosen the adjustment about 20 or 30 notches until the wheel spins freely.

5 Repeat this procedure at all wheels.

6 Install all dust covers, or plugs, in the backing plates or drums.

7 Lower the car to the floor.

8 Make the final adjustment by making numerous forward and reverse stops. Apply the brakes firmly until you obtain a high pedal.

Manual Adjusters. Some cars do not have self-adjusting brakes. These cars may include police cars, taxicabs, and cars with heavy-duty suspension and brake systems. Brakes on these cars are adjusted in the same way as those with self-adjusters. However, manual adjusters do not have a lever to prevent the star wheel from turning backward. Therefore, you do not need a wire to release the adjuster.

As with self-adjusting brakes, you should tighten the star wheel adjusting screw until you feel a heavy drag when you try to turn the wheel. Then you should back off on the adjustment until the wheel turns freely. On cars with manual adjustment you may have to back off from 8 to 12 notches. You should check the manufacturer's manual for the recommended procedure.

Job 5a

ADJUST BRAKES

SATISFACTORY PERFORMANCE

A satisfactory performance on this job requires that you do the following:

1 Adjust the service brakes on the car assigned.
2 Following the steps in the "Performance Outline" and the manufacturer's procedure and specifications, complete the job within 150 percent of the manufacturer's suggested time.
3 Fill in the blanks under "Information."

PERFORMANCE OUTLINE

1 Raise and support the car.
2 Adjust the brakes.
3 Lower the car to the floor.
4 Check the operation of the brakes, making final adjustments if necessary.

INFORMATION

Vehicle identification ______________________________

Reference used ____________________ Page(s) ____________________

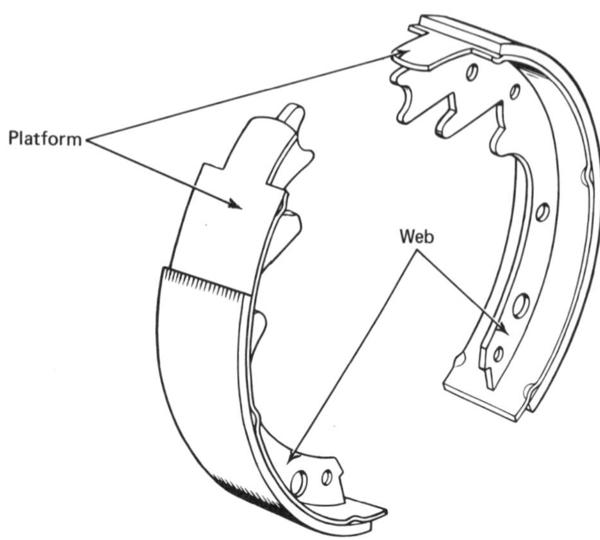

Figure 5.7 A pair of lined primary and secondary shoes showing the platform and web (Chrysler Corporation).

BRAKE SHOE REPLACEMENT—FRONT WHEELS Brake shoe replacement is one of the most commonly performed jobs in an auto shop. Yet, it is one of the most important. A less-than-perfect job may result in a serious accident.

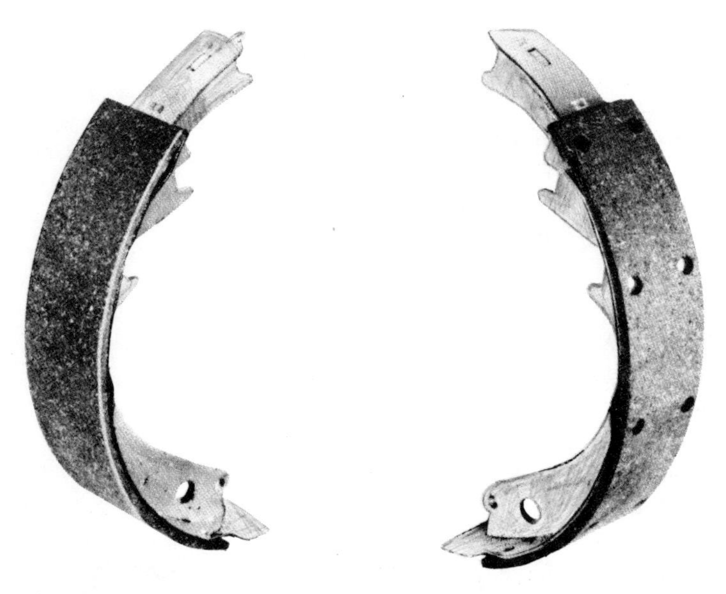

Figure 5.8 Lined brake shoes (Raybestos Division, Raybestos Manhattan, Inc.)

Whenever you work on a brake system, you should inspect each part three times:

1 When you remove it.
2 When you clean it.
3 When you install it.

The discussion that follows will help you to discover what to look for as you examine each part.

Brake Shoes and Lining The most widely used brake shoe is made of two pieces of stamped metal arc-welded together. These pieces are called the platform and the web (Figure 5.7).

The web of the shoe is made of heavy-gauge metal because it must transmit the output force of the wheel cylinder to the platform. The web is shaped and drilled so that the various retracting springs, hold-down springs, and self-adjusting mechanisms can be anchored to it. The platform is formed to the desired curve, centered on the web, and welded to it. The platform provides the surface to which the lining is attached.

There are two methods of holding the lining to the shoe. One method is to use rivets. The other is to use a *bonding agent,* or cement. Linings

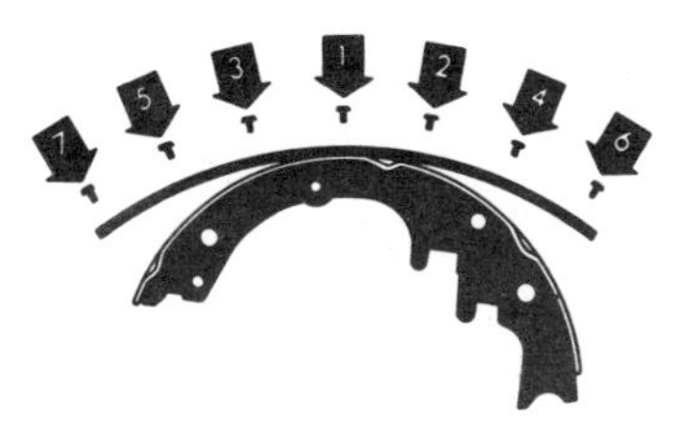

Figure 5.9 Sequence for installing rivets.

held to the shoe by each method are shown in Figure 5.8.

Both methods of holding the lining to the shoe have advantages and disadvantages. Riveted lining is not as rigid as bonded lining. Thus, it is less prone to squeal in use. On the other hand, if it is allowed to wear down to the rivets, the rivets will damage the drum. Since bonded lining has no rivets, it usually has a longer life. But it is more apt to be noisy.

Riveted lining should be replaced if it is worn to less than $\frac{1}{32}$ of an inch (about 1 millimeter) over the heads of the rivets. Bonded lining should be replaced when it is worn to less than $\frac{1}{16}$ of an inch (about 2 millimeters) in thickness.

Worn linings are not the only ones that should be replaced. You should replace any lining that is contaminated by grease or brake fluid.

If the brake lining at one wheel is replaced, the lining at the opposite wheel should also be replaced, even though it appears to be in good condition. Replacing lining at only one wheel will almost always result in a pull to one side when the brakes are applied.

In most cases you will use new or "exchange" brake shoes when you replace worn brake lining. Special equipment is needed to bond new lining on brake shoes, and the work is best handled by a rebuilder of brake parts. It is possible for you to rivet new lining to brake shoes, but this job too is usually best left to the rebuilder.

If you must replace the lining on a riveted brake shoe, you should observe the following precautions:

1 Wear an air purifying respirator.
2 Drill out the old rivets. Punching or pressing them out could distort the shoes.
3 Make sure that the back of the new lining contacts the platform of the shoe smoothly and evenly throughout its entire length. Use a shoe clamp, if possible. If a shoe clamp is not available, start riveting the lining to the shoe at the center pair of holes. Work alternately outward to the ends of the lining, as shown in Figure 5.9. Be sure to draw the lining tight to the shoe before setting the rivets.

Note. Many cars of the same make, year, and model will be fitted with different size brakes. When ordering brake shoes or lining, always mention the shoe dimensions and describe the car on which they are to be installed. The necessary dimensions are (1) size, or diameter (diameter of the drum) and (2) width (width of the shoe).

Retracting Springs

To remove brake shoes, you must first remove the retracting springs. In Chapter 3 you learned that the retracting springs not only return the shoes to the anchor pin but push the pistons back into the wheel cylinders. The pistons are pushed outward by pressure carried through the hydraulic part of the brake system. The retracting springs, then, are part of both the mechanical and the hydraulic systems.

Most hydraulic systems used for drum brakes are designed to have *residual pressure*. That is, there is a slight amount of pressure in the system at all times, even when the brake pedal is released. When you remove brake shoes, this pressure can cause you some problems. When the retracting springs are removed, the residual pressure can sometimes push the pistons out of the wheel cylinder. If this happens, you will have to do extra, unnecessary work. Therefore, before you try to remove the retracting springs, you should make sure that the pistons are held in the cylinder. For this purpose most brake mechanics use a clamp of the type shown in Figure 5.10. You will probably want to add a couple of those clamps to your tool kit.

Figure 5.10 Wheel cylinder clamp (courtesy American Motors Corporation).

Retracting springs are very strong and stiff, but they are easily removed if you use the proper tool. The tool shown in Figure 5.11 is recommended because it can be used to install the springs as well as to remove them.

To remove the springs, place the cupped end of the tool over the anchor pin. Position the tool so that the little hook fits into the open section of the spring (Figure 5.12). Twist the tool, and the little hook will lift the spring clear of the anchor pin and release it.

When installing the springs, use the opposite end of the tool. Hook the spring into the web of the shoe, and place the free end of the spring over the tool. Place the groove or hole at the tip of the tool over the anchor pin (Figure 5.13). Then lever up the tool against the spring tension. The spring will slide up the tool and snap into place on the anchor pin.

Many cars have springs of different tensions on the primary and secondary shoes. In these cars the springs are identified by different colors of paint or by different sizes or shapes. It is extremely important that the springs be installed in their proper place. When you remove the springs, you should be careful to note any differences between them. Check the appropriate shop manual so you can be sure of the correct spring placement.

Whenever you remove a retracting spring, you should also check its condition. You may find that

Figure 5.11 Brake shoe retracting spring tool (courtesy American Motors Corporation).

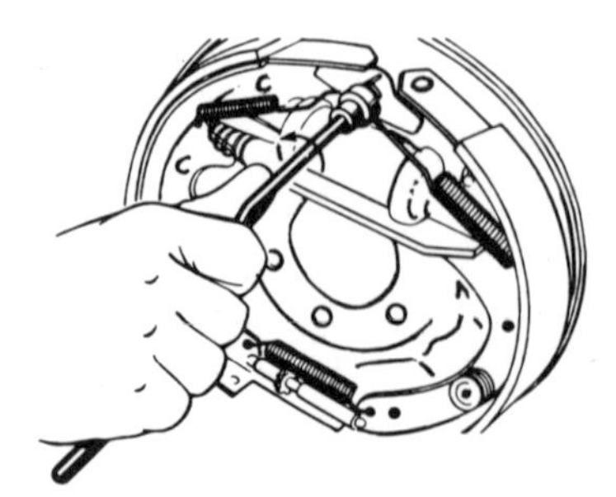

Figure 5.12 Removing a retracting spring (K-D Manufacturing Company).

the paint is burnt or discolored. This indicates that the spring has been overheated. The spring may also be stretched or distorted. The coils of a good spring will be tightly wound (Figure 5.14). One quick method of testing a spring is to drop it on the floor. A good spring will make a dull sound. A bad spring will ring. Any spring showing damage should be replaced. If you must replace a spring or a pair of springs, you should also replace the matching spring or springs on the opposite wheel, even though they may appear to be in good condition. Replacing springs on only one brake assembly will almost always result in a brake pull to one side.

Brake Shoe Hold-Down Assemblies

After the retracting springs have been removed, the hold-down springs should be removed next. These springs too should be inspected for damage and distortion. The hold-down assembly in most common use is called the spring-and-pin type. This type hold-down assembly consists of a pin (nail) with a flattened end, a compression coil spring, and a pair of washers. The washers are cup-shaped, so that they center themselves on the spring. The outer washer is slotted to fit the flattened end of the pin. The hold-down is locked when the outer washer is rotated on the pin a quarter turn. This type of hold-down is shown in Figure 5.15.

You can remove the spring-and-pin hold-down by grasping the outer washer with a pair of slip-joint pliers. Push the washer in, rotate it a quarter turn, and ease it off the pin. However, you will find that there is a much better way of removing these hold-downs. Several toolmakers manufacture

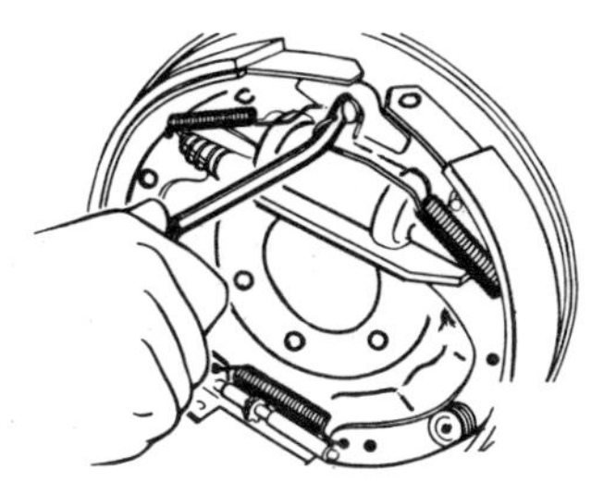

Figure 5.13 Installing a retracting spring (K-D Manufacturing Company).

hold-down spring tools that are designed to save you time and effort. The type shown in Figure 5.16 is used by most mechanics. It would be a worthwhile part of your tool kit.

Star Wheel Adjusters

Though they appear simple, star wheel adjusters require service. Many brake problems are caused by star wheel adjusters that are *frozen*—that is, that are made immovable by rust, corrosion, or improper lubrication.

Whenever you remove a star wheel adjuster, you should take it apart and clean it. The parts are pictured and labeled in Figure 5.17. If the threads on the adjusting screw appear rough or rusted, clean them with a wire brush or a wire wheel. Lubricate them with a thin coating of brake lubri-

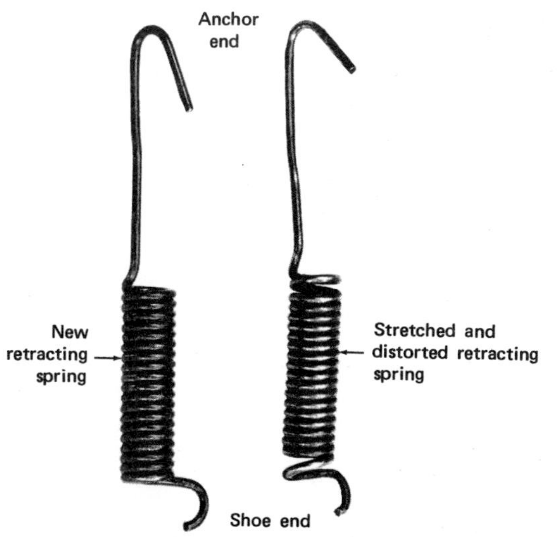

Figure 5.14 Comparison of new and distorted retracting springs (Raybestos Division, Raybestos Manhattan, Inc.)

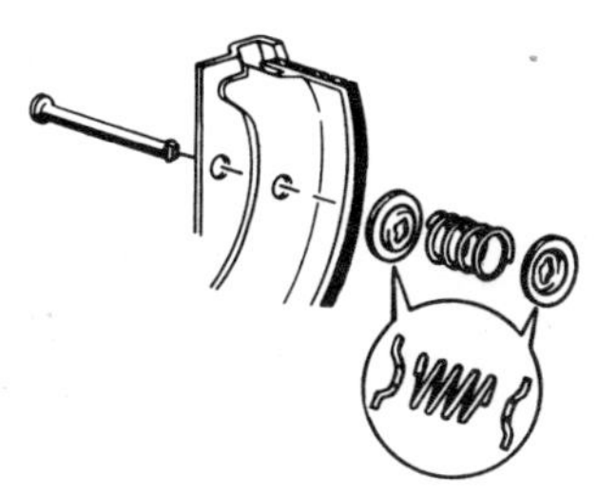

Figure 5.15 Spring-and-pin type of hold-down.

cant. Then turn the adjuster screw all the way into the pivot nut to be sure that it moves smoothly with no sticking or binding. You should then apply a little lube to the inside of the end guide, into which the unthreaded end of the adjuster screw fits. You should also lubricate the washer, which minimizes friction between the screw and the end guide when the screw is turned. Then you should spin the washer on the unthreaded end of the adjusting screw.

On most cars with self-adjusting brakes, the adjusters on the right side of the car have left-hand threads. The adjusters on the left side of the car have right-hand threads. The adjusters should never be interchanged from one side of the car to the other. Interchanging them causes the self-adjusting mechanisms to back off, which loosens the adjustment. This loosening increases the distance between the lining and the drum and results in a very low brake pedal. Sometimes an interchange is made accidentally when brake parts are cleaned, inspected, and lubricated. You can avoid an interchange by keeping all the parts for each wheel in a separate container.

Cable-Type Self-Adjusters Cable-type self-adjusters automatically compensate for any wear on the brake lining. They maintain a minimum clearance between the lining and the drum. The star wheel is rotated by a linkage arrangement that operates after a stop made when the car has been moving in reverse. If the distance between the lining and the drum is excessive, the linkage causes a lever to turn the star wheel one tooth. The parts of a cable-type self-adjuster are shown in Figure 5.18.

The self-adjuster parts are mounted on the secondary shoe. When the brakes are applied as

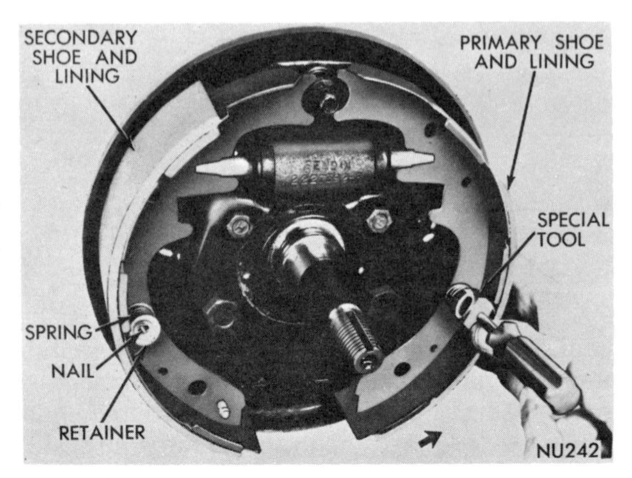

Figure 5.16 Hold-down spring tool and view showing its use (Chrysler Corporation).

the car is moving in reverse, the secondary shoe moves away from the anchor pin. This movement pulls back on the cable guide. Since the upper end of the cable is attached to the anchor pin, it cannot move. Therefore, the lower end of the cable is pulled up.

As the lining wears, the secondary shoe moves farther and farther out to contact the drum. When the shoe moves beyond a set distance, the cable is pulled up enough so that the adjuster lever rises above one of the teeth on the star wheel and engages it. When the brakes are released, the shoe returns to the anchor pin and relaxes its pull on the cable. The adjuster spring then pulls the adjuster lever downward, advancing the star wheel one tooth. If this adjustment restores the proper clearance between the lining and the shoe, the next stop in reverse will not raise the lever far enough to contact the next tooth. But when further wear occurs, the star wheel will be advanced one more tooth.

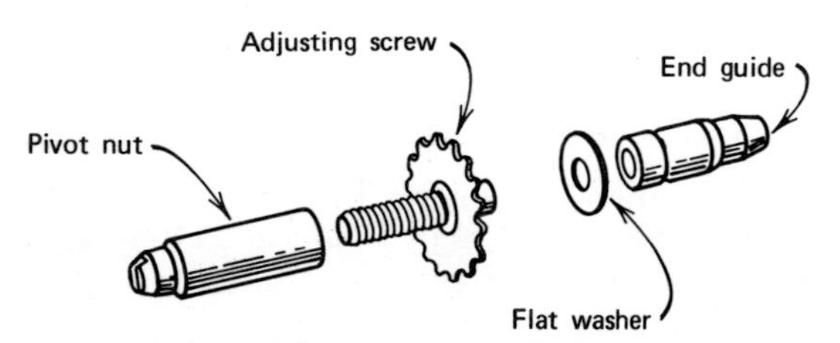

Figure 5.17 Star wheel adjuster assembly.

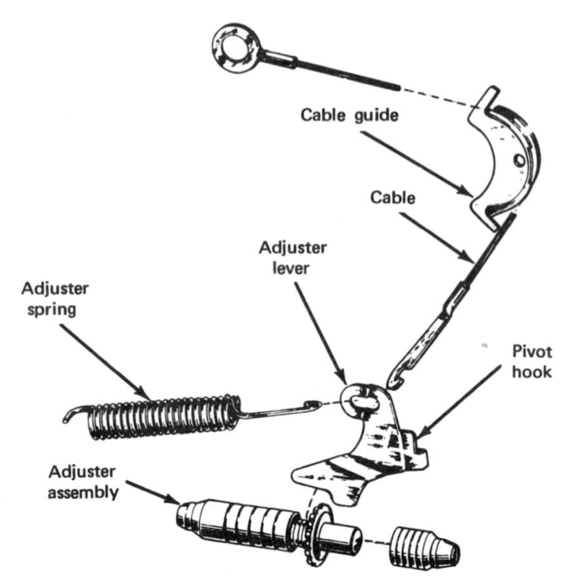

Figure 5.18 Cable-type self-adjuster parts (courtesy American Motors Corporation).

Figure 5.19 Removing retracting springs. Note the use of the wheel-cylinder clamp (Ford Motor Company).

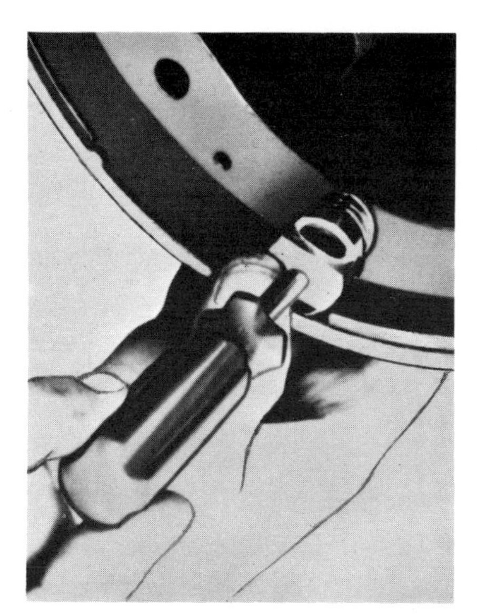

Figure 5.20 Removing a hold-down assembly.

REMOVAL AND INSTALLATION OF FRONT-WHEEL BRAKE SHOES ON A CAR FITTED WITH CABLE-TYPE ADJUSTERS

Here is the procedure for the removal and installation of front-wheel brake shoes on a car fitted with cable-type self-adjusters. You should consult the manufacturer's manuals for specific vehicles. Be sure to wear a respirator.

Removal

1 Raise the car with a jack, and support the car with car stands.

2 Remove a front wheel and brake drum.

3 Install a clamp over the ends of the wheel cylinder.

4 By using a retracting-spring tool, remove the secondary and primary retracting springs. Note any difference in spring size, shape, color, or tension so that the springs can be replaced in their proper position (Figure 5.19).

5 Remove the self-adjuster cable and cable guide. Remove the anchor pin plate if there is one.

6 Using a hold-down spring tool, remove the hold-down washers and springs from both shoes. Remove the hold-down pins (Figure 5.20).

7 Grasp the shoes, one in each hand. Spread them apart slightly to disengage them from the wheel-cylinder push rods, and remove them from the backing plate.

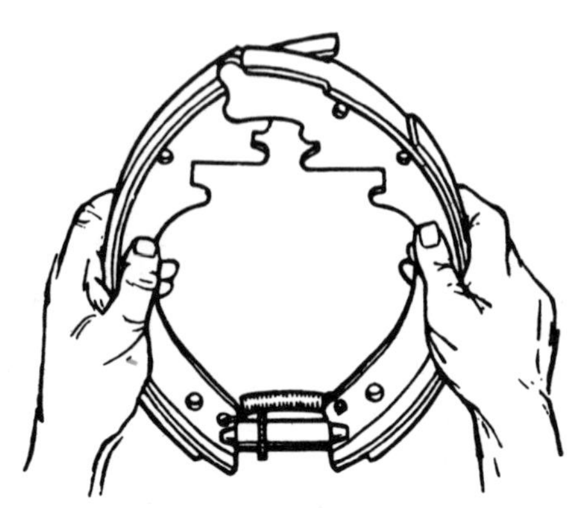

Figure 5.21 Twisting the shoes so that they will overlap (Bendix Automotive Aftermarket Operations).

8 Twist the shoes slightly, and allow them to come together. As they overlap, the spring will relax, and the adjuster, spring, and lever will fall loose (Figure 5.21).

9 Clean all parts, and lubricate the adjuster assembly. Clean the backing plate, and check the bosses for grooves or damage. Check to see if the anchor pin is secure.

10 Check the wheel cylinders for leakage. To do this, carefully peel back the boots at the ends of the cylinder. If fluid seeps or drips out, the cylinder must be replaced or overhauled. Sometimes the inside of the boots will appear wet, but fluid will not drip out. This is a normal condition. Check the wheel-cylinder piston movement by pushing carefully on the push rods. If the pistons appear to be sticky or frozen, overhaul or replace the cylinder. (Procedures for overhauling and replacing wheel cylinders are given in Chapter 8.)

Installation

1 Apply a thin coating of brake lubricant to the backing plate bosses and the anchor pin.

2 Position the shoes on the backing plate and install the hold-downs. Be sure the wheel-cylinder push rods are in place.

3 Install the anchor plate over the anchor pin. Place the eye of the cable over the anchor pin, and let the cable hang over the secondary shoe.

4 Hook the primary retracting spring into its hole in the primary shoe. Use the retracting-spring tool to stretch it up over the anchor pin where it will snap into place.

5 Position the cable guide in its hole in the secondary shoe. Hook the secondary retracting spring through the hole in the cable guide, and align it so that it holds the guide flat against the shoe web.

6 Use the retracting-spring tool to stretch the secondary spring up over the anchor pin and to snap it into place.

Figure 5.22 Using pliers to install an adjuster lever on the secondary shoe (courtesy American Motors Corporation).

7 Install the adjuster assembly, making sure the star wheel is near the secondary shoe.

8 Slip the cable into the groove on the cable guide, and attach the adjuster lever to the cable hook.

9 Hook the adjuster spring into its hole in the primary-shoe web. Attach the free end of the spring to the adjuster lever.

10 Grasp the adjuster lever with a pair of pliers. Then pull the lever over the secondary shoe until it hooks into its mounting hole in the web of the secondary shoe (Figure 5.22).

11 Check the installation of the shoes. Grasp the shoes near the star adjuster, and move them back and forth. Observe the fit of the shoes at the backing plate, anchor pin, and wheel-cylinder push rods. Check to see if the cable is in the groove in the guide and if the guide is flat against the web of the shoe. If necessary, remove and reinstall parts to obtain their proper fit.

12 Repeat all removal and installation steps on the opposite wheel. Remember—: any part you replace on one side of the car must be replaced on the opposite side if even braking is to be obtained.

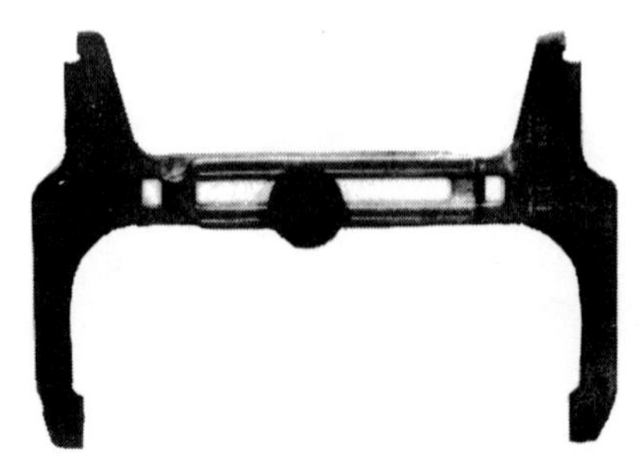

Figure 5.23 Typical drum-to-brake shoe clearance gauge (courtesy American Motors Corporation).

INITIAL ADJUSTMENT

Whenever brake shoes are replaced, they must be adjusted. Most manufacturers recommend the use of a clearance gauge similar to the type shown in Figure 5.23. This tool enables you to adjust the brakes before you install the drums. This can save you a lot of time, because then you can turn the star wheel adjusters with your fingers.

A clearance gauge is actually a pair of inside and outside calipers. When the inside fingers are adjusted to the diameter of the drum, this measurement is duplicated at the outside fingers. The gauge is made so that the proper clearance between the lining and the drum is automatically obtained.

Here are the steps you should take in order to use a clearance gauge properly:

1 Place the clearance gauge inside the brake drum, and expand the gauge to the inner diameter of the drum. Lock the gauge to retain this measurement (Figure 5.24).

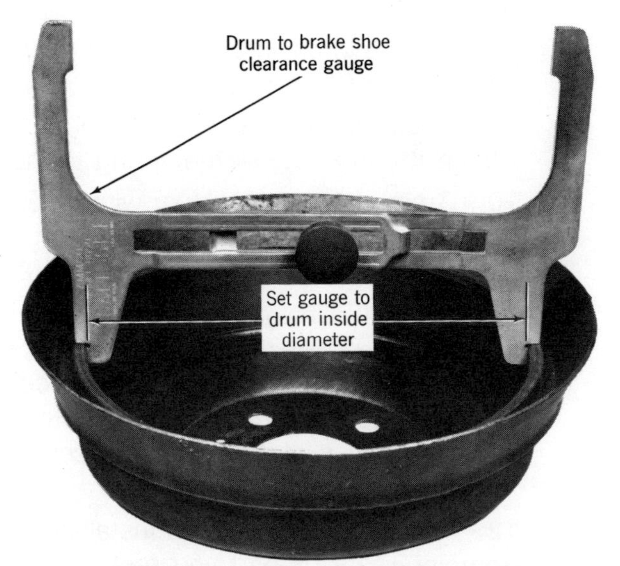

Figure 5.24 Clearance gauge in proper position on brake drum (courtesy American Motors Corporation).

Figure 5.25 Gauge in position for shoe adjustment (courtesy American Motors Corporation).

2 Place the gauge horizontally over the brake assembly, as shown in Figure 5.25. Turn the star wheel to expand the shoes until they make contact with the gauge.

3 Repeat this adjustment procedure at the remaining wheel.

4 Finish the brake job by repacking the wheel bearings and installing the drums.

Make the final adjustment when the wheels are installed and the car is on the floor. To do this, drive the car, making numerous forward and reverse stops and applying the brakes with a firm effort until you obtain a high pedal.

Job 5b

REPLACE FRONT BRAKE SHOES ON A CAR WITH CABLE-TYPE SELF-ADJUSTERS

SATISFACTORY PERFORMANCE

A satisfactory performance on this job requires that you do the following:

1 Replace the front brake shoes on the car assigned.
2 Following the steps in the "Performance Outline" and the manufacturer's procedure and specifications, complete the job within 200 percent of the manufacturer's suggested time.
3 Fill in the blanks under "Information."

PERFORMANCE OUTLINE

1 Raise and support the car, and remove the wheels and drums.
2 Remove the brake shoes and related parts.
3 Clean, inspect, and lubricate the parts.
4 Install the replacement shoes and related parts.
5 Make the initial adjustment.
6 Install the drums and adjust the bearings.
7 Install the wheels and lower the car to the floor.
8 Check the brake fluid level in the master cylinder, and add fluid if needed.
9 Make a final brake adjustment if necessary, and check the operation of the brakes.

INFORMATION

Vehicle identification ____________________

Reference used ____________ Page(s) ____________

Lever-Type Self-Adjusters Lever-type self-adjusters are also mounted on the secondary shoe (Figure 5.26). The lever assembly is attached to the secondary shoe by the hold-down assembly. The inner washer on the spring type hold-down has a tubular projection that passes through holes in the lever and in the shoe web. Thus, the tubular projection becomes a *fulcrum,* or pivot point, for the adjuster lever. The hold-down spring not only holds the shoe against the backing plate, but it holds the lever firmly against the shoe web.

The top of the lever is connected to the anchor pin by a wire link. The bottom of the lever forms a hardened ridge, or *pawl.* The pawl rests against the star wheel and engages its teeth. When the brakes are applied as the car moves in reverse, the secondary shoe moves away from the anchor pin (Figure 5.27). This movement causes the wire link to pull upward on the lever. As the lever pivots, the pawl at the opposite end pushes down on the star wheel. When the brake pedal is released, the secondary shoe returns to the anchor pin, and the lever return spring lifts

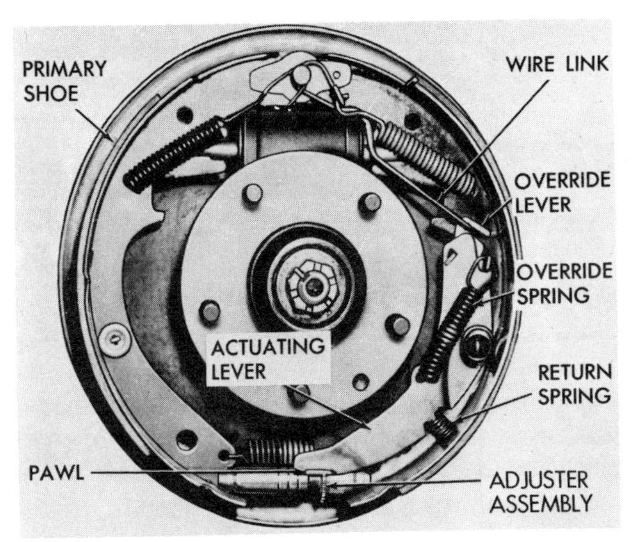

Figure 5.26 A left front brake assembly showing all the parts of a lever-type self-adjuster (courtesy Chevrolet Service Manual, Chevrolet Motor Division).

the lever. If lining wear has allowed the secondary shoe to move out beyond a predetermined distance, the lever will slip up over another tooth on the star wheel. This movement places the lever in a position to advance the star wheel again on the next stop from moving in reverse.

There is an important difference between the operation of the lever-type self-adjuster and the operation of the cable type. On the cable type, the adjuster spring advances the star wheel *after* the brakes have been released following a stop in reverse. On the lever type the lever advances the star wheel *during* the stop. Because of this difference, lever-type adjusters incorporate an override mechanism. This mechanism is a safety device that prevents damage to the linkage and other parts if, for some reason, the mechanism cannot function as it should. For example, if the star adjuster should jam, the adjuster lever would continue to try to turn it. This could

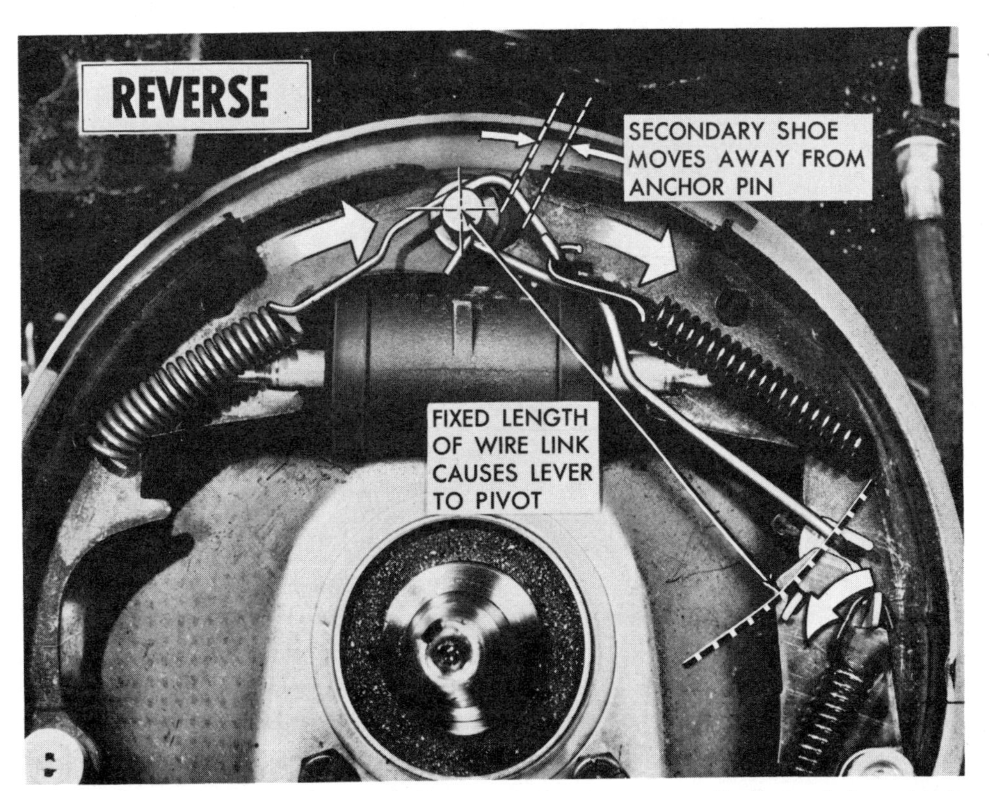

Figure 5.27 Action of a lever-type self-adjuster during a stop in reverse (courtesy Chevrolet Service Manual, Chevrolet Motor Division).

damage both the lever and the adjuster. But because of the override mechanism, this damage is prevented if the star wheel jams. The wire link merely pulls an override lever that stretches the override spring (Figure 5.28). Thus, the override spring absorbs the motion, and the adjuster, or actuating lever, and the star wheel do not move.

Figure 5.28 Operation of the override lever (courtesy Chevrolet Service Manual, Chevrolet Motor Division).

REMOVAL AND INSTALLATION OF FRONT-WHEEL BRAKE SHOES ON A CAR FITTED WITH LEVER-TYPE ADJUSTERS

Replacing brake shoes on a car with lever-type adjusters requires a different procedure than replacing brake shoes on a car with cable-type adjusters. The following steps outline that procedure. Nevertheless be sure to consult the appropriate car manual.

Removal

1 Raise the car with a jack, and support the car with car stands.

2 Remove a front wheel and brake drum.

3 Install a clamp over the ends of the wheel cylinder.

4 Using a retracting-spring tool, remove the secondary and primary retracting springs. Note any difference in spring size, shape, color, or tension so that you can replace the springs in their proper positions.

5 Remove the wire link and the hold-down assemblies.

6 Remove the adjuster-lever assembly.

Note. **The actuating lever, override lever, and override spring are an assembly. They should not be disassembled.**

7 Grasp the shoes, one in each hand. Spread them apart slightly to disengage them from the wheel-cylinder push rods. Then remove them from the backing plate.

8 Twist the shoes slightly, and allow them to come together. As they overlap, the spring will relax, and the star adjuster will fall free.

9 Clean all parts, and lubricate the adjuster assembly. Clean the

backing plate, and check the bosses for grooves or other damage. Also check to see if the anchor pin is secure.

10 Check the wheel cylinders for leakage. (The procedure for this has already been described under "Removal and Installation of Front-wheel Brake Shoes on a Car Fitted with Cable-Type Adjusters.")

Installation 1 Apply a thin coating of brake lubricant to the backing-plate bosses and the anchor pin.

2 Lay the primary and secondary shoes on a clean surface in the position in which they will be installed on the backing plate. Install the adjuster spring, connecting both shoes. Raise one shoe slightly, and overlap the other shoe. Install the adjuster assembly, making sure that the star wheel is closer to the secondry shoe and that it lines up with the long end of the spring.

3 Grasp the shoes, one in each hand, and place them in position on the backing plate. Be sure the wheel-cylinder push rods are in place in the notches on the shoe webs.

4 Install the primary shoe hold-down assembly.

5 Install the adjuster lever assembly on the secondary shoe. Secure the lever and shoe with the hold-down assembly.

6 Install the anchor plate on the anchor pin.

7 Attach the wire link to the adjuster lever assembly, and swing it up over the anchor pin.

8 Install the adjuster lever return spring.

Figure 5.29 Checking the operation of a lever-type self-adjuster (General Motors).

9 Install the primary-shoe retracting spring.

10 Install the secondary-shoe retracting spring.

11 Check the installation of the shoes. To do so, grasp the shoes near the star adjuster, and move them back and forth. Observe the fit of the shoes at the backing plate, anchor pin, and wheel-cylinder push rods.

12 Check the operation of the self-adjuster by pushing down on the lever as shown in Figure 5.29. The pawl should rotate the star wheel easily, and the return spring should push the lever back up when you release it.

Note. **Repeat all removal and installation steps on the opposite wheel. Any part you replace on one side must be replaced on the opposite side if even braking is to be obtained.**

INITIAL ADJUSTMENT

Whenever brake shoes are replaced, they must be adjusted. Although you may know how to adjust brakes by using a clearance gauge, there is another method, which requires no special tools or gauges. The brake drum itself is used as a gauge. To use that method follow these steps:

1 Turn the star wheel adjuster so that the shoes are expanded slightly.

2 Slip the drum over the shoes, and feel for contact with the lining.

3 Repeat steps 1 and 2 until you feel the shoes dragging in the drum as you slide it in place.

4 Remove the drum, and back off on the star wheel one full turn.

5 Install the drum, and check to see that it spins freely.

6 Repeat this procedure on the opposite wheel.

7 Make the final adjustment when the wheels have been installed and the car is on the floor. Drive the car, making numerous forward and reverse stops and applying the brakes with a firm effort until you obtain a high pedal.

Job 5c

REPLACE FRONT BRAKE SHOES ON A CAR WITH LEVER-TYPE SELF-ADJUSTERS

SATISFACTORY PERFORMANCE

A satisfactory performance on this job requires that you do the following:

1 Replace the front brake shoes on the car assigned.
2 Following the steps in the "Performance Outline" and the manufacturer's procedure and specifications, complete the job within 200 percent of the manufacturer's suggested time.
3 Fill in the blanks under "Information."

PERFORMANCE OUTLINE

1 Raise and support the car, and remove the wheels and drums.
2 Remove the brake shoes and related parts.
3 Clean, inspect, and lubricate the parts.
4 Install the replacement shoes and related parts.
5 Make the initial adjustment.
6 Install the drums and adjust the bearings.
7 Install the wheels and lower the car to the floor.
8 Check the brake fluid level in the master cylinder, and add fluid if needed.
9 Make a final brake adjustment if necessary, and check the operation of the brakes.

INFORMATION

Vehicle identification ______________________________

Reference used ____________ Page(s) ____________

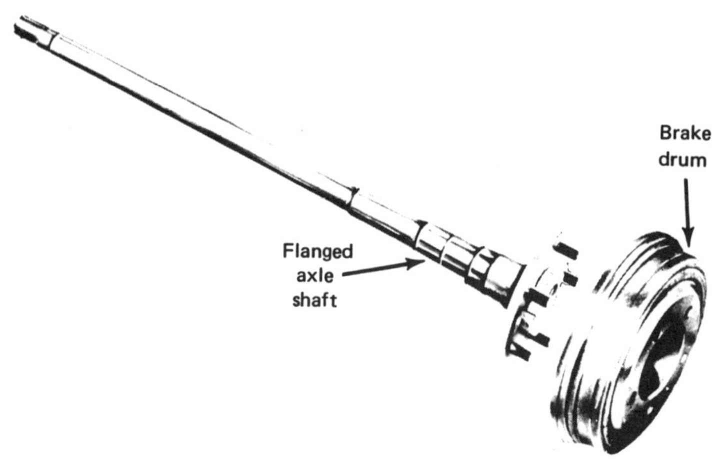

Figure 5.30 View of a floating drum and a flanged axle shaft (courtesy American Motors Corporation).

BRAKE SHOE REPLACEMENT: REAR WHEELS Rear brakes are similar to front brakes, but they have a few extra parts. Also, rear drums are mounted differently from front drums, and they require different removal procedures. The discussion below explains these differences.

Rear Brake Drums There are several types of rear brake drums, but the type you will encounter most often is called a *floating drum.* A floating drum has no hub. It is mounted on a *flange* at the end of the rear axle shaft (Figure 5.30).

The drum is held in place by the wheel. Since the drum has no hub, it has no bearings. The bearings for the rear wheels are on the axle shaft. They do not require repacking as do front wheel bearings.

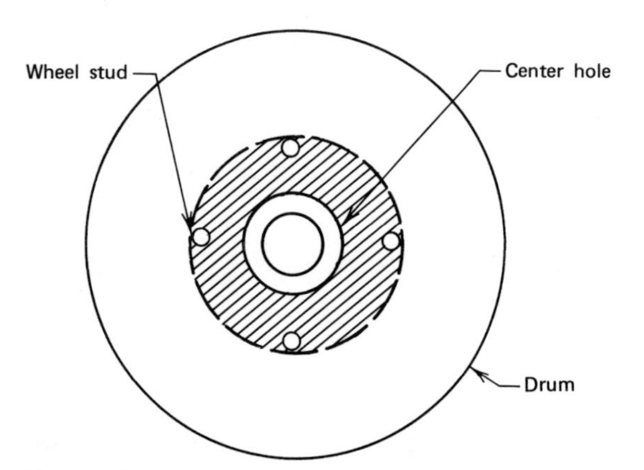

Figure 5.31 Where to apply heat to a frozen floating drum. The shaded area represents the proper place to apply heat.

Floating rear drums usually come off easily. Since they are held on by the rear wheels, you must first remove the wheels. With the parking brake released, you should be able to pull the drums off by hand.

When a rear axle assembly is built at the factory, two or three flat, pressed-steel nuts are usually installed on the rear wheel studs to hold the drums against the axle flanges. These pressed-steel nuts are needed only to keep the drums in place during storage and shipping. Once the wheels have been installed, the pressed-steel nuts serve no purpose. If you find these nuts in place after you remove a rear wheel, you must remove them before the drum will come off. They are not needed and may be discarded when removed.

You will sometimes find floating drums that will not come off the axle flange. As you learned when working with the front drums, you should never force a drum off. There are three reasons why a floating drum may not come off.

1 The parking brake may still be applied. If the drum will not come off, check the parking brake, and be sure it is released. If the brake is released, the drum should turn freely.

2 The brake shoes may have been expanded or adjusted into a badly worn or scored drum. If the

drum will not come off and the parking brake is not set, back off the brake adjustment until the drum slides off.

3 The drum may be frozen on the axle flange. The center hole in the drum fits snugly on a round protrusion on the axle flange. Sometimes rust forms between the drum and the flange, causing the drum to resist normal efforts to remove it. The easiest way of removing a frozen drum is to apply heat to the drum around the center hole. An acetylene torch works best. The heat expands the drum so that it releases its grip on the axle flange (Figure 5.31).

Be sure to clean all traces of rust from the center hole and from the axle flange before you put the drum back on. A light coating of brake lubricant at the center hole will keep this trouble from recurring.

REMOVAL AND INSTALLATION OF A FLOATING REAR DRUM

The following steps outline the procedure for the removal and installation of a typical floating rear drum. Consult the manufacturer's manual for the specific procedure for the car on which you are working.

Removal

1 Raise and support the car.

2 Remove the wheel.

3 Remove and discard any pressed-steel nuts still on the wheel studs.

4 Rotate the drum to find out whether or not the parking brake is released. If necessary, release the brake. Then pull the drum from the axle flange.

Note. **If the drum is loose on the flange but will not slide over the brake lining, back off the brake adjustment. If the drum is frozen to the axle flange, heat the drum around the center hole until it expands enough to be pulled free.**

Installation

1 Clean rust or burrs from the drum center hole. Apply a thin coating of brake lube to the edge of the hole.

2 Position the drum over the brake shoes, and push it into place on the flange. Be sure to align any marks or tabs so that the drum is installed in the same place on the flange.

3 Install the wheel.

4 Adjust the brake if necessary.

5 Lower the car to the floor, and torque the lug nuts.

Parking Brake Cables The rear brakes are part of the parking brake system. The cables that operate the parking brakes pass through holes in the backing plates to pull levers attached to the secondary brake shoes. Whenever the rear brake shoes are replaced, the parking brake cables should be inspected and lubricated.

Before removing rear brake shoes, you should disconnect the parking brake cables so that they can be pulled through their housings for inspection and lubrication. The easiest place to

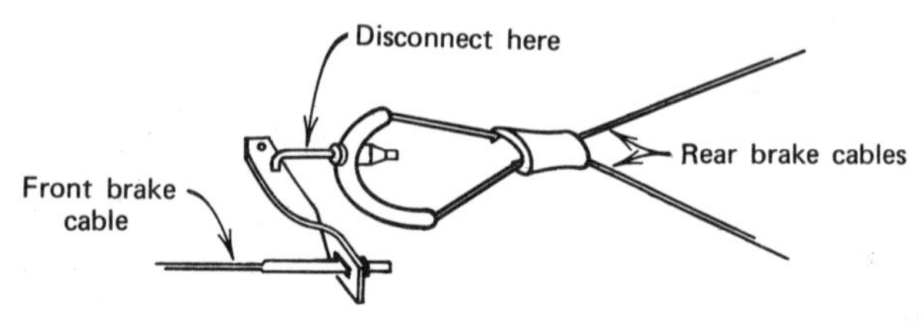

Figure 5.32 Typical parking brake cable equalizer.

disconnect those cables is at the *equalizer.* All cars use an equalizer of some type to equalize the pull of the cables at each wheel. It is usually located where the front cable meets the cables that run to the two rear wheels. Figure 5.32 illustrates a typical equalizer.

REMOVAL AND INSTALLATION OF REAR BRAKE SHOES ON A CAR FITTED WITH CABLE-TYPE SELF-ADJUSTERS

Below are the procedures for the removal and installation of brake shoes on a rear wheel fitted with a cable-type self-adjuster. Consult manufacturer's manuals for specific procedures.

Removal

1 Raise and support the car.

2 Disconnect the parking brake cable.

3 Remove the wheel and the brake drum.

4 Remove the retracting springs and the hold-downs.

5 Remove the self-adjuster cable and the anchor plate.

6 Grasp the shoes, one in each hand. Spread them until they are disengaged from the wheel-cylinder push rods and the parking brake link. Then pull the shoes away from the backing plate.

7 Twist the shoes slightly, the primary over the secondary, and allow them to come together. As they overlap, the adjuster spring will relax and the adjuster will fall loose. Disconnect the adjuster spring and lever.

8 Move the secondary shoe away from the backing plate. This will pull the parking brake cable through the backing plate. Release the cable from the lever.

9 Remove the parking brake lever from the secondary shoe.

Note. **If the lever is held with a C-clip of the type shown in Figure 5.33, carefully pry the clip out of its groove and save it for reuse.**

Figure 5.33 C-clip used to hold the parking brake lever to the secondary shoe.

If the lever is held with a "horseshoe" retainer of the type shown in Figure 5.34, remove it as follows:

Figure 5.34 A horseshoe retainer. This type of retainer is used to hold the parking brake lever to the secondary shoe.

a. Using a sharp, slim punch, indent the top of the horseshoe as shown in Figure 5.35.

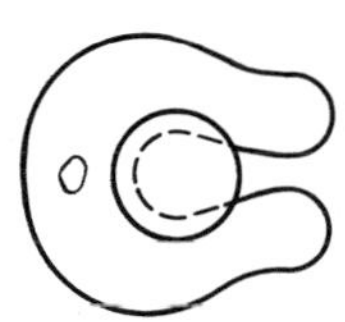

Figure 5.35 Horseshoe retainer in place on parking brake lever. Notice the indentation made by the punch.

b. Tip the punch and drive it across the lever pin as shown in Figure 5.36. This will spread the legs of the horseshoe and slide it out of its groove. Discard the horseshoe retainer.

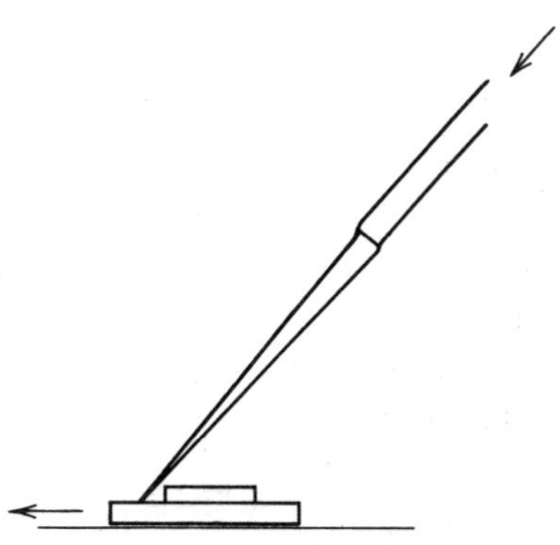

Figure 5.36 Horseshoe retainer being driven out of its groove.

10 Clean all parts. Inspect and lubricate the adjuster. Inspect all springs, and check the wheel cylinder for leaks or sticking pistons.

Installation

1 Apply a thin coating of brake lubricant to the backing plate bosses and the anchor pin.

2 Pull the parking brake cable through its housing, and lubricate the cable. Push the cable back into its housing. Push and pull the

cable repeatedly so the housing is well lubricated. Clean excess lubricant off the exposed cable.

3 Install the parking brake lever on the secondary shoe. Secure the lever with the C-clip or with a new horseshoe retainer. If you use a horseshoe retainer, be sure to squeeze the legs of the retainer so it will not slip off.

4 Connect the parking brake cable to the lever. Position the secondary shoe on the backing plate, and install the hold-down. Check to be sure that the wheel-cylinder push rod is properly positioned.

5 Position the primary shoe on the backing plate and install the hold-down. Check to be sure that the wheel-cylinder push rod is properly positioned.

6 Install the parking brake link and spring by spreading the shoes slightly. Check to see that both ends of the link are in their slots.

7 Install the anchor plate over the anchor pin. Place the eye of the adjuster cable over the anchor pin, and let the cable hang over the secondary shoe.

8 Install the primary retracting spring, cable guide, secondary retracting spring, adjuster, adjuster spring, and adjuster lever in that order.

9 Check the installation of the shoes, and adjust the brake lining to proper drum clearance.

10 Repeat all removal and installation steps on the opposite wheel. Remember, any part you replace on one side of the car must be replaced on the opposite side if even braking is to be obtained.

11 Connect the parking brake cable.

12 Check the pedal height and make final adjustments when the car is on the floor.

13 Check the parking brake adjustment, and correct it if necessary.

Job 5d

REPLACE REAR BRAKE SHOES ON A CAR WITH CABLE-TYPE SELF-ADJUSTERS

SATISFACTORY PERFORMANCE

A satisfactory performance on this job requires that you do the following:

1 Replace the rear brake shoes on the car assigned.
2 Following the steps in the "Performance Outline" and the manufacturer's procedure and specifications, complete the job within 200 percent of the manufacturer's suggested time.
3 Fill in the blanks under "Information."

PERFORMANCE OUTLINE

1 Raise and support the car, and remove the wheels and drums.
2 Remove the brake shoes and related parts.
3 Clean, inspect, and lubricate the parts.
4 Install the replacement shoes and related parts.
5 Make the initial adjustment.
6 Install the drums.
7 Install the wheels and lower the car to the floor.
8 Check the brake fluid level in the master cylinder, and add fluid if needed.
9 Make a final adjustment if necessary, and check the operation of the brakes.

INFORMATION

Vehicle identification ______________________________

Reference used ____________________ Page(s) ____________________

REMOVAL AND INSTALLATION OF REAR BRAKE SHOES ON A CAR FITTED WITH LEVER-TYPE SELF-ADJUSTERS

Here are the procedures for the removal and installation of rear brake shoes on a car fitted with lever-type self-adjusters. Even so, you should consult the manufacturer's manual for specific vehicles.

Removal

1 Raise and support the car.
2 Disconnect the parking brake cables.
3 Remove the wheel and the brake drum.
4 Remove the retracting springs, wire link, and hold-downs.
5 Remove the adjuster lever assembly.

6 Spread the shoes, and remove them from the backing plate.

7 Twist the shoes, and allow them to overlap so the spring will relax and the star adjuster will fall free.

8 Move the secondary shoe away from the backing plate, pulling the parking brake cable. Then release the cable from the lever.

9 Remove the parking brake lever from the secondary shoe.

10 Clean all parts. Inspect and lubricate the adjuster. Inspect all springs, and check the wheel cylinder for leaks or sticking pistons.

Installation

1 Apply a thin coating of brake lubricant to the backing plate bosses and the anchor pin.

2 Lubricate the parking brake cable.

3 Install the parking brake lever on the secondary shoe. Secure the lever with the C-clip or with a new horseshoe retainer. If you use a horseshoe retainer, be sure to squeeze the legs of the retainer so it will not slip off.

4 Lay the primary and secondary shoes on a clean surface in the position in which they will be installed on the backing plate. Install the adjuster spring, connecting both shoes. Raise one shoe slightly, and overlap the other shoe. Install the adjuster assembly, making sure that the star wheel is closer to the secondary shoe and that it lines up with the long end of the spring.

5 Hold both shoes together at the top, and connect the parking brake cable to the lever.

6 Spread the shoes at the top, and put them in position on the backing plate. Be sure the wheel-cylinder push rods are in place in the notches on the shoe webs.

7 Install the parking brake link and spring by spreading the shoes slightly. Check to see that both ends of the link are in their slots.

8 Install the primary shoe hold-down assembly.

9 Install the adjuster lever assembly on the secondary shoe, and secure the lever and shoe with the hold-down assembly.

10 Install the anchor plate on the anchor pin.

11 Attach the wire link to the adjuster lever assembly, and swing it up over the anchor pin.

12 Install the adjuster lever return spring.

13 Install the primary shoe retracting spring.

14 Install the secondary shoe retracting spring.

15 Check the installation of the shoes.

16 Check the operation of the self-adjuster by pushing down on the lever.

17 Repeat all removal and installation steps on the opposite wheel.

18 Adjust the brake lining-to-drum clearance.

19 Connect the parking brake cables.

20 Check pedal height, and make final adjustments when the car is on the floor.

21 Check the parking brake adjustment, and correct it if necessary.

Job 5e

REPLACE REAR BRAKE SHOES ON A CAR WITH LEVER-TYPE SELF-ADJUSTERS

SATISFACTORY PERFORMANCE

A satisfactory performance on this job requires that you do the following:

1 Replace the rear brake shoes on the car assigned.
2 Following the steps in the "Performance Outline" and the manufacturer's procedure and specifications, complete the job within 200 percent of the manufacturer's suggested time.
3 Fill in the blanks under "Information."

PERFORMANCE OUTLINE

1 Raise and support the car, and remove the wheels and drums.
2 Remove the brake shoes and related parts.
3 Clean, inspect, and lubricate the parts.
4 Install the replacement shoes and related parts.
5 Make the initial adjustment.
6 Install the drums.
7 Install the wheels and lower the car to the floor.
8 Check the brake fluid level in the master cylinder, and add fluid if needed.
9 Make a final adjustment if necessary, and check the operation of the brakes.

INFORMATION

Vehicle identification ______________________

Reference used ______________ Page(s) ______________

SUMMARY

By performing the jobs in this chapter, you have developed additional skills. You now can replace the brake shoes on cars with varying types of self-adjusting mechanisms. You are aware of several methods of adjusting brakes, and have learned some of the techniques required.

SELF-TEST

Each incomplete statement below is followed by four suggested completions. In each case, select the *one* that best completes the statement or answers the question.

1 The brake shoe retracting springs are part of the
  I. mechanical system
  II. hydraulic system
  A. I only
  B. II only
  C. Both I and II
  D. Neither I nor II

2 To compensate for lining wear, most drum brakes are fitted with
  A. anchor pins
  B. star wheel adjusters
  C. hold-down springs
  D. retracting springs

3 For a car to have a high brake pedal, its brake lining must be held close to the
  A. backing plate
  B. anchor pin
  C. wheel cylinder
  D. brake drum

4 On cars fitted with self-adjusting brakes, the star wheel is prevented from turning backward by the
  A. adjuster lever
  B. hold-down spring
  C. cable guide
  D. anchor pin

5 On cars fitted with self-adjusting brakes of the cable or lever type, the self-adjuster
  I. operates after a reverse stop
  II. parts are mounted on the secondary shoe
  A. I only
  B. II only
  C. Both I and II
  D. Neither I nor II

6 Access slots in a brake drum should be fitted with a
  I. metal dust cover
  II. rubber dust cover
  A. I only
  B. II only
  C. Both I and II
  D. Neither I nor II

7 When you are installing brake shoes, the sequence for installing the springs is as follows:
  A. primary spring, secondary spring, hold-down springs
  B. secondary spring, hold-down springs, primary spring
  C. hold-down springs, primary spring, secondary spring
  D. primary spring, hold-down springs, secondary spring

8 On brakes fitted with cable-type self-adjusters, the cable guide is held by the
  A. primary shoe retracting spring
  B. secondary shoe retracting spring
  C. primary shoe hold-down assembly
  D. secondary shoe hold-down assembly

9 On self-adjusting brakes, the star wheel adjuster usually
  I. is positioned so that the star wheel is near the secondary shoe
  II. has a left-hand thread when used on the right side of a car
  A. I only
  B. II only
  C. Both I and II
  D. Neither I nor II

10 On brakes fitted with lever-type self-adjusters, the adjuster lever is held by the
  A. primary shoe retracting spring
  B. secondary shoe retracting spring
  C. primary shoe hold-down assembly
  D. secondary shoe hold-down assembly

Chapter 6 Machining Brake Drums and Linings

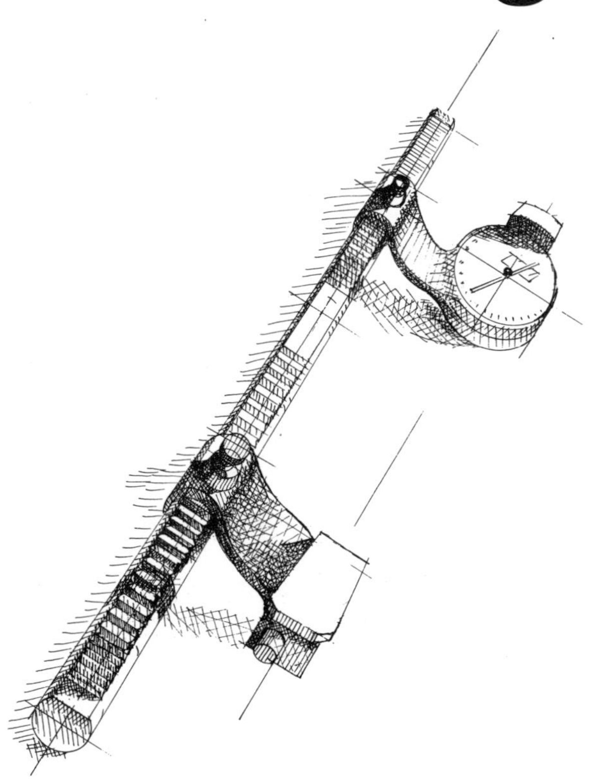

When a brake has been in use long enough for its lining to wear out, its drum usually has some defects too. Often those defects may be corrected by machining the drum. Even if the defects are minor, the drum should be machined when the brake lining is replaced. Machining provides better contact between the drum and the new lining. The better the contact, the better the braking.

Machining and normal wear increase the inside diameter of the drum. Also, they usually increase the diameter of each drum by a different amount. Therefore, if brake shoes are installed in the condition in which they are taken from their box, they almost always fit one or more drums poorly. For this reason, not only drums but new linings too must usually be machined.

After completing this chapter, you should be able to:

1
Check brake drums for wear and other damage.
2
Measure drum wear and judge how serviceable the drum is.
3
Machine drums on a lathe.
4
Arc grind shoes on an arc grinder.

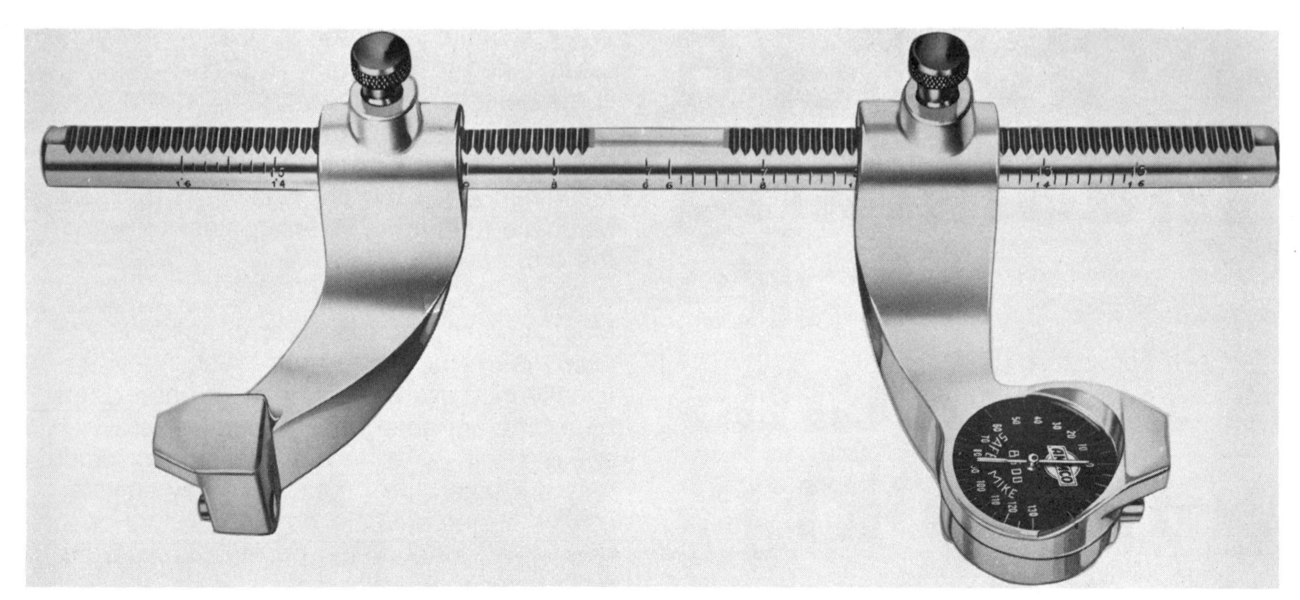

Figure 6.1 A typical brake drum micrometer (courtesy AMMCO TOOLS, Inc.).

BRAKE DRUM INSPECTION A cast-iron brake drum wears at a much slower rate than the lining that contacts it. Therefore, there may be a temptation to ignore the drum unless there is some obvious, serious defect. But to yield to this temptation shows that the importance of the drum is not understood and the result is an imperfect repair job.

If you know what to look for, you will find many common defects while inspecting brake drums. Many of those defects cannot be detected without the use of a *brake drum micrometer.*

The Brake Drum Micrometer A brake drum micrometer is an instrument that measures the diameter of a brake drum in thousandths of an inch. There are many types of drum micrometers, or "mikes," in use. But most are similar to the one shown in Figure 6.1.

The scale on a drum micrometer indicates the amount of *oversize* there is in a drum. Oversize is any increase in the original drum diameter. Brake drums are made in standard diameters. American auto manufacturers commonly use drums of even-inch or half-inch sizes. You will find drums that measure 9, 10, and 11 in. You will also find some that measure $9\frac{1}{2}$, $10\frac{1}{2}$, and other half-inch sizes.

Measuring a Drum with a Drum Micrometer
A drum micrometer is easy to use. Just follow the steps described here.

1 Set the mike to the original diameter of the drum you wish to measure. Since this measurement is in even inches or half inches, you can use a ruler to find it.

2 Insert the mike into the drum, and push it down until its depth stops hit the edge of the drum. The stops on most mikes position the mikes so that they measure the drum diameter at a point about one inch from the edge of the drum. Figure 6.2 shows a drum mike correctly positioned in a drum.

3 With the mike held in place, move one end back and forth in a slight arc until the dial shows the highest reading you are going to obtain. This reading, in thousandths of an inch, *plus the diameter to which you set the mike,* is the exact diameter of the drum.

Here is how to interpret your readings. Suppose that you want to measure a 10-in. drum. Set the

Figure 6.2 A drum micrometer in place in a brake drum. Note that the micrometer indicates that the drum is oversize (courtesy AMMCO TOOLS, Inc.).

Figure 6.3 Maximum diameter marking on a 10-inch drum (Chrysler Corporation).

mike at 10 in., place it in the drum, and rock it to obtain the highest reading. If this reading is 0.015 in., it means that the actual diameter of the drum is 10.015 in. As the drum measured 10.000 in. when it left the factory, the mike reading means that 0.015 in. has been worn away. The drum is 0.015 in. oversize.

Now suppose that you mike another 10-in. drum and that the mike reads 0.045 in. Though it is possible for a drum to wear that much and even more, such a high reading usually indicates that the drum has been machined. Another mechanic has probably machined the drum to 0.030 in. oversize to eliminate a defect. The additional 0.015 in. has probably worn away since that time. But knowing how the drum got to be oversize is less important than discovering exactly how much it is oversize. You cannot discover that without using a drum mike.

Drum Defects There are many drum defects that you must be aware of. The most common of them are listed in this section.

Oversize Any drum that has been in use will show some wear. When drums are manufactured, they are made thick enough to allow for a certain amount of wear and machining. There is, however, a limit to how much material can be worn and machined away. Most passenger car drums may be safely machined to 0.060 in. over their original diameter. Most manufacturers warn against machining a drum to more than 0.060 in. oversize. Some state motor vehicle agencies make it illegal.

Even so, some allowance must be made for wear after a drum has been machined to 0.060 in. oversize. Therefore, it is generally accepted that a drum can wear to 0.090 in. over its original diameter. Many manufacturers now mark their drums with the maximum allowable wear diameter (Figure 6.3).

Cracks Any drum having a crack should be immediately replaced. A cracked drum could fail, causing a serious accident.

Scoring Scoring is most commonly caused by dirt trapped between the lining and the drum. Cars fitted with riveted lining will score their drums if dirt becomes trapped in the rivet holes. The rivets themselves will score the drums if the

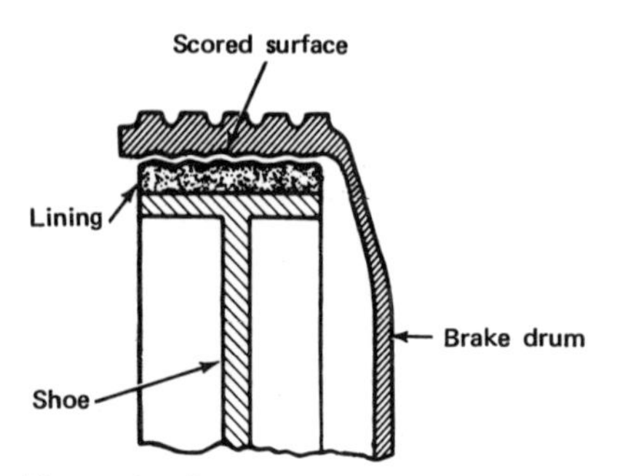

Figure 6.4 Scored brake drum (Ford Motor Company).

lining is allowed to wear down to the heads of the rivets (Figure 6.4).

Barrel Shape Sometimes a drum shows excessive wear at the center of its friction surface. In most cases this wear is caused by extreme braking pressure. Such pressure distorts the platform so that the pressure is concentrated at the center of the lining (Figure 6.5).

Out-of-Round You can recognize an out-of-round drum by measuring the diameter of the drum at several places. Out-of-round drums usually cause most of the complaints of a *pulsating brake pedal.* A pulsating brake pedal is one that moves up and down when the brakes are applied while the car is in motion. As the drum turns, the brake shoes move in and out, following the warped shape of the drum. This causes pulsations that are transmitted to the brake pedal (Figure 6.6).

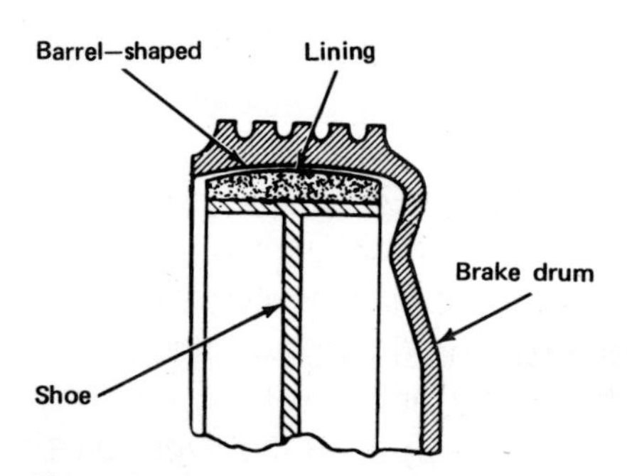

Figure 6.5 Barrel-shaped brake drum (Ford Motor Company).

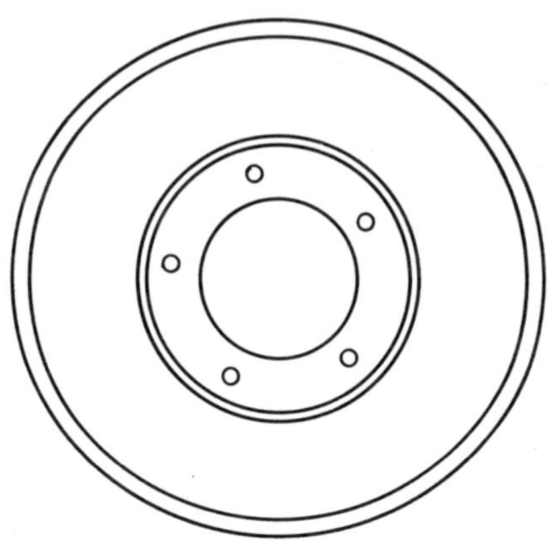
Figure 6.6 Out-of-round brake drum.

Eccentric Eccentric, or off-center, drums can also cause a pulsating brake pedal. Eccentric and out-of-round drums can be caused by the driver's applying the parking brakes when the drums are overheated. Then as the drums cool, they settle into a distorted position (Figure 6.7). Other causes are improper lug nut torque and tightening lug nuts in an improper sequence.

Bell-Mouth High braking temperatures and high braking pressures may cause a drum to distort, or bend. The distortion will often exceed the elastic limit of the material from which the drum is made. When this occurs, the drum may not return to its original shape as it cools. The drum has less support and dissipates less heat

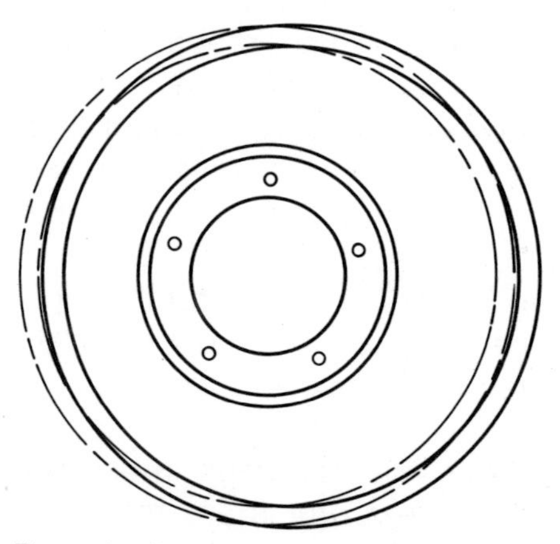
Figure 6.7 Eccentric brake drum.

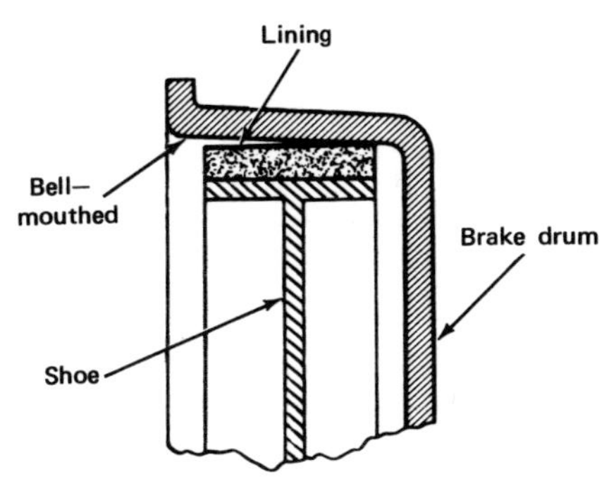

Figure 6.8 Bell-mouthed brake drum (Ford Motor Company).

at its open end, or mouth, than at its closed end. Thus, it may settle into the bell-mouthed shape shown in Figure 6.8.

Taper Taper wear is the presence of more wear near the closed end of a drum than near the open end. It is caused by a heated drum which expands more at its open end than at its closed end. Therefore, the pressure of the shoe near the closed end is greater than the pressure near the open end. This results in more wear near the closed end. When the drum cools and returns to its original diameter, the extra wear at the inner edge becomes obvious. Taper wear is shown in Figure 6.9.

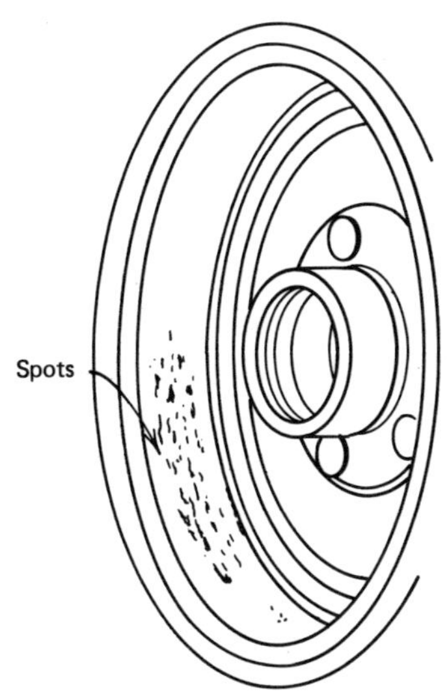

Figure 6.10 Hard spots on the drum surface.

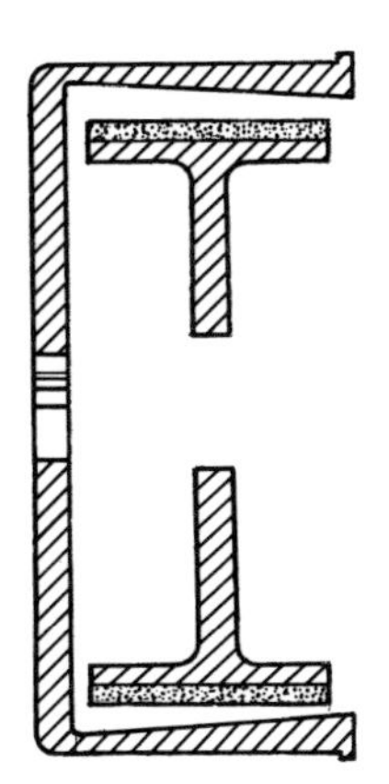

Figure 6.9 Taper drum wear.

Hard Spots Hard spots are found on the friction surface of many drums. In most cases, such spots do not become obvious until the drum is being machined. They are formed by extreme heat and pressure. After the drum has been machined, the hard spots show up as small bumps on the friction surface. The bumps have to be removed by additional grinding (Figure 6.10).

Detecting and Remedying Drum Defects Of all the defects listed, only a cracked or scored drum can be detected visually. The other defects can be found only with a drum mike or by an attempt to machine the drum.

With the exception of a cracked drum, all of the listed defects can be remedied by machining the drum on a drum lathe. But remember, defects that cannot be eliminated by machining to within 0.060 in. oversize require that the drum be replaced.

Cast-iron drums can become distorted in storage. For this reason, most drum manufacturers suggest that a light machining cut be made on all new drums. This cut ensures that the drums will be perfectly round when you install them. In many cases, however, you should machine new drums anyway. This may seem strange, but there is a good reason for it.

For equal braking, the drums on each axle should have the same diameter. This means that the two front drums should be of the same diameter and that the two rear drums should also have the same diameter. Suppose you are doing a brake job and you find that both front drums are badly scored. You try to save the drums by machining them. The left front drum presents a good surface after you have machined it to 0.050 in. oversize. Since this diameter is within the limits allowed by the manufacturer, the drum can be re-installed. But suppose the right front drum still has some deep scores after you have machined it to 0.060 in. oversize. You cannot safely remove any more metal, which means that the drum must be replaced.

Now suppose that a new drum of standard diameter is placed on the right front. The car will pull to the left when the brakes are applied. This is because the left drum is 0.050 in. larger in diameter than the right drum and thus provides more leverage in braking. Ideally, both drums on the same axle should have exactly the same diameter. Usually, however, a difference of 0.015 in. between the drums will not cause a noticeable difference in braking. Any difference greater than 0.015 in. in diameter will usually result in the car's pulling to the side of the larger drum when the brakes are applied. For this reason, you should always machine the drums on the same axle to within 0.015 in. of each other. So, when you place a new drum on the right-hand side of an axle that has a 0.050 in. oversize drum on the left, you should machine the new drum to at least .035 in. oversize. Then the car will stop without pulling to one side.

Note. **To save yourself time and trouble, always mike both drums sharing an axle. Always machine the more badly worn or more oversized drum first. Mike the drum when finished. Then machine the second drum to within 0.015 in. of the first.**

New drums are usually coated with an antirust compound. This coating must be removed before the drum is installed. If not removed, it reduces the friction between the brake shoes and the drum. You should not use gasoline, kerosene, or other petroleum-based solvents to remove it. They can penetrate the pores of the cast iron and carry some of the coating with them. Then as the drum becomes heated in operation, traces of the coating can come to the surface and contaminate the new lining. To get rid of all the coating, you should clean drums with alcohol or some other greaseless solvent.

MACHINING BRAKE DRUMS The friction surface of a brake drum is most often refinished by machining it on a brake drum lathe. Many different types of brake drum lathes are used in auto shops. The specific operating procedures vary with each design. For this reason, each manufacturer furnishes an instruction manual with its lathes. You should study the manual to become familiar with the lathe controls. Figure 6.11 shows one type of brake drum lathe.

All modern drum lathes have a rotating arbor, or shaft. The drum is mounted on this arbor. The arbor turns the drum while a cutting tool, called a *bit,* is moved across the inside surface of the drum. The drum, of course, must be perfectly centered and securely held on the arbor. The cutting bit is adjusted so that it removes a few thousandths of an inch of metal from the drum per cut. Several cuts may be necessary, each cut removing a thin layer of metal until all imperfections have been removed. The depth and speed of each cut can be adjusted. This helps you make sure that the finished drum will have a smooth, flat surface to contact the full width of the brake shoe. It also helps you make sure that the drum will be perfectly round so that there will be no pedal bounce, or pulsation.

Mounting Drums on a Lathe As you work on different makes and models of cars, you will find that these three types of brake drums are commonly used:

1 Hubbed drums with bearing races.
2 Floating drums.
3 Hubbed drums with splined or tapered holes.

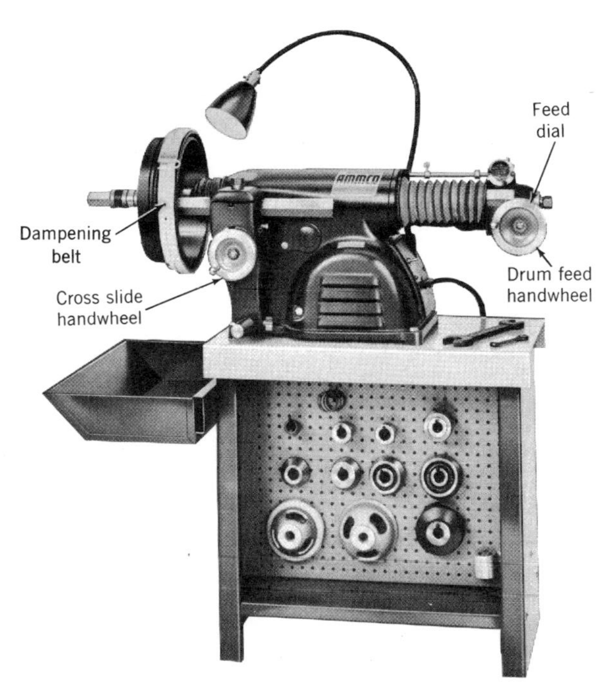

Figure 6.11 A typical brake drum lathe (courtesy AMMCO TOOLS, Inc.)

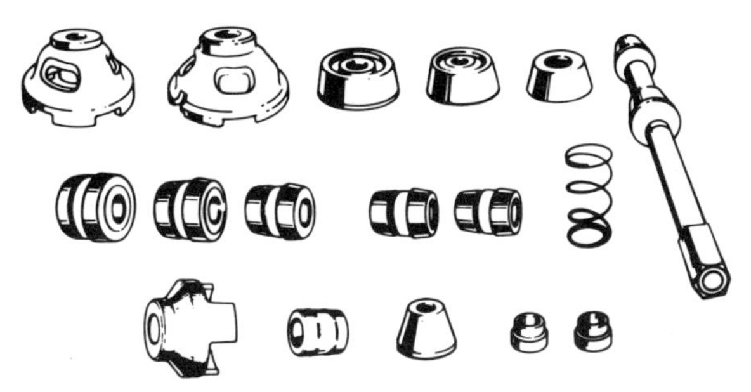

Figure 6.12 A set of accessories furnished with a brake drum lathe (courtesy AMMCO TOOLS, Inc.)

Each of these drums must be mounted on the arbor of a drum lathe in a different manner.

The quality of the work done by any brake drum lathe depends largely upon the care with which you mount the drum on the arbor. Each lathe is furnished with a complete set of bushings, fittings, and adapters, as shown in Figure 6.12. By properly choosing these accessories, you can accurately mount any drum you are likely to encounter.

HUBBED DRUMS WITH BEARING RACES

These drums are usually found at the front wheels of a car. They rotate on bearings, and the bearing cups are pressed into the hub. The inner, tapered surfaces of the bearing cups are perfectly round, so they are used to position the drum on the arbor.

Included among the accessories furnished with a drum lathe are an assortment of *taper adapters* similar to those shown in Figure 6.13. The adapters are designed to fit into the bearing cups and to make contact with the center of the inner, tapered surface. With the bearings removed from the hub, you should select a taper adapter that fits into the inner cup. This adapter should then be pushed completely onto the arbor. Next, select a taper adapter that fits the outer cup. Slide the drum over the arbor, and hold it in place on the large adapter. Now slide the smaller adapter in place. Slide a spacer or two over the arbor so that the arbor nut, when installed, will push the

Figure 6.13 Taper adapters of various sizes (courtesy AMMCO TOOLS, Inc.)

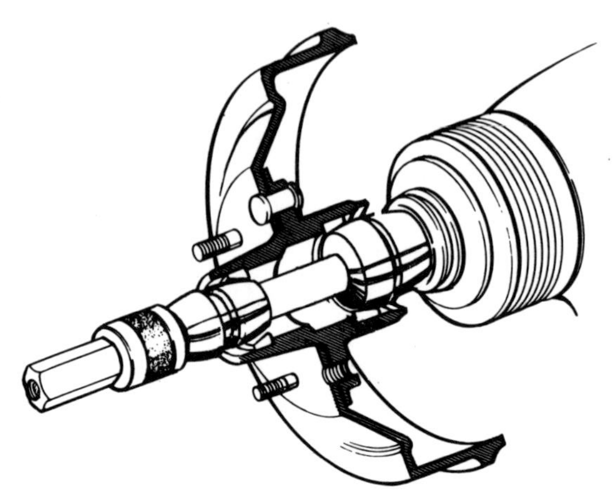

Figure 6.14 Sectioned view of a hubbed drum with bearing races mounted on a lathe arbor with taper adapters (courtesy AMMCO TOOLS, Inc.)

adapters tightly into the bearing cups. This automatically centers the drum while securing it to the arbor. Figure 6.14 illustrates the correct mounting of a hubbed drum with bearing races.

Each time you mount a drum on a lathe, you must check the accuracy of your mounting. The following steps show the best way to do it.

1 Turn the cross slide handwheel inward (clockwise) to move the tool bit away from the surface of the drum.

2 Turn the feed handwheel so that the bit is positioned inside the drum about one inch from its outer edge.

3 Make sure the feed lever is in neutral. Turn on the lathe.

4 While the drum is rotating, slowly turn the cross slide handwheel outward until the tool bit barely touches the surface of the drum. Turn off the lathe.

5 Examine the drum surface. If the drum is perfectly round and accurately mounted, the tool bit will make a mark completely around the drum (360°). Usually, however, the tool mark will not run completely around the drum. When it does not, the drum is out-of-round, eccentric, or improperly mounted.

6 Loosen the arbor nut. Rotate the drum on the arbor one-half turn. Tighten the arbor nut. Move the feed handwheel slightly so that the tool bit moves to one side of the original mark.

7 Turn on the lathe. The tool bit will make another mark. If the second mark is along side the first, your mounting was accurate and the drum is out-of-round or eccentric. If the marks are not side-by-side, the drum was inaccurately mounted. Repeat steps 6 and 7 until you make a tool mark that is along side the one made just before it.

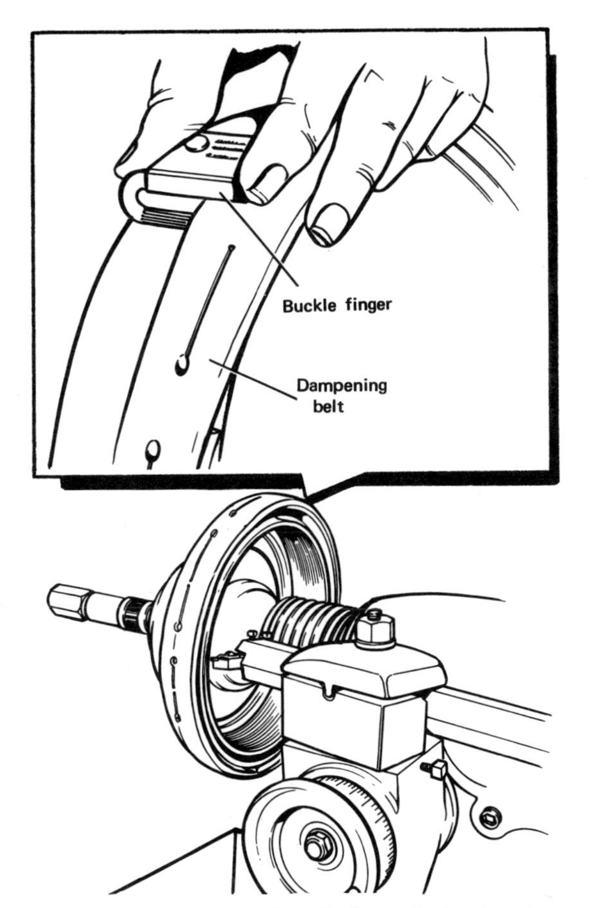

Figure 6.15 Rubber belt installed on a brake drum to dampen vibrations during machining (courtesy AMMCO TOOLS, Inc.)

After a drum has been mounted, it will ring like a bell if you strike it gently with a metal tool. A bell rings by vibrating. A brake drum is shaped like a bell, and it will vibrate when it is supported at the center. If the drum vibrates when you are machining it, its vibrations will cause it to move in and out while the bit is cutting the drum surface. The finished surface will be quite rough and will resemble the diagonal teeth of a file. This surface is usually called a *herringbone pattern.* A herringbone pattern on the finished surface of a drum causes rapid lining wear as it actually files away the lining. It will also cause brake pull and erratic braking.

A large rubber belt is supplied with all brake drum lathes. This belt is called a silencing belt or a dampening belt. It is used to dampen, or silence, vibrations that may be set up in the drum when it is being machined. Wrap the drum with the rubber belt (Figure 6.15). The belt should be installed with a slight stretch, and it should contact all parts of the outer surface of the drum. You can test the effectiveness of the belt by striking the drum again after it is wrapped. The ringing sound should be dampened, or silenced. Always remember to install the rubber dampening belt before machining a drum.

Job 6a

MOUNT A HUBBED DRUM WITH BEARING RACES ON A LATHE

SATISFACTORY PERFORMANCE

A satisfactory performance on this job requires that you do the following:

1 Mount a hubbed drum with bearing races on a brake drum lathe arbor.
2 Following the steps in the "Performance Outline," complete the job within 15 minutes.

PERFORMANCE OUTLINE

1 Select the proper adapters.
2 Install the adapters and drum on the arbor.
3 Install the belt.
4 Check accuracy of mounting and adjust if necessary.

FLOATING DRUMS

Floating drums do not have a hub and are not supported by bearings. These drums are used on rear axles. They are mounted on the axle flanges and are held in place by the wheel and its lug nuts. Some manufacturers also use floating drums at the front wheels. There, such drums are held against a flange on the front hub. Front wheel floating drums have alignment tabs so you can always install them on the hub in the same position.

Since floating drums have no hubs or bearing cups, you cannot use taper adapters for mounting them on a drum lathe. Floating drums are mounted with aligning cups and a centering cone. These lathe accessories are shown in Figure 6.16. The drum is sandwiched between the

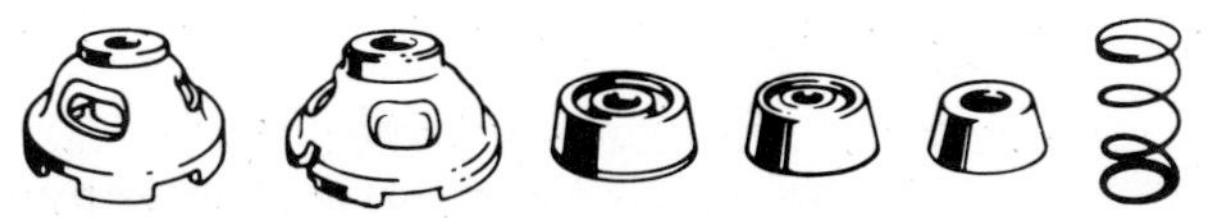

Figure 6.16 Aligning cups, centering cones, and a spring. These brake drum lathe accessories are used to mount floating drums on a lathe arbor (courtesy AMMCO TOOLS, Inc.)

aligning cups while the centering cone, pushed out by a spring, accurately centers the drum.

To mount a floating drum, follow these steps:

1 Slide the inside aligning cup over the arbor.

2 Slide on the coil spring.

3 Select a centering cone that fits the hole in the center of the drum.

4 Slide this cone, small side out, onto the arbor.

5 Slide the drum onto the arbor until the cone protrudes from the center hole.

6 Push the drum against the inside cup, and hold it there against the tension of the spring.

7 Slide the outside retaining cup onto the arbor, and push it against the drum.

8 Slide a spacer or two over the arbor so that the nut, when tightened, will push the cups together and will hold the drum. (Figure 6.17 shows the position of the drum and the mounting parts when they are correctly installed on the arbor.)

9 Check your mounting by taking two light cuts with the tool bit, as you did when you mounted the hubbed drum.

Some manufacturers suggest another method of mounting front wheel floating drums on a brake drum lathe arbor, as follows:

1 Remove the hub from the spindle.

2 Remove the inner bearing.

3 Bolt the drum to the hub, using the lug nuts. If you put the nuts on backward, with the tapered sides facing out, they will hold the drum tight against the flange on the hub.

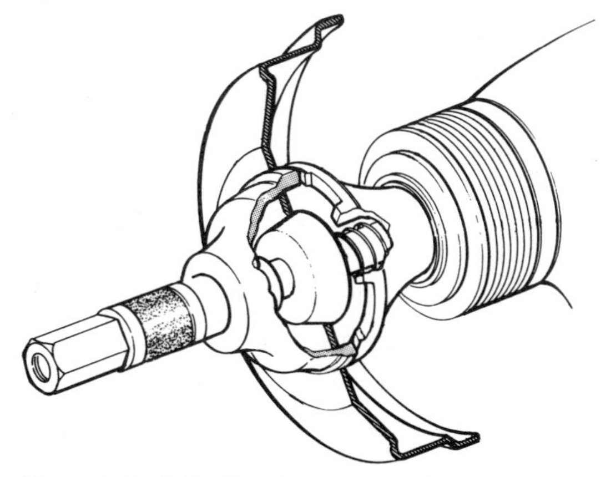

Figure 6.17 A floating drum mounted on a lathe arbor. Note the use of aligning cups and a centering cone (courtesy AMMCO TOOLS, Inc.)

4 Mount the hub-and-drum assembly on the lathe arbor, using taper adapters as if it were a hubbed drum.

Job 6b

MOUNT A FLOATING DRUM ON A LATHE

SATISFACTORY PERFORMANCE

A satisfactory performance on this job requires that you do the following:

1 Mount a floating brake drum on the arbor of a brake drum lathe.
2 Following the steps in the "Performance Outline," complete the job within 15 minutes.

PERFORMANCE OUTLINE

1 Select the proper centering cone and aligning cups.
2 Install the cone, cups, and drum on the arbor.
3 Install the belt.
4 Check the accuracy of the mounting and adjust if necessary.

HUBBED DRUMS WITH SPLINED OR TAPERED HOLES

Drums of this type are found on cars that have splined or tapered ends on their rear axle shafts. The splined or tapered hole in the hub provides an accurate mounting surface, therefore, these drums are easy to mount on a lathe arbor. Here are the steps:

1 Select a centering cone that fits the inner hole in the hub, and slide it onto the arbor.

2 Select a cone that fits the outer hole.

3 Slide the drum on the arbor.

4 Slide on the remaining cone.

5 Install any necessary spacers.

6 Tighten the arbor nut.

Note. **When the arbor nut is tightened, the cones will automatically center the drum and hold it in that position. Figure 6.18 illustrates this type of mounting.**

7 Check the accuracy of your mounting.

TAPERED REAR AXLES (SMALL)

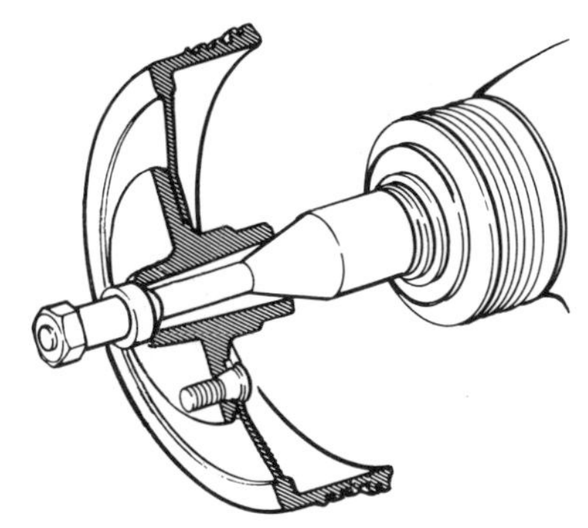

Figure 6.18 View showing a drum mounted on a lathe arbor by means of centering cones (courtesy AMMCO TOOLS, Inc.)

Job 6c

MOUNT A HUBBED DRUM WITH A SPLINED OR TAPERED HOLE ON A LATHE

SATISFACTORY PERFORMANCE

A satisfactory performance on this job requires that you do the following:

1 Mount a hubbed drum with a splined or tapered hole on the arbor of a brake drum lathe.
2 Following the steps in the "Performance Outline," complete the job within 15 minutes.

PERFORMANCE OUTLINE

1 Select the proper centering cones.
2 Install the cones and the drum on the arbor.
3 Install the belt.
4 Check the accuracy of the mounting and adjust if necessary.

OPERATING THE DRUM LATHE

Now that you have mounted several different types of drums and have checked the accuracy of your mounting, you are ready to machine drums.

Because you operated the lathe to check your drum mountings, you are familiar with the main controls: the cross slide handwheel and the feed handwheel. The cross slide handwheel moves the tool bit into contact with the drum and is used to vary the depth of the cut. On some lathes the feed handwheel moves the drum back and forth so that the tool bit contacts the entire drum surface. On other lathes the feed handwheel moves the tool bit back and forth.

There is one other control that you must learn about before you machine a drum. This control sets the feed speed, which is the speed with which the tool bit moves across the drum surface. The faster the bit moves, the rougher the cut it makes. A rough cut is all right for removing material in a hurry, but it will not leave a smooth braking surface. A fine cut will leave a smooth surface, but it takes too much time if repeated cuts are needed. Therefore, you will normally use one or more rough cuts to machine away the defects. Then you will use a slow, fine cut to obtain a smooth surface.

Some lathes have a broad range of feed speeds. Their speeds are set by turning a numbered dial. The higher numbers give you higher speeds, and the lower numbers give you lower speeds. With experience, you can dial the correct speed for any job. If your lathe is fitted with a feed speed dial, it also has a feed lever that engages the lathe feed. Lathes that do not have a feed speed dial usually provide just two speeds. These speeds are for a coarse cut and a fine cut. A lever allows you to shift into the speed you want for the job. It not only selects the speed, but also starts the feed.

Before you start the lathe study the controls for a few minutes. If you are unsure of anything, recheck the instruction manual.

The following procedure will enable you to operate most lathes.

1 Mount the drum and check the mounting.

2 Turn the feed handwheel so that the tool bit is drawn out toward the edge of the drum. The tool bit will cut into and remove the "rust ridge" that is present at the open end of the drum.

3 Turn the handwheel back so that the bit moves back into the drum. Continue turning until the bit starts to enter the inner corner of the drum. Proceed slowly here because there is usually another ridge in the corner. Keep turning the handwheel slowly until the bit cuts away the inner ridge and reaches the inner edge of the drum surface.

4 Note the setting on the dial of the cross slide handwheel. Advance it 0.005 in. This will push the bit 0.005 in. into the drum surface. Lock the cross slide so that your adjustment will not change.

5 If your lathe has a feed speed dial, adjust it to *16*. This will give you a relatively fast cut. Lock your adjustment and engage the feed. If your lathe has two speeds, shift the feed lever into the fast-cut, or rough-cut, position.

6 When the cut is finished and the tool bit is clear of the drum, disengage the feed lever and shut off the lathe.

7 Inspect the drum surface. If defects still exist, start the lathe and turn the feed handwheel to run the tool bit back into the corner. Loosen the lock on the cross slide and advance the handwheel another 0.005 in. Engage the feed lever for another fast cut. Repeat the fast cuts of 0.005 in. until all defects are eliminated.

8 Run the bit back into the corner of the drum. Advance the cross feed handwheel 0.002 in. If your lathe has a feed speed dial, adjust it to 2. This will give you a very slow cut. Lock your adjustment and engage the feed lever. If your lathe has two speeds, shift the feed lever into the slow-cut, or fine-cut, position.

9 When the cut is finished and the tool bit is clear of the drum, disengage the feed lever and shut off the lathe. Check the drum surface. It should be smooth to the eye and to the touch.

10 Remove the drum from the arbor, and remove the belt from the drum. Mike the drum to be sure its diameter does not exceed the greatest allowable oversize after machining (usually 0.060 in.).

Job 6d

MACHINE A BRAKE DRUM

SATISFACTORY PERFORMANCE

A satisfactory performance on this job requires that you do the following:

1 Machine a brake drum to within the specifications of the manufacturer.
2 Following the steps in the "Performance Outline," complete the job in 30 minutes.
3 Fill in the blanks under "Information."

PERFORMANCE OUTLINE

1 Mount the drum on the lathe arbor.
2 Eliminate the defects with one or more rough cuts.
3 Machine to a smooth surface with a fine cut.

INFORMATION

Drum size (original) ________________ Width ________________

Actual drum size before machining ________________

Actual drum size after machining ________________

Maximum allowable diameter after machining ________________

Recommendations ________________

BRAKE SHOE GRINDING The actual diameter of a brake drum after machining may vary from 0.005 to 0.060 in. oversize. New or relined brake shoes cannot possibly fit all the drums when the drums vary in diameter. Of course, the shoes will eventually wear in, but they will do so only after thousands of miles of use. If you want to turn out a perfect brake job, you must ma-

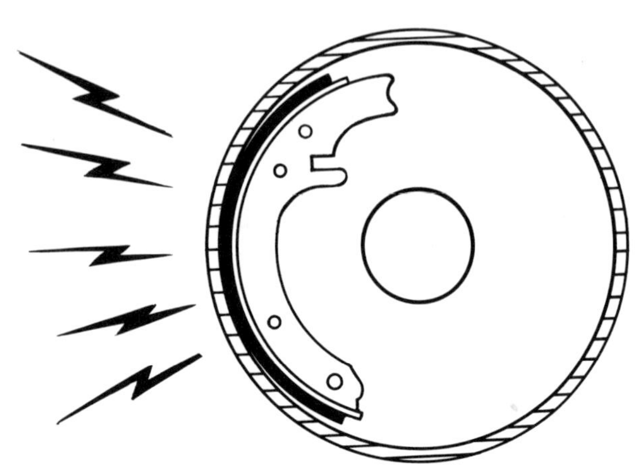

Figure 6.19 Standard lining installed in an oversize drum (courtesy AMMCO TOOLS, Inc.)

chine the brake lining to fit the friction surface of the drum. This machining is called *arc grinding.*

Arc grinding is necessary for several reasons:

1 To fit the lining to a drum that has been machined to an oversize diameter.

2 To allow for slight misalignment and a possible stackup of manufacturing tolerances.

3 To allow a slight shoe flexing to minimize the possibility of brake noise.

Drum Machined Oversize Because machining a drum increases its diameter, lining manufacturers provide lining in two thicknesses. When you buy lining, you can request the lining that is the proper thickness for the job. Standard brake lining is made to fit drums of original diameter. It has the same thickness as the lining that was installed on the car when it was new. Most manufacturers recommend that standard lining be installed when the actual drum diameter does not exceed 0.030 in. oversize. This procedure usually results in a satisfactory lining-to-drum contact that will improve with wear in a short time. Oversize lining is made to fit drums that have been machined to 0.030 in. or more oversize. Though the two sizes of lining provide a satisfactory fit in most instances, they may also lead to problems with the completed job.

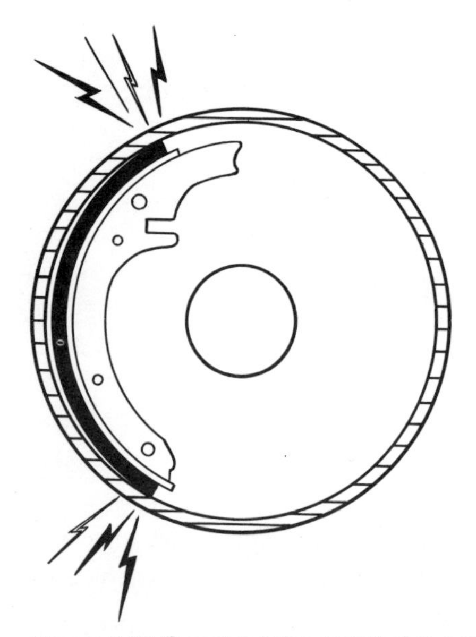

Figure 6.20 Oversize lining installed in a near-standard drum (courtesy AMMCO TOOLS, Inc.)

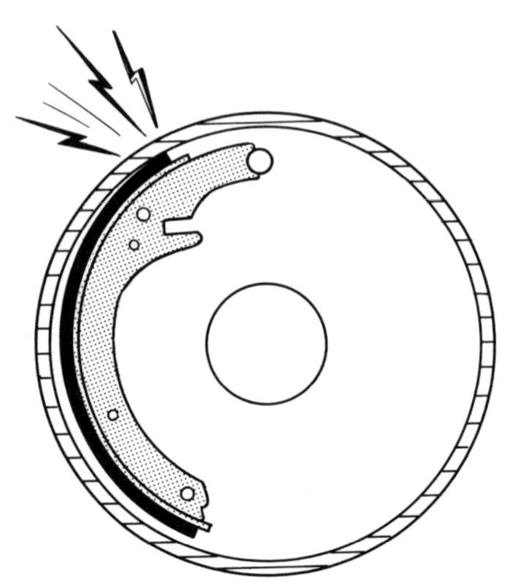

Figure 6.21 Shoe too high in drum (courtesy AMMCO TOOLS, Inc.)

Suppose you were to install a set of standard linings in an oversize drum. As Figure 6.19 shows, this would result in initial contact at the middle of the lining and an excessive clearance at the ends. The lining in the middle of the shoe would be doing all the work and would be subjected to extreme temperatures and severe wear. In addition, the pedal would feel spongy because the shoe would flex, or bend, excessively. The lining at the center of the shoe would eventually wear enough so that more lining surface would contact the drum. But until then, the brakes would not feel or function properly.

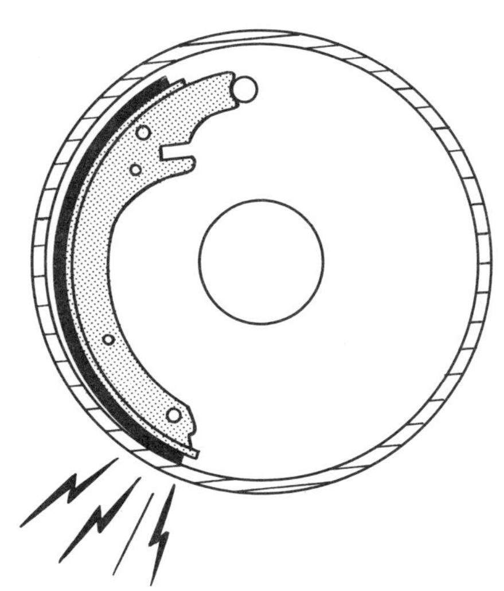

Figure 6.22 Shoe too low in drum (courtesy AMMCO TOOLS, Inc.)

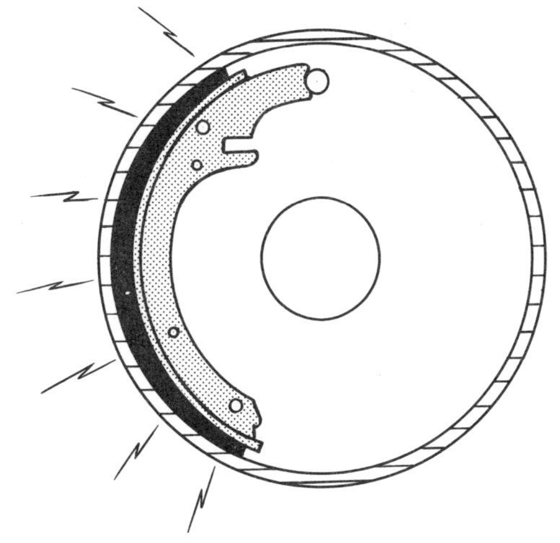

Figure 6.23 Ideal lining-to-drum contact (courtesy AMMCO TOOLS, Inc.)

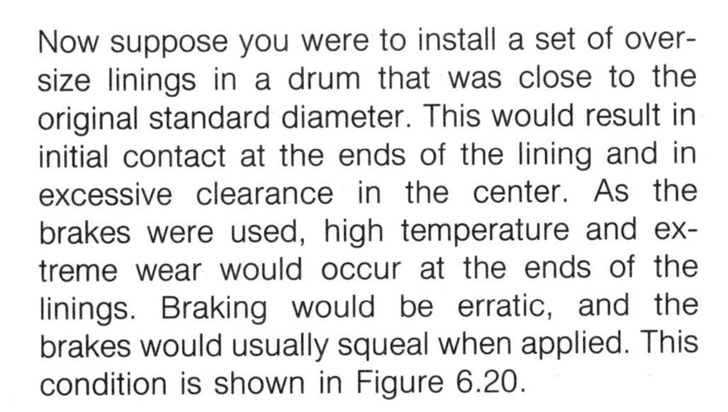

Now suppose you were to install a set of oversize linings in a drum that was close to the original standard diameter. This would result in initial contact at the ends of the lining and in excessive clearance in the center. As the brakes were used, high temperature and extreme wear would occur at the ends of the linings. Braking would be erratic, and the brakes would usually squeal when applied. This condition is shown in Figure 6.20.

Misalignment and Stack-Up of Manufacturing Tolerances In the mass production of auto parts, normal wear and slight misalignment of assembly machinery are allowed for and are held to certain tolerances. In the production of backing plates, the anchor pins can be attached slightly too high or too low. Likewise, brake shoes can have slight variations in the position of the anchor pin radius. Usually these variations are slight. However, a buildup of allowable tolerances at times results in shoes that are positioned too high or too low in relation to the drum. Figures 6.21 and 6.22 illustrate the types of lining-to-drum contact that occur when the shoes are not properly positioned.

Since the position of the anchor pin is not adjustable, there is no way to correct improper shoe positioning except by machining the lining. Failure to check and correct these conditions usually results in brake squeal and erratic braking.

Allowance for Flexing It may seem that the best fit between the lining and the drum is obtained by grinding the lining to the exact radius of the drum. But this is not so. If the lining is ground to the exact radius of the drum, no allowance is made for any anchor pin misalignment. Also, a perfect fit results in a brake shoe that is too rigid in operation. A brake shoe must be allowed to flex slightly. You can allow for flexing by grinding the lining to a radius slightly smaller than that of the drum. Such grinding provides a slight clearance at the ends of the lining. This clearance compensates for any misalignment of the anchor pin. It also allows a little flex in the shoes for quiet operation (Figure 6.23).

Each brake job involves a different set of variables. Drum size, shoe and lining size, and the position of the anchor pin must always be considered. Because of these variables, each shoe must be specially fitted to the drum it will contact.

Figure 6.24 A brake shoe grinder (courtesy AMMCO TOOLS, Inc.)

Determining the Proper Radius of Brake Shoes The best initial braking is obtained when the lining is ground to a diameter of 0.030 in. less than the actual diameter of the drum. Most manufacturers recommend this difference in diameter. Most brake shoe arc grinders automatically provide that difference. The actual diameter of each drum is found by measurement. Then the grinder can be adjusted to grind each pair of shoes to fit the drum they will be used with. Two adjustments are provided. The first adjustment sets the grinder to the original diameter of the drum. This adjustment is usually made in increments of one-half inch. The second adjustment consists of a micrometer dial that is set to the actual oversize of each drum.

Arc Grinding Brake Lining

All brake shoe grinders operate on a common principle. The brake shoe is clamped into a fixture that allows the shoe to be moved in an arc across an abrasive disc or drum. The arc is adjustable so that the lining can be ground to fit any size drum. Figure 6.24 illustrates a typical brake shoe grinder.

There are many types of brake shoe grinders. Specific operating instructions are given with each machine. Therefore study the instruction manual that is furnished with the machine you will use.

Since most of the brake systems on which you will be working use a fixed anchor, the brake shoes should be *fixed anchor ground.* In other words, the shoes should be held in the grinder in a position that duplicates their position on the backing plate. Also, the shoes should be ground in the brakes-applied position. To do this, you must hold the shoes in the position they are in when forced against the brake drum.

Most shoe grinders incorporate a fixed-anchor device that enables you to duplicate on the grinder the exact position the shoes take on the backing plate. The fixed-anchor device does this by duplicating three dimensions. These dimensions are identified on Figure 6.25 by the letters *A, B,* and *C.* Dimension *A* is the distance from the center of the backing plate to the center of the anchor pin. Dimension *B* is the radius, or one-half

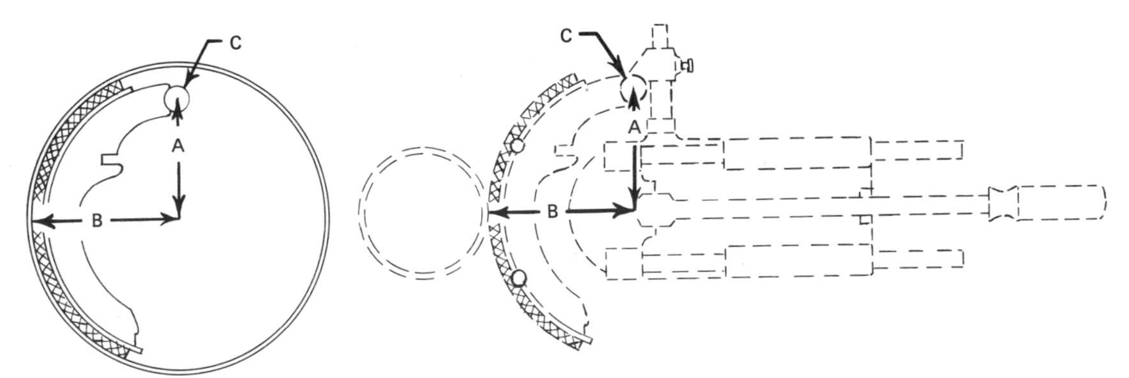

Figure 6.25 The three dimensions considered in arc grinding brake lining (courtesy AMMCO TOOLS, Inc.)

the diameter, of the brake drum. Dimension *C* is the diameter of the anchor pin. These dimensions are transferred to the grinder before the shoes are mounted.

In most instances you will duplicate dimension *A* by using a gauge sleeve or a gauge plate selected with the help of a chart supplied with the grinder (Figure 6.26).

To set dimension *B*, adjust the grinder to the actual diameter of the drum, as shown in Figure 6.27.

To obtain dimension *C*, place a sleeve of the proper diameter over a pin on the shoe clamp. Select this sleeve by the use of a chart or manual supplied by the maker of the grinder (Figure 6.28).

When all three dimensions have been set, the shoe is placed in the clamp so that it is in contact with the anchor sleeve. When the shoe is advanced toward the abrasive, it will pivot on the anchor sleeve. In this way it duplicates its action when the star adjuster is expanded as shown in Figure 6.29.

Since it is difficult to observe the area of the lining that is being ground, you should mark it with a number of chalk lines. Since the grinding removes the chalk, you can easily follow the extent of the grinding. Arc grinding releases a considerable amount of asbestos dust. You should always wear an air respirator during the grinding operation.

With the shoe held in position, you should grind as follows:

1 Turn on the grinder.

2 Turn the feed handle inward until the lining barely contacts the abrasive disc or drum.

3 Swing the handle from side to side so that the entire length of the lining passes over the abrasive.

4 Check the chalk marks on the lining to determine the extent of the grinding.

A

Figure 6.26 A gauge sleeve and gauge plate. Use these tools to obtain dimension A when setting up to arc grind brake lining (courtesy AMMCO TOOLS, Inc.)

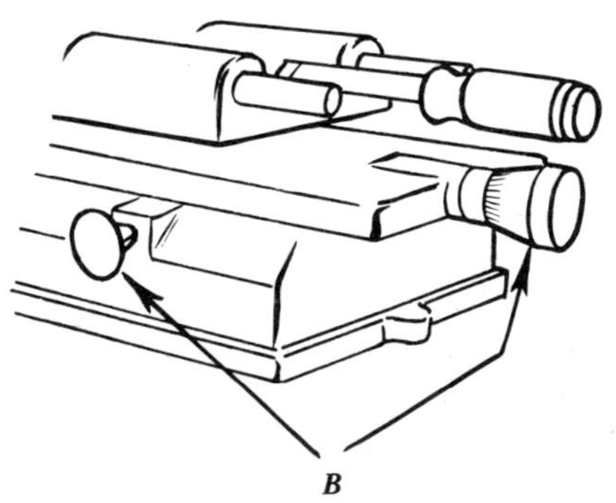

Figure 6.27 The slide lock pin and micrometer adjustment. Use these controls to adjust the grinder to dimension B (courtesy AMMCO TOOLS, Inc.)

Figure 6.28 Anchor sleeves used to duplicate dimension C when arc grinding brake lining (courtesy AMMCO TOOLS, Inc.)

5 Advance the feed handle slightly, and repeat steps 3 and 4 until all chalk marks have been ground away.

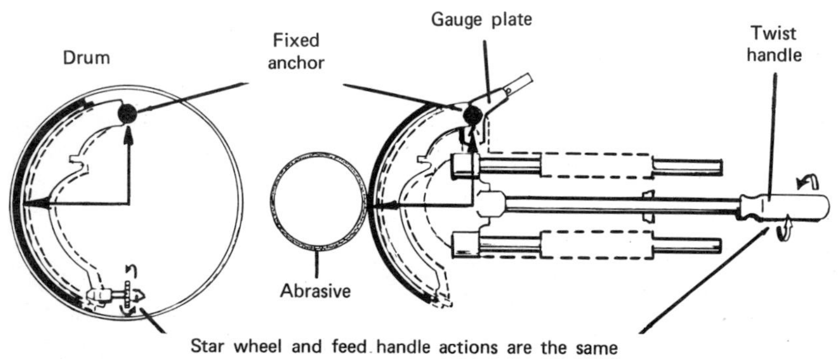

Figure 6.29 The motion of the brake shoe on the grinder duplicates the motion of the shoe on the backing plate when the star wheel adjuster is turned (courtesy AMMCO TOOLS, Inc.)

Job 6e

ARC GRIND BRAKE SHOES

SATISFACTORY PERFORMANCE

A satisfactory performance on this job requires that you do the following:

1 Arc grind a pair of lined brake shoes to fit an oversize drum.
2 Following the steps in the "Performance Outline," complete the job in 15 minutes.
3 Fill in the blanks under "Information."

PERFORMANCE OUTLINE

1 Determine the correct settings of the grinder adjustments.

2 Adjust the grinder.
3 Mount each shoe in the grinder.
4 Grind each shoe.

INFORMATION
Drum size (original) ____________ Width ____________

Actual size of drum ____________________

Grinder settings: inches ____________________

thousandths ____________________

SUMMARY

Here are some of the important facts stressed in this chapter. First, safe and efficient braking depends on many factors. The braking surfaces of the drum and lining are very important, and correcting any defects in them requires specific diagnostic and repair skills. Second, drums may exhibit many faults, which can usually be corrected by machining. Third, machining increases the diameter of the drum. Brake shoes should be fitted to this new diameter. They should also be ground slightly undersize to allow quiet operation. Fourth, machining the drums and shoes should be part of every brake job.

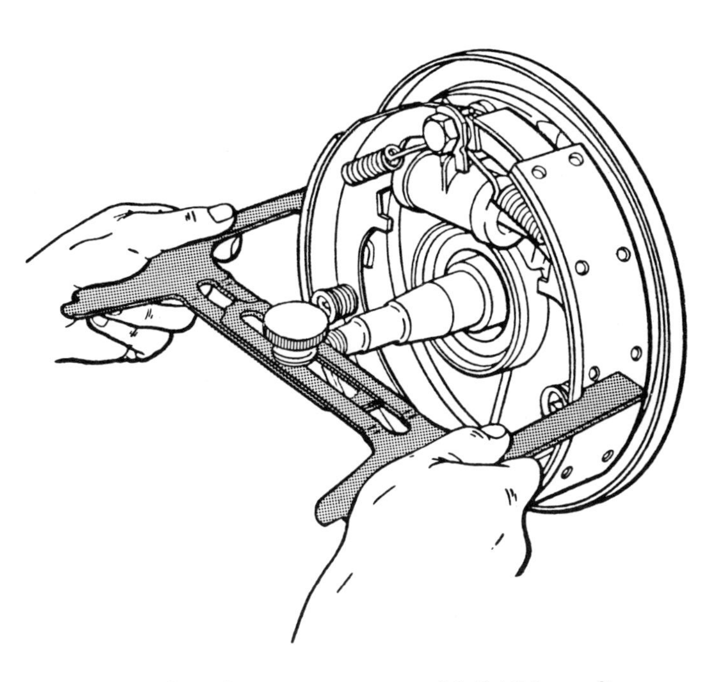

SELF-TEST

Each incomplete statement or question in this test is followed by four suggested completions or answers. In each case select the *one* that best completes the statement or answers the question.

1 Which of the following statements about the tool shown above do you consider true?
   I. The tool is a brake drum micrometer.
   II. The tool is being used to obtain the correct measurement to set an arc grinder.
   A. I only
   B. II only
   C. Both I and II
   D. Neither I nor II

2 A brake drum measures 10.040 in. after machining. To arc grind a set of brake shoes to fit this drum, you should adjust the grinder to
   A. 10.010 in.
   B. 10.020 in.
   C. 10.030 in.
   D. 10.040 in.

3 Road testing a car reveals that the brake pedal bounces or pulsates when the brakes are applied. What could cause this condition?
   I. Out-of-round brake drums.
   II. Eccentric brake drums.

A. I only
B. II only
C. Either I or II
D. Neither I nor II

4 Hard spots in a brake drum are caused by
A. excessive heat and pressure
B. improper mounting of the drum on the lathe
C. machining a drum without using a rubber belt to dampen out the vibrations
D. machining a drum at too high a speed

5 Which of the following statements regarding brake drums are considered proper machining practice?
I. Brake drums should never be machined more than 0.060 in. oversize
II. Brake drums on a common axle should be machined to within 0.015 in. of each other
A. I only
B. II only
C. Both I and II
D. Neither I nor II

6 When mounting a hubbed drum with bearing races on a drum lathe, you should use
A. a centering cone and a pair of aligning cups
B. taper adapters fitted to the bearing cups
C. a pair of centering cones fitted to the hub
D. a taper adapter and a pair of centering cups

7 After machining a brake drum, you notice a herringbone pattern on the friction surface. The most probable cause of this pattern is
A. the use of an improperly fitted radii adapter
B. an improperly sharpened tool bit
C. cutting the drum at high feed speed
D. improper drum vibration dampening

8 The most accurate method of machining floating drums that are used on front hubs is to
A. mount the drum on the arbor using a centering cone and a pair of aligning cups
B. secure the drum to its hub and mount the assembly as if it were a hubbed drum with bearing cups
C. mount the drum on the arbor with taper adapters
D. center the drum with a pair of centering cones and mount the assembly with an aligning cup

9 Two mechanics are discussing the arc grinding of brake shoes.
Mechanic A says arc grinding is necessary to fit the shoes to the new, oversize drum diameter.
Mechanic B says arc grinding is necessary to allow a slight shoe flexing to minimize the possibility of brake noise.
Who is right?
A. A only
B. B only
C. Both A and B
D. Neither A nor B

10 Oversize brake lining should be used if drum oversize exceeds
A. 0.010 in.
B. 0.020 in.
C. 0.030 in.
D. 0.040 in.

Chapter 7 Hydraulic System Basics

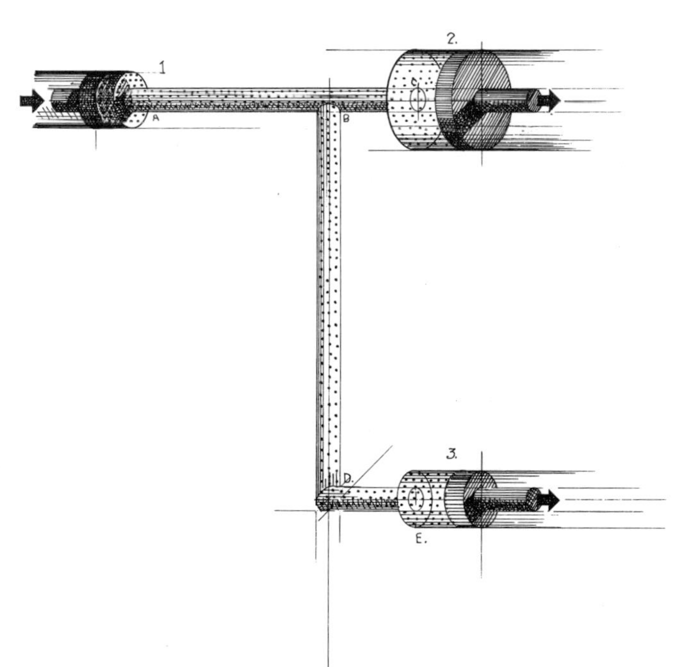

Automotive service brake systems multiply the force the driver applies to the brake pedal and transmit it to the wheels. They achieve this change through *hydraulics.* Hydraulics is the use of fluids, or liquids, to do work. Service brake systems use a hydraulic system to slow or stop a moving car.

In this chapter you will learn the basic principles of hydraulics. You will learn how those principles are used to provide a safe, efficient braking system. Your objectives will be to:

1
Understand the basic rules of hydraulics.

2
Determine pressures and forces in hydraulic systems.

3
Identify the properties of brake fluid.

4
Check brake fluid level.

5
Identify wheel cylinder parts.

6
Identify the parts of a master cylinder.

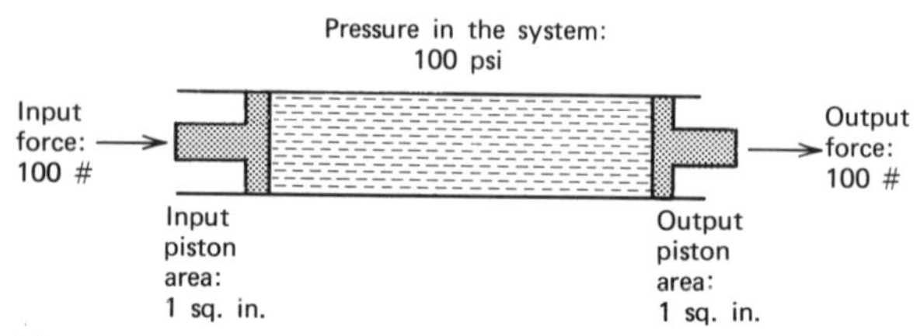

Figure 7.1 A simple hydraulic system.

BASIC RULES OF HYDRAULICS The first step toward understanding how a hydraulic system operates is to learn the three basic rules of hydraulics:

1 *A fluid cannot be compressed.* The fluid in a hydraulic system acts as a solid and, therefore, it can be used to transmit pressure.
2 *A fluid takes the shape of its container.* If there is nothing else in a hydraulic system, the fluid in the system tends to fill it completely. Pressure applied to the fluid is then transmitted to all parts of the system.
3 *Pressure applied to a fluid is transmitted equally in all directions.* When pressure is applied to the fluid in one part of a hydraulic system, equal pressure is transmitted to all other parts.

These three basic rules must always be considered in any hydraulic system. In working with them, you will use two words that are familiar to you, but they must be used according to a strict definition. These two words are *force* and *pressure.*

Force is a push, or pushing effort. It is measured in pounds and is at times indicated by the symbol #. The driver of a car applies force to the hydraulic system by pushing on the brake pedal. This force, multiplied by the leverage of the brake pedal lever, is applied to the piston of the master cylinder.

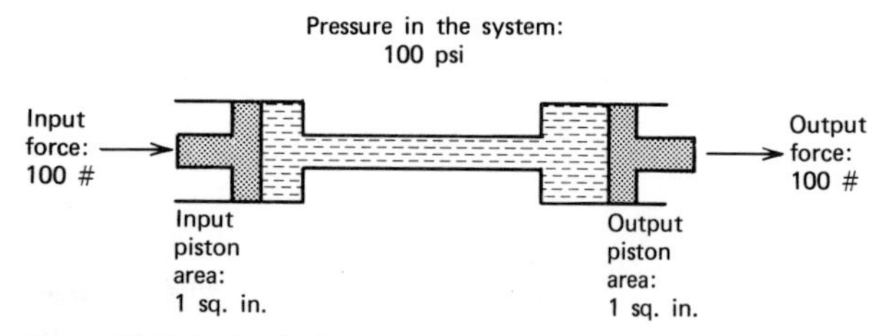

Figure 7.2 A simple hydraulic system with one input cylinder and one output cylinder connected by a tube.

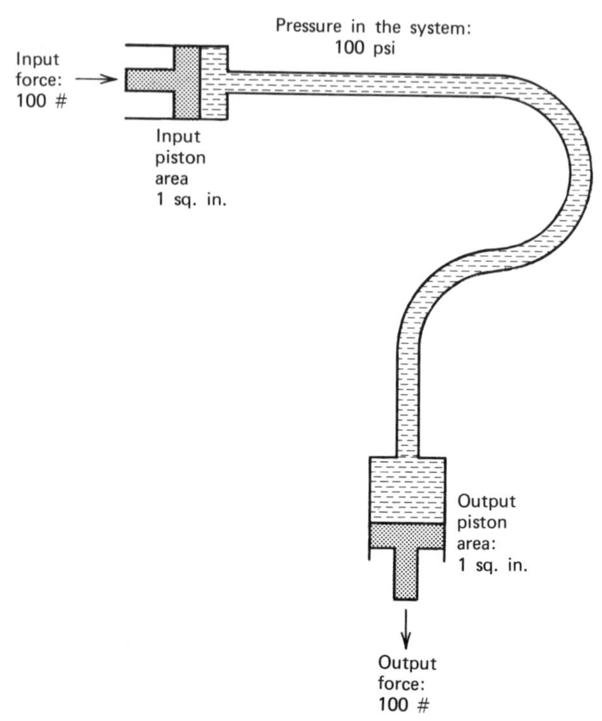

Figure 7.3 A simple hydraulic system in which pressure is transmitted through a curved tube or hose.

Pressure is a quantity, or amount of force applied to a definite area. It is measured in pounds per square inch, usually written psi. Pressure in a service brake hydraulic system pushes the wheel cylinder pistons outward. This motion forces the brake shoes against the drum.

A force applied to a hydraulic system is called an *input force.* The input force is applied to an input piston which builds up a pressure within the system. This pressure is transmitted to the various parts of the system, where it can be used to move another piston or a group of pistons. The force applied by these pistons is called the *output force,* and the pistons themselves are called output pistons.

Hydraulic Systems The first rule of hydraulics, you may recall, makes it possible to transmit pressure. Some information about how pressure is transmitted is both interesting and useful to an auto mechanic.

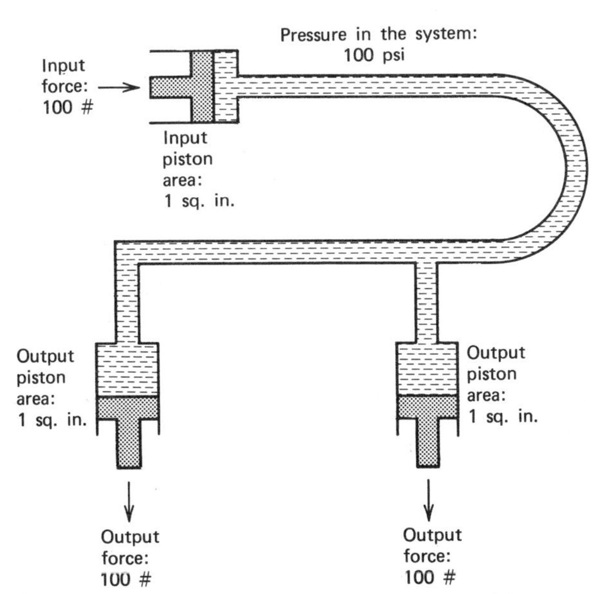

Figure 7.4 A hydraulic system with two output pistons.

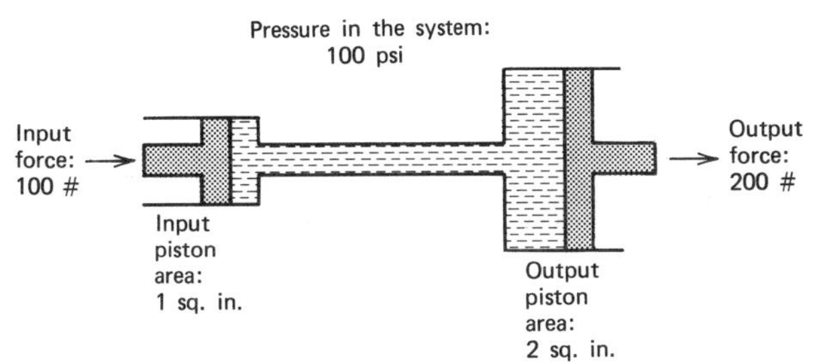

Figure 7.5 A simple hydraulic system with an output piston twice the area of the input piston.

Transmitting Pressure A simple hydraulic system is shown in Figure 7.1. A cylinder with a cross-sectional area of one square inch is fitted with two pistons, one at each end. An input force of 100# (pounds) applied to an input piston that has an area of one square inch builds up a pressure of 100 psi in the system. This pressure is transmitted to the output piston. Since the output piston also has an area of one square inch, the pressure transmitted to it causes it to move outward with a force of 100#.

The action shown in Figure 7.1 takes place in strict accordance with the three basic rules of hydraulics; first, the fluid in the cylinder cannot be compressed; second, it completely fills the cylinder; and third, it transmits pressure equally in all directions. Nevertheless, since the walls of the cylinder will not move, the pressure can move only the piston.

Because of the second rule of hydraulics, Figure 7.1 can be modified to resemble more closely the hydraulic system used to apply brakes. Since the fluid in a hydraulic system takes the shape of its container, the single, long cylinder can be replaced with two short cylinders connected by a tube (Figure 7.2). The output force still matches the input force because the piston areas remain the same.

The basic system can be further changed as shown in Figure 7.3—the connecting tube is curved as it might be if it were used on a car. This system functions in the same manner as those shown previously.

Because of the third rule of hydraulics, the pressure in a hydraulic system can be applied to more than one output piston. Figure 7.4 shows how the pressure can be applied to two pistons, both of which deliver the same output force.

Controlling Output Force Figure 7.4 also shows the way in which hydraulic pressure can be applied to two pistons. It shows that when each output piston has the same area as the input piston, the output force of each output piston is the same as the input force. Therefore, the combined output force of both output pistons is double the input force. What would happen if the two pistons were replaced by one piston whose area was double the area of the input piston? Figure 7.5 answers this question.

In Figure 7.5 we see a system that has an output piston whose area is twice the area of the input piston. Since the area over which the pressure is exerted has been doubled, the output force has also been doubled.

In a system that has two or more output pistons, the output force of each piston depends upon the area of the piston. If the areas of the

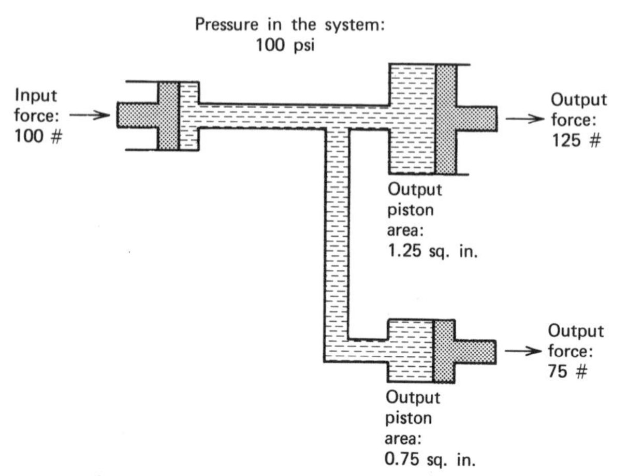

Figure 7.6 A simple hydraulic system with output pistons that have different areas. Note that the output forces are proportionate to the areas of the output pistons.

output pistons are different, the output force of each piston is different. A system with two output pistons having different areas is shown in Figure 7.6.

When used in an automotive brake system, the parts of a hydraulic system have different names: the input cylinder is called a master cylinder; in drum brake systems the output cylinders are called wheel cylinders, in disc brake systems, they are called calipers. You will learn about calipers in a later chapter.

Since the output forces of a hydraulic system can be altered by changing the size of the output pistons, car manufacturers select wheel cylinder sizes so that the brakes are applied with just the right amount of force. Whenever a car is brought to a stop, the front of the car has a tendency to dip down. This tendency is caused by a weight shift to the front of the car. During this shift, the front of the car is heavier than normal, and the rear of the car is lighter. Therefore, in stopping the car, the front brakes must do much more work than the rear brakes.

This being the case, the front brakes should be applied with more force. For this reason, most cars have larger wheel cylinders for their front brakes than for their rear brakes. The cylinder sizes actually used on a car are chosen to provide the proper braking for that particular car.

Figure 7.7 pictures a hydraulic system that could be used in an automotive brake system. Note how the force that the driver applies to the master cylinder is increased at the front brakes but decreased at the rear brakes.

Computing Output Force In Figure 7.7 the input force of 400# is obtained by means of the leverage of the brake pedal, which multiplies the force exerted by the driver. Since this force is applied to a master cylinder piston that has an area of one square inch, the force is converted to a pressure of 400 psi in the hydraulic system. The pressure in a hydraulic brake system can be computed by dividing the input force by the area of the input piston. This method is expressed in the following simple formula:

$$\text{Pressure} = \frac{\text{Input force}}{\text{Area of input piston}}$$

The pressure in the system shown in Figure 7.7 is found by using this formula as follows:

$$\frac{400\#}{1 \text{ sq. in.}} = 400 \text{ psi}$$

This pressure is transmitted throughout the entire system, and it exerts a push on the pistons in the wheel cylinders. Since the wheel cylinder pistons have different areas, they exert different forces on the brake shoes. Knowing the pressure in the system and knowing the area of the wheel cylinder pistons, you can determine the output forces by using the following formula:

$$\text{Output force} = \text{Pressure} \times \text{Output piston area}$$

The output force of each front wheel cylinder piston is computed in this way:

$$400 \times 1.25 = 500$$

The output force of each rear wheel cylinder piston is computed as:

$$400 \times 0.75 = 300$$

As a mechanic, you will not be required to compute pressure and forces, but you must have an understanding of the relationships of force, pressure, and piston areas. This understanding will aid you in diagnosing and correcting brake problems.

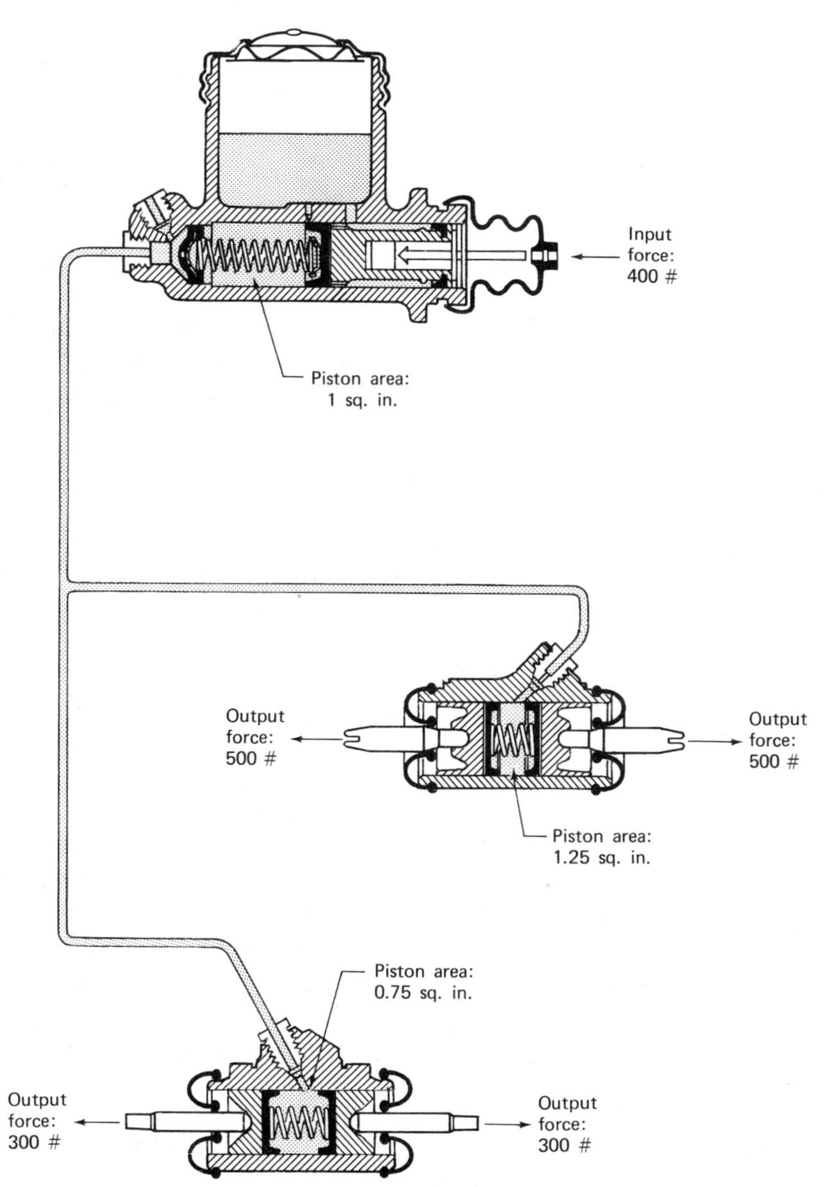

Figure 7.7 Output forces in a typical hydraulic brake system.

Job 7a

COMPLETE STATEMENTS OF THE BASIC RULES OF HYDRAULICS

SATISFACTORY PERFORMANCE

A satisfactory performance on this job requires that you do the following:

1 Complete the statements of the basic rules of hydraulics by writing on the lines provided the correct words chosen from the list below the statements. (Some words are used more than once.)
2 Correctly complete all the statements within 10 minutes.

PERFORMANCE SITUATION

1 A fluid cannot be ____________. The ____________ in a hydraulic system acts as a ____________. Therefore, it can be used to transmit ____________.

2 A fluid takes the shape of its ____________. If there is nothing else in a hydraulic system, the ____________ in the system tends to fill it completely. ____________ applied to the fluid is transmitted to all parts of the system.

3 ____________ applied to a fluid is transmitted ____________ in all directions. When ____________ is applied to the fluid in one part of a hydraulic system, an equal ____________ is transmitted to all other parts.

Pressure	Equality	Compressed	Fluid
Hydraulic	Solid	Container	Brake

Job 7b

DETERMINE PRESSURES AND FORCES IN A HYDRAULIC SYSTEM

SATISFACTORY PERFORMANCE

A satisfactory performance on this job requires that you do the following:

1 Using the measurements given, determine the missing pressure, area, or force in each of five systems.

2 Correctly supply 8 of the 10 required answers within 30 minutes.

PERFORMANCE SITUATION

	Input Force	Input Piston Area	Pressure	Output Piston Area	Output Force
1	100#	1 sq. in.	______ psi	2 sq. in.	______ #
2	100#	______ sq. in.	50 psi	______ sq. in.	100#
3	200#	2 sq. in.	______ psi	1 sq. in.	______ #
4	______ #	1 sq. in.	200 psi	______ sq. in.	400#
5	100#	______ sq. in.	200 psi	______ sq. in.	200#

BRAKE FLUID Because fluid is so important to any hydraulic system, there are certain things about the fluid used in brake svstems that you must know. If you do not know them, you may do more harm to a brake system than good.

Standards Since the pressure in a hydraulic system is transmitted by a fluid, the fluid must meet certain standards. These standards have been set by the Society of Automotive Engineers (SAE) and the Department of Transportation (DOT). Before using a brake fluid, check the container to be sure it has SAE and DOT approval.

All brake fluids sold by reputable manufacturers meet or exceed SAE and DOT specifications. You should never use a brake fluid of questionable origin, and you should always check the container to see that it is marked DOT 3 or DOT 4.

Fluid meeting DOT 3 specifications is suitable for use in drum or disc brake systems. Fluid meeting DOT 4 specifications is also suitable for drum and disc systems. However, it is recommended for extra-heavy duty. It will withstand higher temperatures than fluid meeting DOT 3 specifications.

A brake fluid must not rust or corrode metals such as iron, brass, and aluminum. Many parts of the brake system are made of these metals. Furthermore, brake fluid must not cause rubber to swell or dissolve. Brake hoses and certain other parts of the brake system are made of rubber and, for this reason, you should never use oil or any petroleum-based fluid in a hydraulic brake system. If the fluid in a brake system is found to be contaminated with a petroleum product, the system will have to be taken apart and thoroughly flushed out. Moreover, when the system is put back together, all its rubber parts will have to be replaced with new ones.

Brake fluid must also have a low freezing point

and a high boiling point. A low freezing point is necessary because a vehicle may be operated in very cold temperatures. A high boiling point is necessary because much of the heat of braking is transmitted to the fluid. Brake fluid must be chemically stable within the ranges of temperature and pressure to which it is subjected. The fluid must not break down, and it must not form sludge deposits that may gum up the system. Also, the fluid must provide lubrication for the pistons and other moving parts in the system.

Finally, brake fluid must be *miscible.* That is, it must mix with other brake fluids. As the brake lining wears, the adjusters move the shoes closer to the drum. This action causes the wheel cylinder pistons to move outward in their bores, creating more space in the system. To fill this space you add more fluid to the system, but you have no way of knowing the brand of fluid that is already in the system. Any fluid you add must mix well with the fluid already there, and it must also mix without causing any chemical action that may damage the system.

Brake Fluid and Water Brake fluid is *hydroscopic.* In other words, it tends to absorb water, and it can even absorb moisture from the air. For this reason, brake fluid should always be stored in a closed, air-tight container. Even small amounts of water in brake fluid will lower the boiling point of the fluid. In severe usage, such as panic stops or repeated braking, the temperature of the fluid is raised to well above the boiling point of water. When the water in the system boils, a steam pocket forms in the system.

Unlike a fluid, steam can be compressed. Therefore, when there is steam in the brake system, the pressure in the system is wasted in compressing the steam. This results in a lack of pressure at the wheel cylinders and, consequently, the wheel cylinder pistons cannot push the brake shoes out tight enough against the drum. To stop the car, the driver must push down the brake pedal farther than usual. In extreme cases, the pedal goes all the way to the floor without developing enough pressure to stop the car.

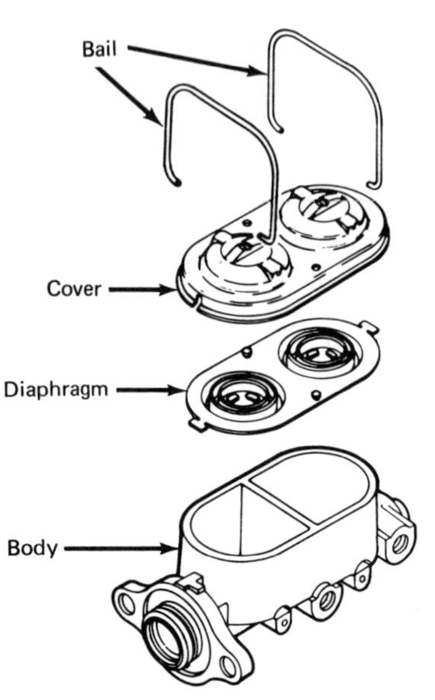

Figure 7.8 A typical dual master cylinder. The cover has been removed to show the position of the diaphragm (courtesy Chevrolet Service Manual, Chevrolet Motor Division).

Water in a brake system can cause problems other than loss of pressure. Since water is heavier than brake fluid, it settles in the lowest parts of the systems. Many of the wheel cylinders you attempt to rebuild will be found ruined beyond further use. Water will have caused deep rust pits at the bottom of the cylinders, and these pits can allow fluid to leak out past the pistons. Many car makers use aluminum pistons which do not rust. However, you will find aluminum pistons that are frozen in their bores by a whitish corrosion. Such corrosion is caused by water between the pistons and their bores.

Most master cylinder reservoirs have a diaphragm, or bellows seal, under the cylinder cover (Figure 7.8). This seal is folded so that it will move up or down with changes in the fluid level and yet will seal out air. The seal cuts down on the amount of moisture the fluid can absorb.

Job 7c

IDENTIFY TERMS RELATING TO BRAKE FLUID

SATISFACTORY PERFORMANCE

A satisfactory performance on this job requires that you do the following:

1 Identify each word or abbreviation by matching its number with the correct definition.
2 Complete part 1 in 10 minutes.

PERFORMANCE SITUATION

1 Hydroscopic	_______ Society of Automotive Engineers
2 DOT	_______ Mixes well with other fluids
3 Miscible	_______ Department of Transportation
4 SAE	_______ Helps prevent air contact with brake fluid
5 Bellows seal	_______ Tends to absorb water
	_______ Prevents leakage past pistons

CHECKING THE LEVEL OF BRAKE FLUID The master cylinder reservoir should be checked regularly to maintain the reserve level of fluid. The procedure for checking the fluid level and adding fluid to a typical master cylinder is outlined below. You should refer to an appropriate manual for specific procedures that may be necessary on particular cars.

1 Clean the reservoir cover thoroughly so that no dirt can fall into the reservoir when the cover is opened.

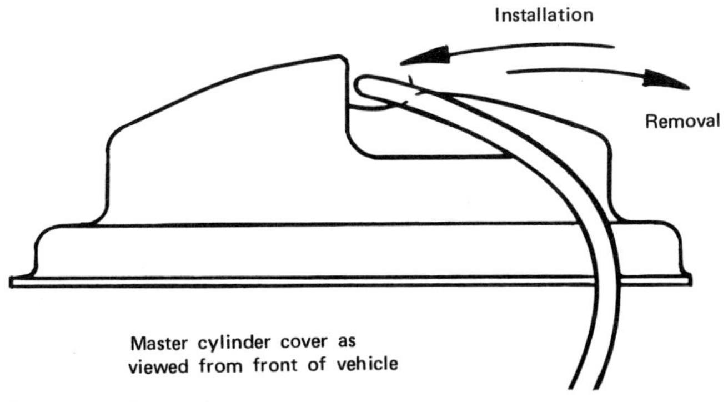

Figure 7.9 Bail or clasp used to secure the cover on a master cylinder reservoir (courtesy Chevrolet Service Manual, Chevrolet Motor Division).

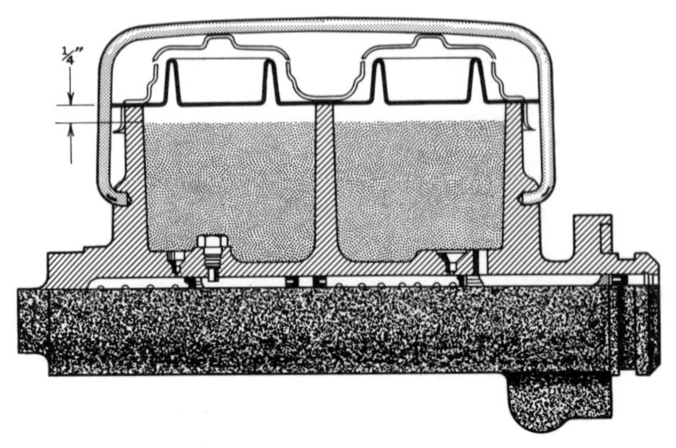

Figure 7.10 Sectioned view of a master cylinder showing the correct level of brake fluid (courtesy Chevrolet Service Manual, Chevrolet Motor Division).

2 Use a screwdriver to pry the bail, or clasp, to one side as shown in Figure 7.9. Hold the cover with your free hand to avoid dropping it.

3 Lift the cover, turn it over, and check the diaphragm seal. If the seal is unfolded, refold it, so that it lies flat again.

4 Check the fluid level. Add fluid, if necessary, to raise it to within $\frac{1}{4}$ in. (about 6 mm) of the top of the casting, as shown in Figure 7.10.

5 Carefully set the cover back in place so the seal is not distorted.

6 Using a screwdriver, carefully pry the bail up and into place in its retaining groove on the cover.

Job 7d

CHECK BRAKE FLUID LEVEL

SATISFACTORY PERFORMANCE

A satisfactory performance on this job requires that you do the following:

1 Check the level of brake fluid in a master cylinder reservoir, and add fluid if necessary.
2 Following the steps in the "Performance Outline," complete the job within 10 minutes.
3 Fill in the blanks under "Information."

PERFORMANCE OUTLINE

1 Protect the fender with a fender cover.
2 Remove the reservoir cover.
3 Check the seal, and refold it if necessary.
4 Add fluid to the correct level if it is needed.
5 Install the cover.
6 Remove the fender cover.

INFORMATION

Vehicle identification ______________________________

Was fluid needed? ____________ Was fluid added? ____________

WHEEL CYLINDERS Though many wheel cylinder designs exist, the most common design is the twin piston type (Figure 7.11). In this design, the fluid enters the cylinder at the center and transmits its pressure equally to the two pistons. The pressure pushes the pistons out toward the ends of the cylinder and, since the cylinders are connected to the brake shoes by means of push rods, the shoes are pushed outward into contact with the drum. Wheel cylinders have only a few simple parts, but each part has an important function. You must understand the function of each part before you can conduct a proper brake inspection or can properly diagnose any brake problem.

Boot Cup Bleeder screw Cup Boot

Piston Cylinder Return spring Piston

Figure 7.11 Exploded view of a twin piston wheel cylinder showing all its component parts (Ford Motor Company).

The Cylinder Body The cylinder body is made of cast iron and has a highly polished cylinder bore. The cylinder is attached to the backing plate, usually by two bolts. A threaded inlet port is provided so that a hose or steel tube can be connected. A bleeder valve is located near the top of the cylinder so that air can be purged from the system.

The Pistons The twin pistons are made of aluminum or of sintered iron. Their inner end is flat. Their outer end is usually concave, or dished, to provide a self-centering seat for the push rods. Many aluminum pistons are anod-

Figure 7.12 Wheel cylinder pistons. Note that the flat ends of the pistons face the center of the wheel cylinder whereas the concave ends face outward (Ford Motor Company).

ized; anodizing provides better wear resistance and minimizes the effects of corrosion (Figure 7.12).

The Piston Cups Piston cups provide a leak-proof seal at the cylinder walls. They are designed with sharp edges, or lips, which enables the fluid to press them tighter against the cylinder walls as the pressure increases (Figure 7.13).

The Spring The spring holds the cups firmly against the pistons when the system is at rest. This action prevents the cups from sliding or cocking in the cylinder, either of which would result in leakage.

Dust Boots Dust boots keep dirt, water, and other foreign matter from entering the cylinder. They are not meant as fluid seals; the fluid should be kept in the cylinder by the piston cups. When inspecting a brake system you should always peel back the dust boots to check for fluid behind them. If fluid leaks out when a dust boot is peeled back, the piston cup is leaking, and, therefore, the cylinder should be either rebuilt or replaced.

Figure 7.13 Wheel cylinder piston cups. Note that the lips of the cups face inward, toward the center of the wheel cylinder (Ford Motor Company).

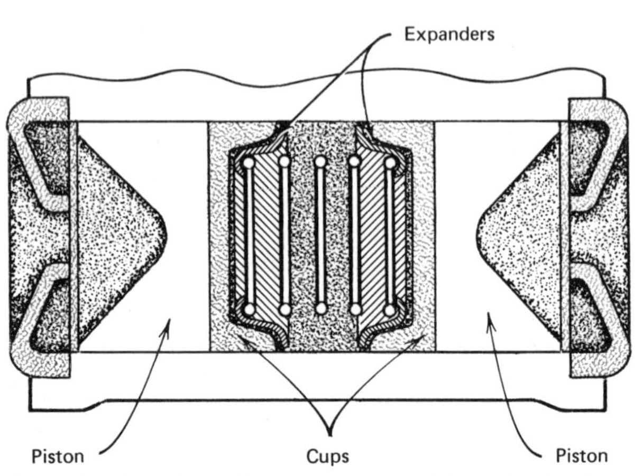

Figure 7.14 Sectioned view of a twin piston wheel cylinder fitted with cup expanders. Note how the expanders direct the spring pressure to the lips of the cups.

The Push Rods The push rods are steel links that connect the pistons to the shoes. They transmit the movement of the pistons to the shoes, pushing the shoes into contact with the drum.

Cup Expanders While not universally used, cup expanders are provided on each end of the spring by some manufacturers (Figure 7.14). The expanders direct the spring pressure to the lips of the cups. This pressure holds the lips tight against the walls of the cylinder. The spring pressure is maintained even when the brakes are released. Thus, the cup expanders help prevent air from entering the cylinder when the pistons move back in.

Job 7e

IDENTIFY WHEEL CYLINDER PARTS

SATISFACTORY PERFORMANCE

A satisfactory performance on this job requires that you do the following.

1 Identify the numbered parts on the drawing under "Performance Situation" by placing each number in front of the correct part name below the drawing.
2 Correctly identify all the parts within 10 minutes.

PERFORMANCE SITUATION

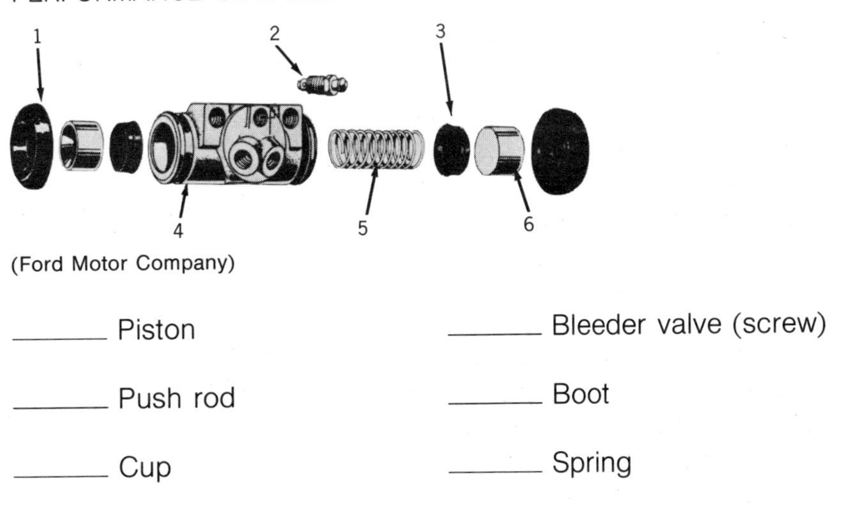

(Ford Motor Company)

______ Piston	______ Bleeder valve (screw)
______ Push rod	______ Boot
______ Cup	______ Spring
______ Cylinder	______ Retainer

THE MASTER CYLINDER To fully understand the operation of the master cylinder, you must know its parts. In diagnosing brake problems, you must understand how the failure of any of these parts can affect the brake system.

All modern cars have a dual hydraulic system. In most instances, the hydraulic system for the front brakes is completely independent of the hydraulic system for the rear brakes. A dual system, of course, requires a dual master cylinder. A dual master cylinder can be compared to two single master cylinders connected in tandem (one in front of the other). The easiest way to learn their parts and the function of those parts is to study a single master cylinder, such as the one shown in Figure 7.15. When you understand how a single unit works, you will progress to the dual master cylinder. Study the parts and their position carefully. Refer back to Figure 7.15 as you study the description of the parts on the pages that follow.

The Cylinder and Reservoir Body The master cylinder body is made of cast iron. The top of the casting is a reservoir that stores extra brake fluid that may be needed by the system. The reservoir also provides an area into which the fluid can expand when its temperature rises. The lower part of the body contains the cylinder, which is a long, smooth hole in which a piston

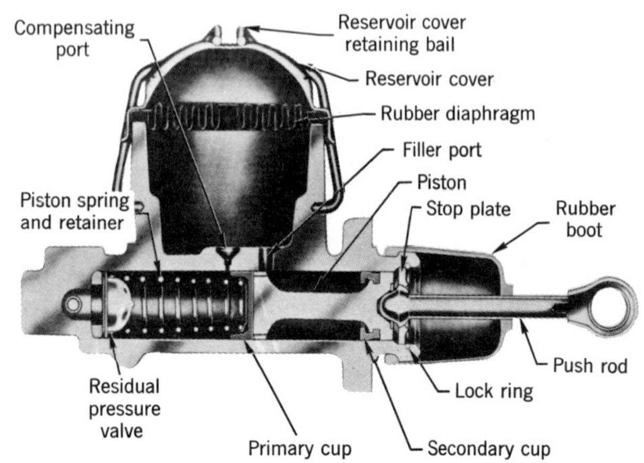

Figure 7.15 A cut-away view of a single master cylinder (General Motors).

can slide back and forth. Two holes, or ports, connect the reservoir to the cylinder.

The Filler Port The filler port is a large hole connecting the reservoir with the cylinder. It allows fluid to flow from the reservoir into the cylinder.

The Compensating Port The compensating port is a small hole connecting the reservoir to the cylinder. This hole allows for the expansion and contraction of the brake fluid caused by changes in temperature. The heat produced by braking causes the fluid to expand, and this expansion creates an excess of fluid in the system. When the brake pedal is released, the compensating port is uncovered, and the excess fluid escapes to the reservoir. When the temperature of the fluid in the system drops, the fluid can return to the system. A plugged or restricted compensating port causes a pressure buildup in the system. Such a buildup means brakes that drag or fail to release. A compensating port can be plugged by foreign matter, blocked by a swollen primary cup, or covered by the primary cup if the pushrod is improperly adjusted.

The Piston The piston used in the master cylinder is spool-shaped and is fitted with two cups or seals. It is pushed forward in the cylinder when the brake pedal is depressed. Then the force of the piston applies pressure to the fluid in the system.

The Primary Cup The forward cup, or seal, on the piston is called the primary cup. It seals in all the pressure ahead of the piston. Any leakage here allows the brake fluid to return to the reservoir through the compensating port and the filler port. A failure of the primary cup is usually indicated by a loss of pedal. In other words, the brake pedal sinks to the floor under the steady pressure by the driver.

The Secondary Cup The secondary cup provides a sealed chamber behind the primary cup. Fluid flows from the reservoir, through the filler port, and into this chamber. Since the sealing edges of both the primary and secondary cups face forward, the fluid in this chamber can flow under the edges of the primary cup and into the cylinder. However, the fluid cannot flow past the secondary cup; and, as a result, it cannot leak out of the cylinder. If the secondary cup is worn or damaged, fluid will leak out of the push rod opening. The leak will be evident either at the firewall or where the master cylinder is fastened to the booster unit. In many instances, a leaking secondary cup will allow fluid to seep out around the boot and to drip onto the carpet or floor mat under the brake pedal.

The Residual Pressure Valve (Check Valve) The residual pressure valve is found in master cylinders used in drum brake systems. This valve maintains a slight static pressure in the hydraulic system at all times, even when the brake pedal is released. The pressure varies from 5 psi to 20 psi, depending on the particular system. This slight pressure holds the lips of the wheel-cylinder piston cups tightly against their cylinder walls. Thus it aids in keeping air from entering the system when the system is at rest. A failure of the residual pressure valve is usually indicated when a brake system requires frequent bleeding to maintain a firm pedal.

The Push Rod The push rod transmits to the master cylinder piston the force that the driver applies to the brake pedal. As the piston must return completely to uncover the compensating port after each brake application, the length of the push rod is very critical. The push rod must be adjusted so that there is a slight bit of free play, or travel, between the rod and the piston when the brake pedal is released. A push rod that is adjusted too long will not allow the primary cup to uncover the compensating port. The result will be a failure of the brakes to release after several brake applications. A push rod that is adjusted too short will allow excessive pedal travel before the brakes start to aply.

Reservoir Cover and Seal The reservoir cover is vented to the atmosphere but is fitted with a diaphragm-type seal. This seal acts as a gasket to prevent leakage. It also serves to separate the surface of the fluid from the atmosphere. As brake fluid is hydroscopic, this seal helps prevent contamination of the fluid with water absorbed from the air.

Job 7f

IDENTIFY MASTER CYLINDER PARTS

SATISFACTORY PERFORMANCE

A satisfactory performance on this job requires that you do the following:

1 Identify the numbered parts on the drawing under "Performance Situation" by placing the number for each part in front of the correct part name under the drawing.

2 Correctly identify all the parts within 15 minutes.

PERFORMANCE SITUATION

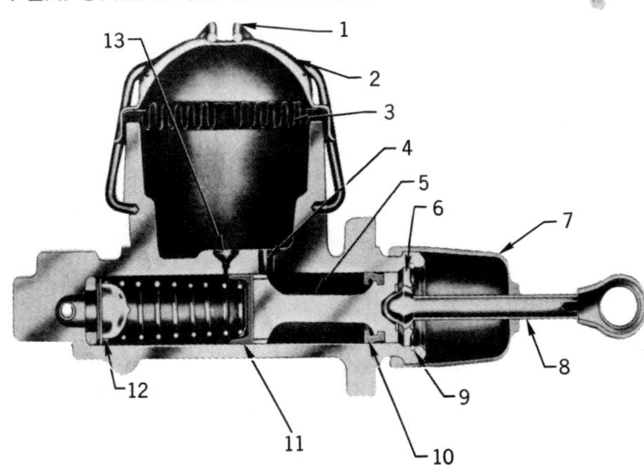

______ Filler port

______ Reservoir cover

______ Secondary cup

______ Residual pressure valve

______ Primary cup

______ Cover retaining bail

______ Piston

______ Compensating port

______ Stop plate

______ Piston pin

______ Push rod

______ Diaphragm

______ Boot

______ Lock ring

SUMMARY

In this chapter you have learned the basic rules of hydraulics and how they apply to brake systems. You have learned the standards that brake fluid must meet and the properties it must have to meet them. Moreover, you can now identify all the parts of wheel cylinders and master cylinders, and you understand how each part functions.

SELF-TEST

Each incomplete statement or question in this test is followed by four suggested completions or answers. In each case select the *one* that best completes the statement or answers the question.

1 If a wheel cylinder is replaced with one having larger pistons
   I. the pressure in the system will be increased
   II. the brakes at that wheel will be applied with a greater force
   A. I only
   B. II only
   C. Both I and II
   D. Neither I nor II

2 The front brakes do more work than the rear brakes because the
   I. front wheel cylinders are closer to the master cylinder than are the rear wheel cylinders
   II. weight of a car shifts to the front during a stop
   A. I only
   B. II only
   C. Both I and II
   D. Neither I nor II

3 The standards for brake fluid are determined by the
   I. Society of Automotive Engineers (SAE)
   II. Department of Transportation (DOT)
   A. I only
   B. II only
   C. Both I and II
   D. Neither I nor II

4 Cars equipped with disc brakes and subjected to severe braking should use brake fluid meeting which of the following specifications?
   A. DOT 1
   B. DOT 2
   C. DOT 3
   D. DOT 4

5 Most master cylinders have a diaphragm seal under the cover of the reservoir. This seal is used to
   A. maintain pressure in the system
   B. prevent fluid evaporation
   C. prevent the entrance of moisture-laden air
   D. prevent fluid leakage past the primary cup

6 If a can of brake fluid is left open, the fluid will
   A. absorb moisture from the air
   B. evaporate
   C. become miscible
   D. corrode the container

7 The spring in a wheel cylinder

A. pulls the pistons back when the brake pedal is released
B. holds the brake shoes out close to the drum
C. holds the cups against the pistons
D. maintains a slight pressure in the system

8 Brakes that drag or fail to release can be caused by a
A. leaking primary cup
B. plugged compensating port
C. failure of the residual pressure valve (check valve)
D. leaking secondary cup

9 A master cylinder push rod must be adjusted so that
I. there is a slight bit of free play present before the push rod moves the piston
II. the compensating port is uncovered by the primary cup
A. I only
B. II only
C. Both I and II
D. Neither I nor II

10 A drum brake system requires frequent bleeding to maintain a firm pedal. The most probable cause of this problem is a
A. restricted compensating port
B. leaking primary cup
C. leaking residual pressure valve (check valve)
D. leaking diaphragm seal

Chapter 8 Basic Hydraulic System Service

Now that you know some of the basic principles of hydraulics, you are ready to put your knowledge to work in servicing some of the hydraulic components of a brake system. In this chapter you will learn how to purge air from a hydraulic system. A hydraulic system works properly only if it is completely filled with fluid. Air or a leak in the system results in loss of pressure, which means loss of braking power. Any condition causing a loss of braking power should be repaired immediately.

In this chapter you will learn how to do all the things necessary for servicing three major parts of a hydraulic system: (1) lines, (2) hoses, and (3) wheel cylinders. Your objectives will be to:

1
Bleed a brake system.
2
Replace brake hoses.
3
Fabricate and replace brake lines.
4
Replace wheel cylinders.
5
Overhaul wheel cylinders.

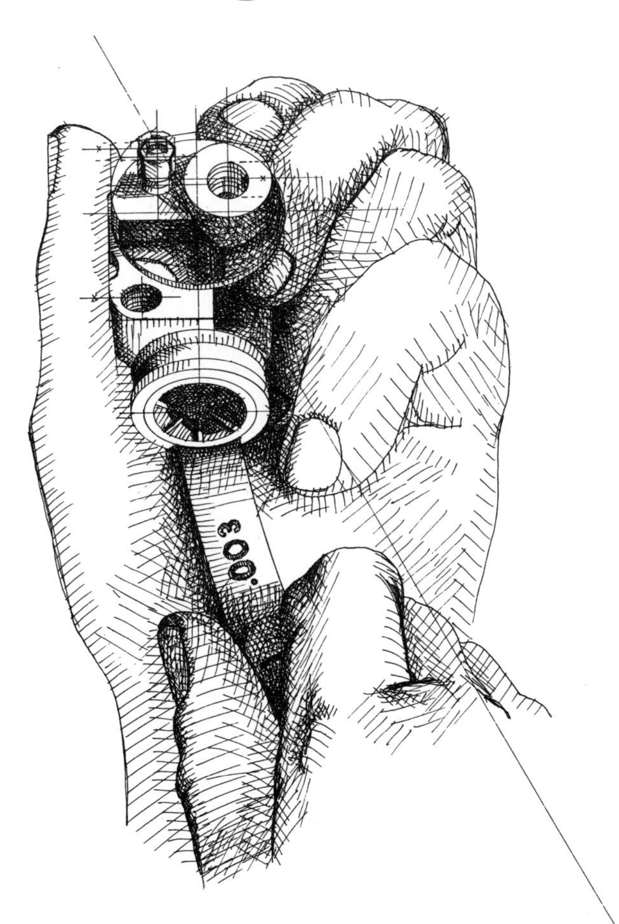

BLEEDING As stated before, a hydraulic system will function properly only when there is no air in the system. Air, like steam, can be compressed and can therefore cause the same problem that steam causes. For this reason, car makers provide for *bleeding,* or removing, all air from the hydraulic system.

The need for bleeding is usually indicated by a soft, spongy feeling when the brake pedal is depressed. This means that air is present in the system. Air may enter the system during any repair that requires the removal of a part of the system. Air can enter the system for other reasons, too. It can enter because the reservoir is not filled with fluid and a residual pressure check valve in the master cylinder is defective.

Bleeding a system means forcing fluid through the system while allowing the air to escape. To make bleeding possible, bleeder valves are provided on all wheel cylinders and at other places where air might be trapped. Two methods of bleeding are commonly used: *manual bleeding* and *pressure bleeding.*

Manual Bleeding Manual bleeding requires no special equipment, but it is a two-mechanic operation. One mechanic pumps the brake pedal slowly and firmly to push the fluid through the system. The other mechanic opens and closes the bleeder valves and watches for the escape of air bubbles. As the brake system is bled, much of the fluid is forced out. If the fluid level in the master cylinder reservoir is allowed to drop below the filler port, more air will be pulled into the system. Therefore, the fluid level in the reservoir must be continuously watched and maintained.

Pressure Bleeding This method of bleeding requires the use of a *brake bleeder tank,* an example of which is shown in Figure 8.1. A brake bleeder tank is a container in which brake fluid is held under pressure. Because brake fluid is hydroscopic, bleeder tanks have two chambers separated by a flexible diaphragm. One chamber contains the fluid. The other contains air. The air chamber can be pressurized with compressed air from the shop air system.

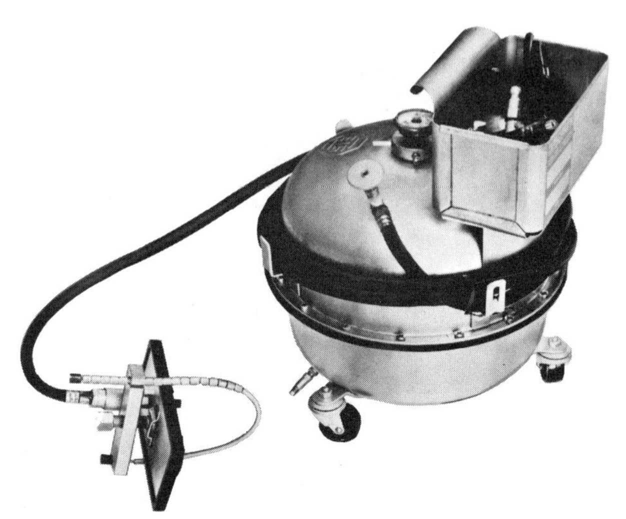

Figure 8.1 A brake bleeder tank (Chrysler Corporation).

When the air chamber is pressurized, it exerts pressure through the flexible diaphragm against the fluid in the fluid chamber. A valve on the fluid side of the tank allows the fluid to escape through a hose. The hose is tightly connected to the top of the master cylinder reservoir by an adapter (Figure 8.2). There are various adapters to fit master cylinder castings of different shapes.

The use of a brake bleeder allows you to bleed brakes by yourself. The bleeder pressurizes the system and maintains the level of fluid in the reservoir. You can open and close the bleeder valves without depending on another mechanic.

Other Methods of Bleeding Other methods of bleeding are sometimes required to purge air from certain hydraulic systems. One of those methods is *gravity bleeding.* It allows the fluid, by its own weight, to push the air out through a bleeder valve located low in the system. Another method is *reverse bleeding.* It involves pumping fluid into the system through a low point, such as a wheel cylinder, to force air out through the master cylinder. These methods are seldom used. When required, they are

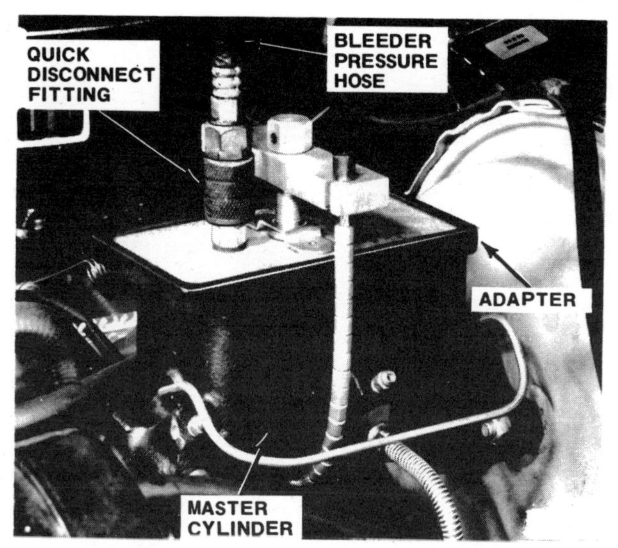

Figure 8.2 A brake bleeder adapter cover installed on a master cylinder (courtesy American Motors Corporation).

specified and described in the manufacturer's service manuals.

Special Bleeding Tools Most bleeder valves have *hexagonal* (six-sided) heads that can be gripped with a box wrench. However, those valves are usually located where they are hard to reach and are hard to turn with conventional wrenches. You will also find many bleeder valves that are rusted in place. An improperly fitted wrench or an improperly applied force may round off the corners of the hex-head or break off the valve. For these reasons, you should use special wrenches designed for opening bleeder valves. These wrenches are called bleeder wrenches. They have specially formed hexagonal openings, and are available in many sizes and shapes. Figure 8.3 shows one of the more commonly used bleeder wrenches.

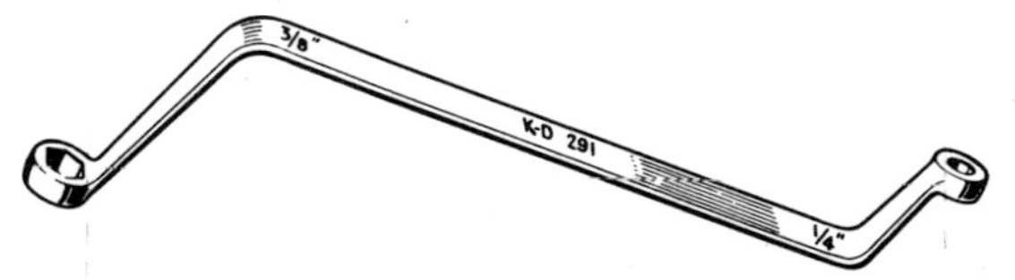

Figure 8.3 A typical bleeder wrench (K-D Manufacturing Company).

Figure 8.4 A bleeder hose and jar in use (courtesy Chevrolet Service Manual, Chevrolet Motor Division).

Regardless of the method you use in bleeding a hydraulic system, you should always use a bleeder hose and jar. One end of the hose is pushed over the end of the bleeder valve. The other end is placed in a jar that contains enough brake fluid to cover the end of the hose. The fluid in the jar helps you see any air bubbles that might be forced out of the system. The use of the bleeder hose and jar also frees you from having to clean up the mess you would have made if you did not use them. Figure 8.4 shows a bleeder hose and jar in use.

***Note.* Never reuse brake fluid even though a considerable amount of it may be expelled**

from a system while brakes are bled. After bleeding brakes, discard all the fluid in the bleeder jar.

Bleeding Procedure

In the usual procedure for bleeding a hydraulic brake system the valve that is farthest from the master cylinder is bled first and the valve that is closest to it is bled last. However, auto manufacturers do not always recommend this procedure. Their recommendations vary with the type of system (drum, disc, or combination), the type of master cylinder (single or dual), the type of external valves with which the system is fitted and, finally, the type of antiskid system used. The proper procedure for bleeding any system is given in the manufacturer's service manual for each car.

Job 8a

BLEED BRAKES

SATISFACTORY PERFORMANCE

A satisfactory performance on this job requires that you do the following:

1 Bleed the brake system on the car assigned.
2 Following the steps in the "Performance Outline" and the manufacturer's procedure and specifications, complete the job and obtain a high, hard brake pedal within 150 percent of the manufacturer's suggested time.
3 Fill in the blanks under "Information."

PERFORMANCE OUTLINE

1 Raise and support the car.
2 Inspect the system for leaks.
3 Fill the master cylinder.
4 Adjust the service brakes if necessary.
5 Bleed the system.
6 Check the pedal height and feel.
7 Lower the car.
8 Check brake operation.

INFORMATION

Vehicle identification ______________________

Reference used ______________ Page(s) ______________

BRAKE HOSES The hydraulic system contains several flexible lines, or hoses. They must be flexible to allow for the turning of the wheels, and are made of alternating layers of rubber and fabric (Figure 8.5). The threaded metal fittings on the hose ends grip the hose internally as well as externally. They are compressed on with special equipment. This type of construction enables the hose to withstand high pressures without swelling or bursting.

Brake hoses should be replaced whenever the outer jacket becomes cracked or worn. Fluid stains visible at the end fittings indicate leaks. Although there may not yet be any fluid loss, the

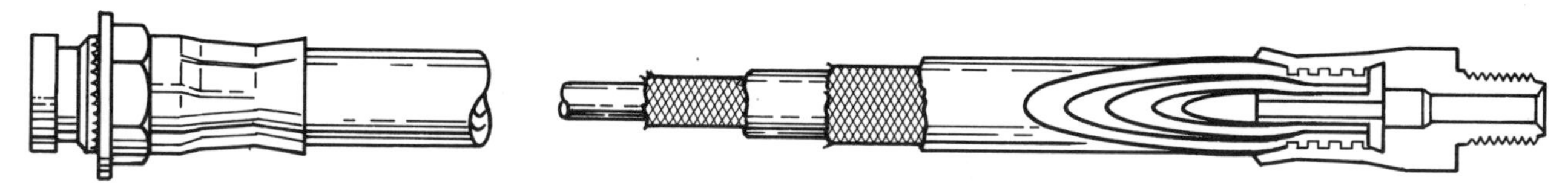

Figure 8.5 A typical two-ply hose made of rubber and fabric. Note how the fittings on the ends are crimped so that the hose is gripped internally and externally (Bendix Automotive Aftermarket Operations).

hose should be replaced before it fails. Bubbles or blisters in the outer jacket, as shown in Figure 8.6, also mean that a hose should be replaced.

Whenever you replace a brake hose, check during and after the installation to see that the hose is not in contact with any part of the car body or chassis. Any contact can result in rubbing that can wear through the hose. Allowance must also be made for the movement of the suspension system, the loads the car may carry, and the complete range of steering angles. The outer jacket of most brake hoses usually has parallel ribs in its surface. These ribs provide a means of checking for a twisted hose. When installing a brake hose, avoid any kinking or twisting of the hose. If a kink or a twist does occur, loosen the hose at the tubing nut end, remove the lock clip, and reposition the hose to eliminate the kink or twist. Install the lock clip and tighten the tubing nut to maintain the new position.

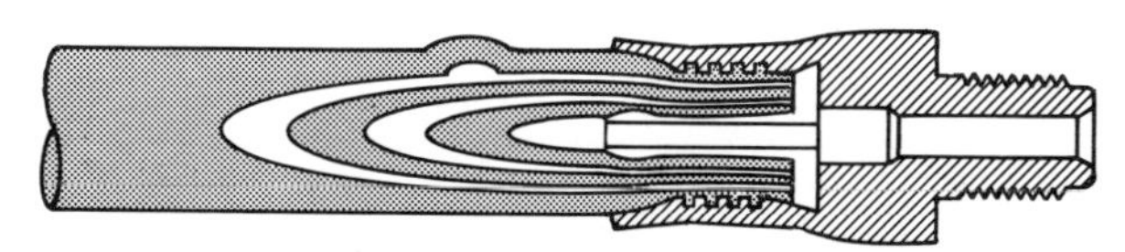

Figure 8.6 A sectioned view of a brake hose showing how fitting leakage will cause fluid stains and the formation of bubbles in the jacket (Bendix Automotive Aftermarket Operations).

REPLACING BRAKE HOSES

For removing and installing a front brake hose, check the shop manual for specific procedure that may apply to a particular car. The general procedure is given below.

Removal

1 Clean all dirt from the ends of the hose. This reduces the possibility of getting dirt into the system during assembly.

2 By using pliers as shown in Figure 8.7, remove the brake hose lock clip that holds the hose to the chassis bracket.

3 Hold the end of the hose with an open end wrench. Use a tubing wrench or flare nut wrench to disconnect the tubing nut (Figure 8.8).

4 Use an open end wrench to loosen the brake hose from the wheel cylinder (Figure 8.9).

Installation

1 Check the area near the threaded hole in the wheel cylinder, the threaded end of the tubing, and the area near the chassis bracket for dirt. Carefully clean away any dirt that you find.

2 Place a new copper washer over the threaded end of the hose, and thread the hose into the wheel cylinder.

3 Tighten the hose with an open end wrench.

4 Hold the hose in place at the chassis bracket, and connect the

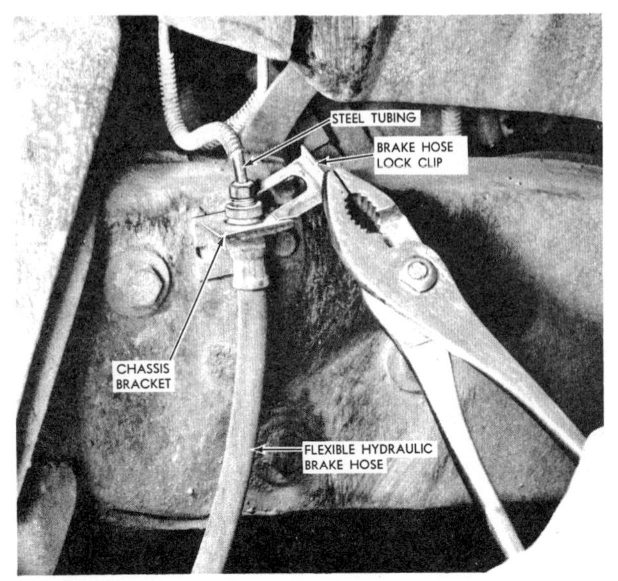

Figure 8.7 Removing a brake hose lock clip (Raybestos Division, Raybestos Manhattan, Inc.)

Figure 8.8 Disconnecting a tubing nut. Note that the tubing nut is turned with a tubing wrench while the hose is held with an open end wrench (Raybestos Division, Raybestos Manhattan, Inc.)

tubing to the hose. To reduce the possibility of thread damage, use your fingers to start the threads.

5 Position the hose so that it is not kinked or twisted, and tighten the tubing nut. Hold the hose with an open end wrench, and turn the tubing nut with a tubing wrench or a flare nut wrench.

6 Install the brake hose lock clip to secure the hose to the chassis bracket.

7 Check the hose position to be sure the hose will not contact anything when the wheels are turned and when they move through the limits of the suspension system. Reposition the hose if necessary.

8 Bleed the brake system.

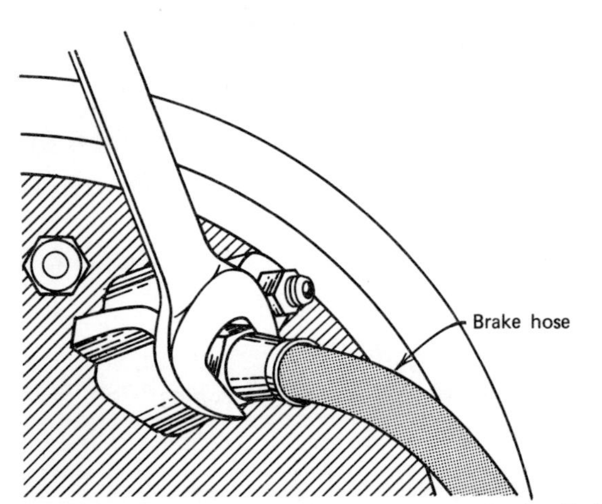

Figure 8.9 Removing a brake hose from a wheel cylinder.

Job 8b

REPLACE A BRAKE HOSE

SATISFACTORY PERFORMANCE

A satisfactory performance on this job requires that you do the following:

1 Replace a brake hose on the car assigned.
2 Following the steps in the "Performance Outline" and the manufacturer's procedure and specifications, complete the job within 150 percent of the manufacturer's suggested time.
3 Fill in the blanks under "Information."

PERFORMANCE OUTLINE

1 Raise and support the car.
2 Disconnect the brake line from the hose.
3 Remove the hose retaining clamp.
4 Remove the hose from the wheel cylinder or caliper.
5 Install the hose at the wheel cylinder or caliper.
6 Attach the brake line to the hose.
7 Align the hose, tighten the fitting, and install the clamp.
8 Bleed the brakes.
9 Check the pedal height and feel.
10 Lower the car.
11 Check brake operation.

INFORMATION

Vehicle identification ______________________________

Location of hose replaced ______________________________

Reference used ________________ Page(s) ________________

BRAKE LINES Brake lines are subjected to the same high pressures as other parts of the hydraulic system. Therefore, such lines are always made of steel tubing. If any section of a brake line is damaged or weakened by rust or corrosion, that entire section must be replaced with steel tubing of the same size, type, and length. Copper tubing should never be used in a hydraulic brake system because it does not have the strength of steel tubing and can rupture under high braking pressures.

The fittings and connections in the hydraulic system are as important as the lines. The ends of the steel tubing must be *double flared,* which means that they are expanded and then folded over themselves to provide a strong, uniform surface. The flared end can then be held tightly in a fitting. Tubing can also be single flared, but a single flare will usually split and cause a failure in the hydraulic system. Figure 8.10 illustrates both single and double flares.

You should purchase steel brake lines "ready made" whenever possible. Most automotive supply houses stock various sizes and lengths

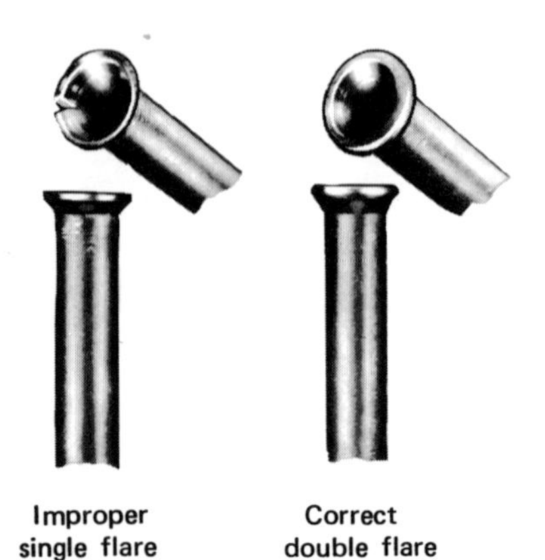

Figure 8.10 Flared ends of brake tubing (General Motors).

of brake lines that are complete with the necessary nuts and fittings installed. These lines are all double flared, and they are furnished in straight lengths as shown in Figure 8.11.

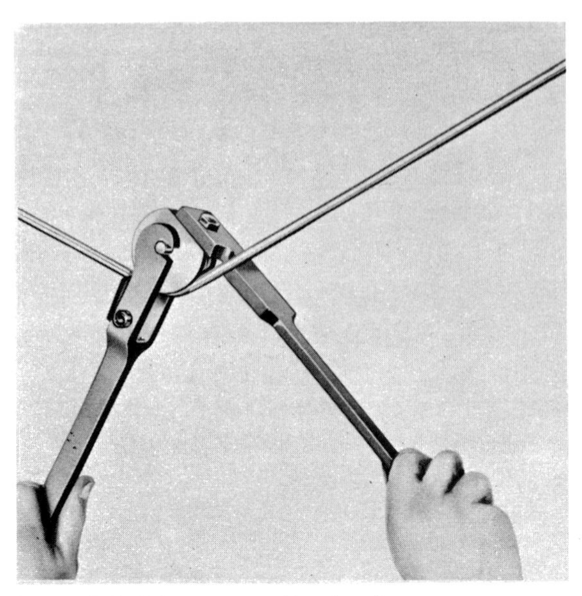

Figure 8.13 A lever-type tubing bender.

Figure 8.11 A length of stock steel brake line (Raybestos Division, Raybestos Manhattan, Inc.)

Bending Steel Tubing To replace a defective brake line, bend the new line to duplicate the shape of the one you remove. Figure 8.12 shows a brake line bent to clear a suspension part of the frame. Note that the bends are smooth and not kinked. Although steel tubing

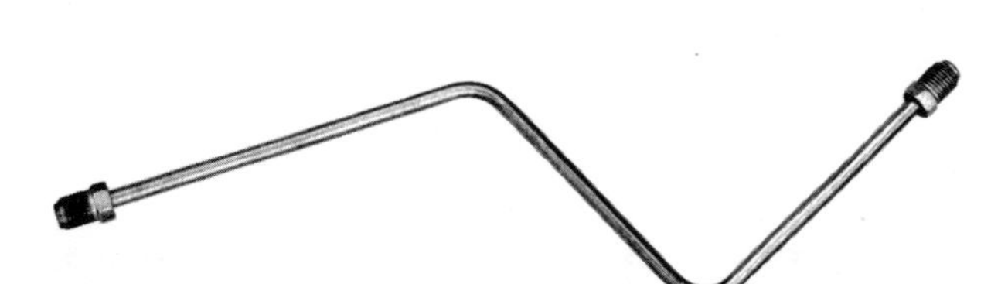

Figure 8.12 A length of stock steel brake line bent to clear chassis components (Raybestos Division, Raybestos Manhattan, Inc.)

may be bent by hand to obtain a gentle curve, any attempt to bend a tight curve by hand will usually kink the tubing. Since a kink in a brake line will weaken the line, a kinked line should never be used. To avoid kinking, always use a bending tool. There are two types of bending tools. The first type uses levers to roll the tubing around a wheel that is shaped like a pulley. This tool, shown in Figure 8.13, can be attached to the tubing at any place along its length, even if the tubing ends are flared.

The second type of tubing bender is a tightly wound coil spring that is slipped over the tub-

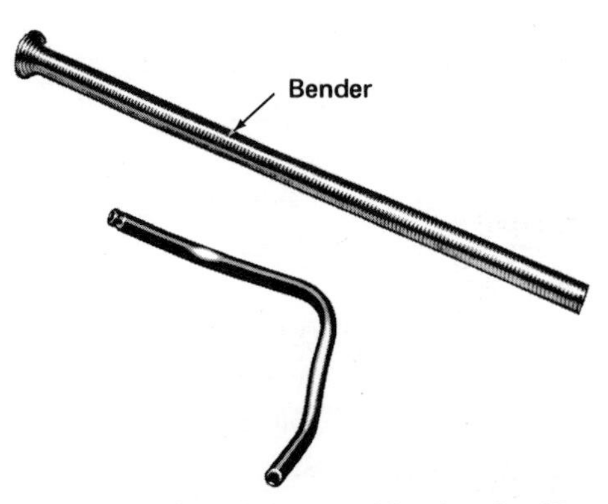

Figure 8.14 A spring-type tubing bender (General Motors).

ing. With the spring in place, you can bend the tubing by hand without danger of kinking it. This type tool is shown in Figure 8.14. Its use, however, is limited to unflared tubing because it will not slide over a flared end.

Cutting Steel Tubing When you have to make your own brake lines, you will have to cut the tubing to the length you need. A special tool, called a *tubing cutter,* must be used to cut the ends of the tubing perfectly square and to keep the tubing itself perfectly round. A tubing cutter does this by supporting the tubing with a pair of rollers while a sharp cutting wheel is rolled around the tubing. A typical tubing cutter is shown in Figure 8.15.

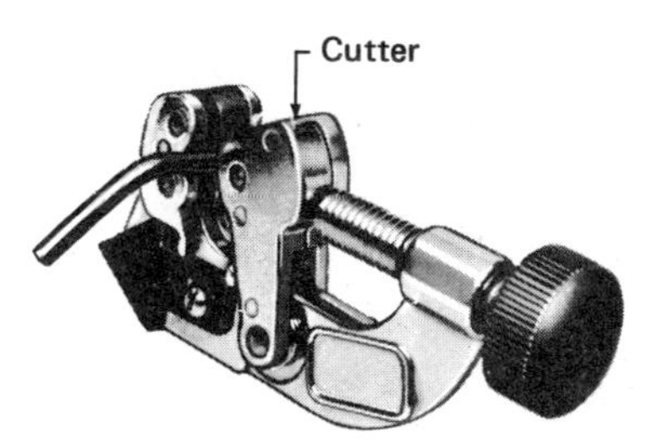

Figure 8.15 A tubing cutter (General Motors).

After the tubing has been cut, any burrs that may have formed inside the edge of the tubing must be removed (Figure 8.16).

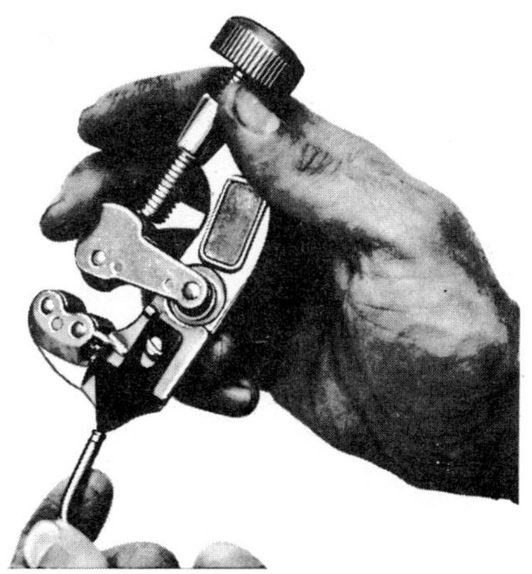

Figure 8.16 Removing burrs from the inner edge of the tubing (General Motors).

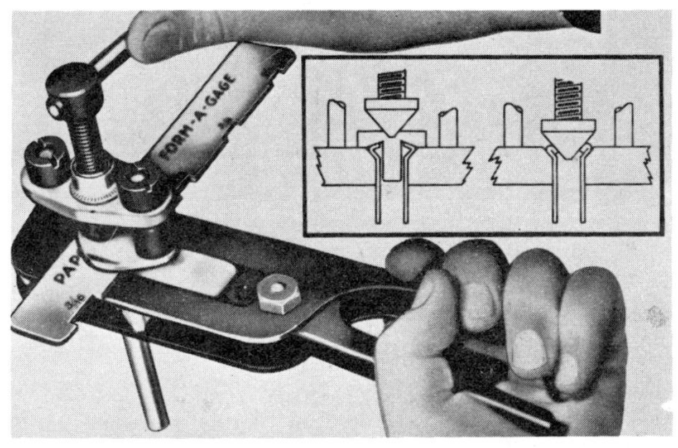

Figure 8.17 A double flaring tool in use. The insert shows the two steps necessary to obtain the proper double flare (Chrysler Corporation).

Double Flaring Steel Tubing To obtain a double flare, you must use a special double flaring tool similar to the one shown in Figure 8.17. There are several types of double flaring tools available. Their use varies with each type. Be sure to follow the instructions furnished with the tool you use.

Regardless of the type tool you use, double flaring is accomplished in the two steps shown in Figure 8.18. The size and angle of the flare are determined by the tool. Careful use of the

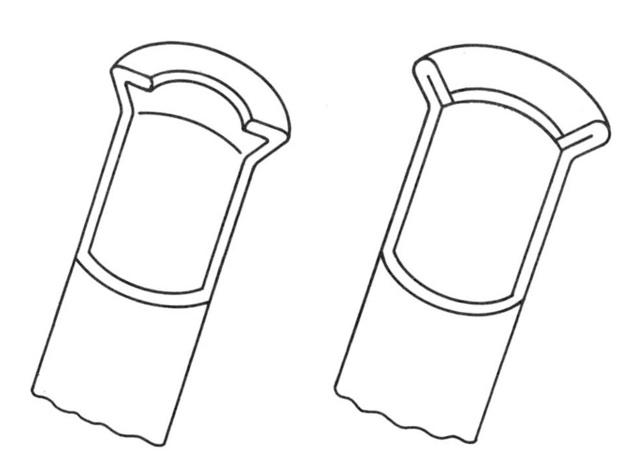

Figure 8.18 Flaring operations (courtesy Chevrolet Service Manual, Chevrolet Motor Division).

double flaring tool will enable you to produce strong, leakproof connections. Figure 8.19

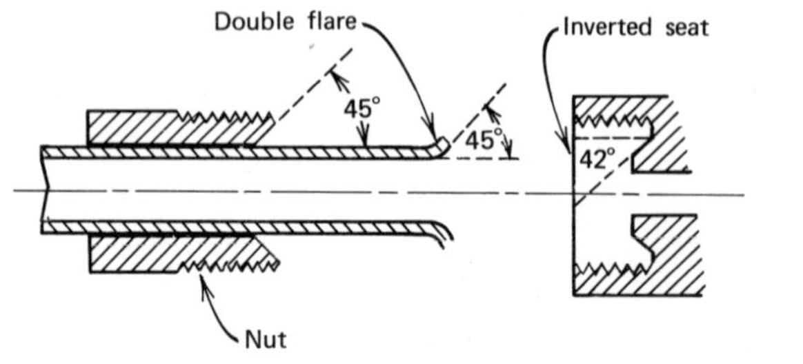

Figure 8.19 Cross-sectional view of brake tubing and fittings.

shows the importance of the correct flare. Notice that the angle of the flare and the nut are both 45° while the angle of the seat is 42°. When the nut is tightened into the fitting, the difference in angles, called an *interference angle,* causes both the seat and the flared end of the tubing to wedge together. Properly assembled, brake lines connected in this way provide joints that can withstand the high pressures in the hydraulic system.

FABRICATING A STEEL BRAKE LINE

Here is the procedure you should follow in making a brake line from steel tubing:

1 Using a tubing cutter, cut the tubing to the desired length.

2 With the reamer attached to the tubing cutter, remove any burrs from the inner edges of the tubing.

3 Use a tube bender to shape the tubing as required.

4 Install a tubing nut or the required fitting on one end of the tubing (Figure 8.20).

Figure 8.20 Nut installed on end of tubing (General Motors).

5 Using a double flaring tool, position the tubing so that its end protrudes a distance equal to the height of the forming die (Figure 8.21).

6 Force the forming die down to buckle the tubing. To avoid cocking the press, use both hands (Figure 8.22).

7 Remove the forming die and again turn the press down to fold the metal into a double flare (Figure 8.23).

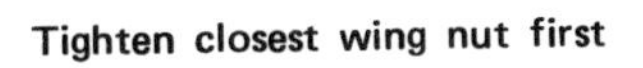

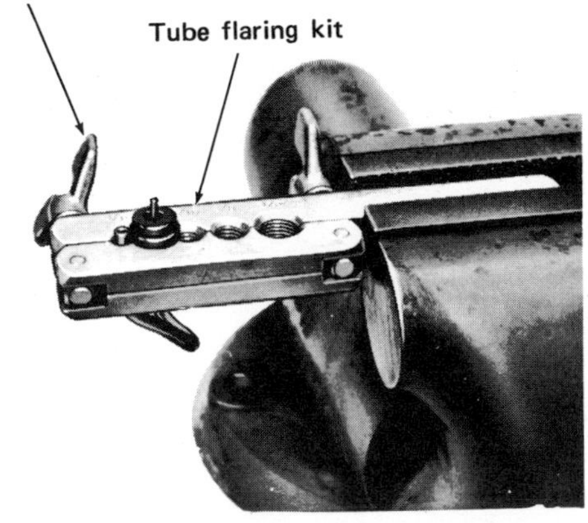

Figure 8.21 Using the forming die to gauge the height of the tubing in the flaring tool clamp (General Motors).

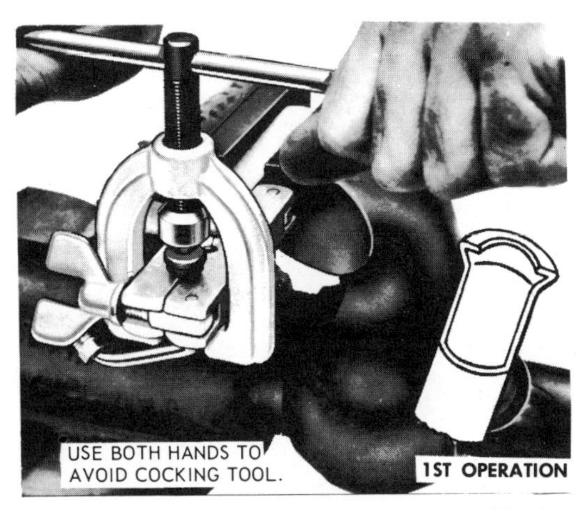

Figure 8.23 The second flaring operation. The tapered end of the flaring tool completes the fold in the tubing (General Motors).

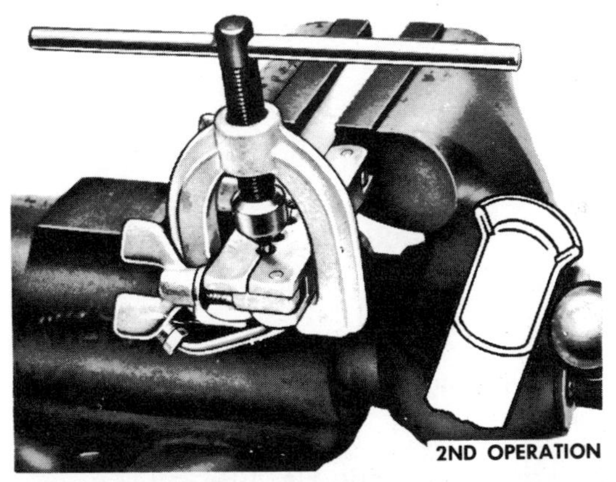

Figure 8.22 The first flaring operation. The forming die is used to obtain the buckle crimp in the tubing (General Motors).

8 Repeat steps 4 through 7 on the other end of the tubing.

Job 8c

REPLACE A BRAKE LINE

SATISFACTORY PERFORMANCE

A satisfactory performance on this job requires that you do the following:

1 Replace a brake line on the car assigned.
2 Following the steps in the "Performance Outline" and the manufacturer's procedure and specifications complete the job within 150 percent of the manufacturer's suggested time.
3 Fill in the blanks under "Information."

PERFORMANCE OUTLINE

1 Raise and support the car.
2 Disconnect and remove the brake line.
3 Fabricate a new brake line from steel tubing.
4 Install the brake line.
5 Bleed the system.
6 Check the pedal height and feel.
7 Lower the car.
8 Check brake operation.

INFORMATION

Vehicle identification ______________________

Location of line replaced ______________________

Reference used ______________ Page(s) ______________

WHEEL CYLINDER SERVICE

Replacing and overhauling wheel cylinders are two of the most commonly performed services for the hydraulic system. Wheel cylinders should be replaced or overhauled whenever the brake shoes are replaced, even though the cylinders show no signs of leakage.

Removal and Installation

To remove and install a front wheel cylinder, you should check the service manual of the car on which you are working. Here is the general procedure:

Removal

1 Raise and support the car. Remove the wheel, drum, and brake shoes to gain access to the wheel cylinder.

2 Using an open end wrench, loosen the brake hose but do not remove it.

3 Using a socket wrench or a box wrench, remove the wheel cylinder mounting bolts.

4 Pull the wheel cylinder away from the backing plate, and unscrew the wheel cylinder from the hose.

Installation 1 Clean the backing plate and the end of the hose so that dirt will not be carried into the system.

2 Install a new copper seal washer on the threaded end of the brake hose.

3 Thread the new wheel cylinder on the hose, turning the wheel cylinder until it contacts the washer.

4 Hold the wheel cylinder in position on the backing plate, and install the retaining bolts finger tight.

5 Using a socket or box wrench, tighten the wheel cylinder retaining bolts.

Note. Some backing plates are made with formed metal tabs called piston stops. These tabs project out at the ends of the wheel cylinders (Figure 8.24). They prevent the wheel cylinder pistons from accidentally coming out of their bores when the brake shoes are removed. Other backing plates are formed so that the wheel cylinder fits between two protrusions. These protrusions act as piston stops also. When installing wheel cylinders on backing plates with piston stops, you must be careful that the boots and pistons are not jammed against the stops. The boots and pistons should be squeezed into the cylinder and held there while the attaching bolts are tightened.

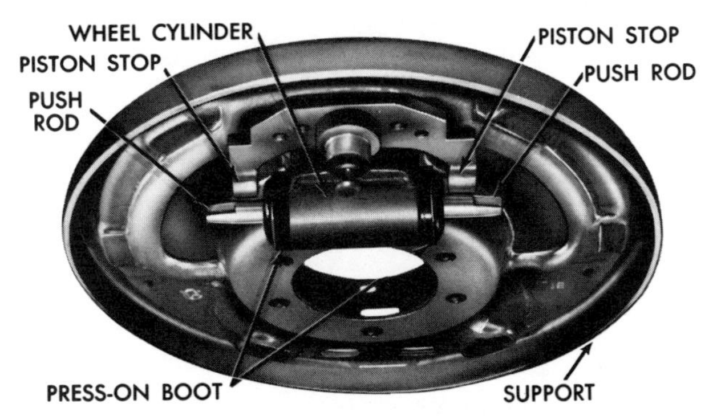

Figure 8.24 View of a backing plate showing the location of the piston stops (Chrysler Corporation).

6 Using an open end wrench, tighten the brake hose.

7 Check to be sure the hose is not kinked or twisted. Be sure the hose will not come in contact with anything when the wheels are turned and when they move through the limits of the suspension system.

8 Install the brake shoes and attached hardware.

9 Install the brake drum, and adjust the bearings.

10 Bleed the system.

11 Install the wheel and lower the car to the floor.
12 Check the pedal height and brake fluid level.

Job 8d

REPLACE A WHEEL CYLINDER

SATISFACTORY PERFORMANCE

A satisfactory performance on this job requires that you do the following:

1 Replace a wheel cylinder on the car assigned.
2 Following the steps in the "Performance Outline" and the manufacturer's procedure and specifications, complete the job within 150 percent of the manufacturer's suggested time.
3 Fill in the blanks under "Information."

PERFORMANCE OUTLINE

1 Raise and support the car.
2 Remove the wheel, the drum, and the brake shoes.
3 Remove the wheel cylinder.
4 Install the wheel cylinder.
5 Install the brake shoes and the drum.
6 Adjust the bearings.
7 Bleed the system.
8 Check the pedal height and feel.
9 Install the wheel.
10 Lower the car.
11 Check brake operation.

INFORMATION

Vehicle identification ________________________

Location of wheel cylinder replaced ________________________

Reference used ________________ Page(s) ________________

OVERHAULING A WHEEL CYLINDER Before you attempt to overhaul a wheel cylinder, you should open the bleeder valve. In many instances the bleeder valve will be rusted in place and may break off when you attempt to turn it. If the bleeder valve will not open or if it breaks off when you attempt to turn it, replace the cylinder.

The following steps outline the procedure for overhauling a common twin-piston wheel cylinder. Though the procedure shows a wheel cylinder that has been removed from a vehicle, in many cases a wheel cylinder can be overhauled without being removed.

1 Remove the rubber boots. Push out the pistons, cups, and spring (Figure 8.25). Discard the cups and boots. They should never be reused.

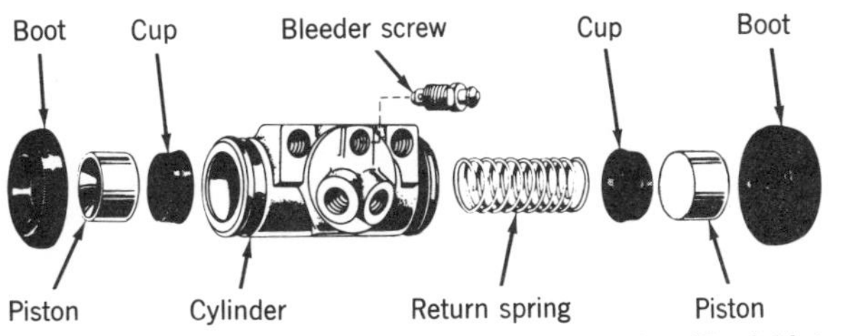

Figure 8.25 Exploded view of a wheel cylinder (Ford Motor Company).

2 Inspect the cylinder bore by holding the cylinder in front of a light. Discard the cylinder if it is deeply pitted or scored.

3 Wash all parts in clean brake fluid.

4 Using a $\frac{1}{4}$ in. drill motor and a wheel cylinder hone, of the type shown in Figure 8.26, hone the cylinder by moving the hone back and forth in the cylinder bore. The hone stones should be kept wet with brake fluid, and the hone should never be pushed or pulled out of the cylinder ends while it is still turning (Figure 8.27).

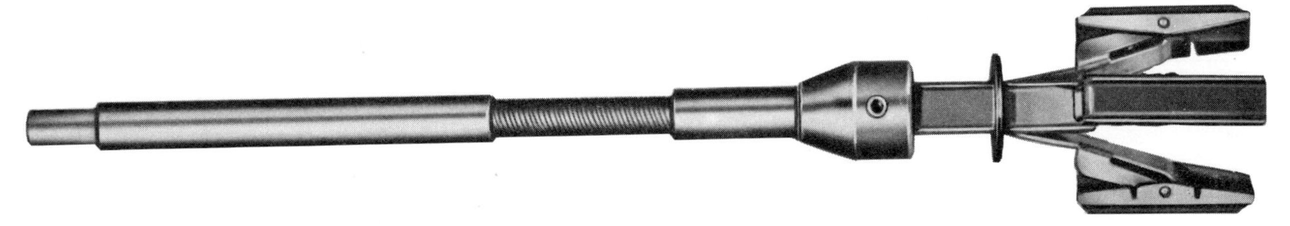

Figure 8.26 A wheel cylinder hone (courtesy AMMCO TOOLS, Inc.)

5 Clean the cylinder and reinspect the bore. Discard the cylinder if there are any surface defects.

6 Using a clean piston and a feeler gauge, check the fit of the piston to the bore, as shown in Figure 8.28. Consult the appropriate service manual for the allowable clearance. Discard the cylinder if the allowable clearance is exceeded.

7 Wet the pistons, the cylinder bore, and the new rubber cups with clean brake fluid. With the spring in the center of the cylinder, install the rubber cups in each end, making sure that the lips of the cups are facing inward toward the spring.

Figure 8.27 Honing a cylinder (courtesy AMMCO TOOLS, Inc.)

Note. Some manufacturers use cup expanders in their wheel cylinders. If cup expanders are present, install them with their concave side facing the spring before installing the cups.

8 Install the pistons, making sure the flat sides of the pistons are resting against the flat sides of the cups.

9 Install the boots, making sure their outer edges are secure in the retaining grooves.

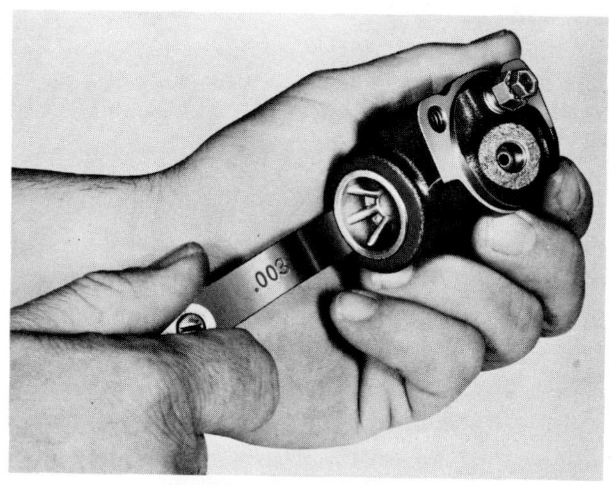

Figure 8.28 Checking wheel cylinder piston fit (courtesy Chevrolet Service Manual, Chevrolet Motor Division).

Job 8e

OVERHAUL A WHEEL CYLINDER

SATISFACTORY PERFORMANCE

A satisfactory performance on this job requires that you do the following:

1 Overhaul a wheel cylinder according to the specifications of the manufacturer.
2 Following the steps in the "Performance Outline," complete the job within 150 percent of the manufacturer's suggested time.
3 Fill in the blanks under "Information."

PERFORMANCE OUTLINE

1 Disassemble a wheel cylinder.
2 Inspect the cylinder.
3 Clean the parts.
4 Inspect the parts.
5 Hone the cylinder.
6 Clean the cylinder.
7 Inspect the cylinder.
8 Reassemble the cylinder.

INFORMATION

Cylinder identification ______________________

Bore diameter ______________ Piston clearance ______________

SUMMARY

In this chapter you have learned how to perform many of the operations needed in repairing the hydraulic system. You can bleed brake systems manually and with a pressure bleeder. You can replace brake hoses. You can fabricate steel brake lines and install them in a system. You can also remove wheel cylinders, overhaul them, and replace them.

SELF-TEST

Each incomplete statement or question in this test is followed by four suggested completions or answers. In each case select the *one* that best completes the statement or answers the question.

1 Two mechanics are discussing brake lines. Mechanic A says that brake lines are made of steel tubing with single flared ends. Mechanic B says that brake lines are made of copper tubing with double flared ends. Who is right?
   A. A only
   B. B only
   C. Both A and B
   D. Neither A nor B

2 Brake bleeding is an operation commonly performed to
   A. purge the hydraulic system of air
   B. drain the reservoir of excessive fluid
   C. relieve excessive pressure at the wheel cylinders
   D. remove water from the hydraulic system

3 Air can enter the hydraulic system by all the following means except
   A. a defective residual pressure valve (check valve)

B. a blocked compensating port
C. replacement of a hydraulic system part
D. failure to maintain fluid in the reservoir

4 When using a bleeder tank to bleed brakes, the hose from the tank must be securely connected to the
A. filler port of the master cylinder
B. bleeder valve of the wheel cylinder
C. top of the master cylinder reservoir
D. residual pressure valve (check valve)

5 A bleeder hose and jar are used to
A. fill the master cylinder reservoir
B. check for air discharged while bleeding
C. drain excess fluid from the master cylinder
D. pressurize the system for bleeding

6 Before attempting to overhaul a wheel cylinder, you should
A. measure the piston diameter
B. open the bleeder valve
C. remove the piston stops
D. install a piston clamp

7 After disassembling a wheel cylinder, a mechanic finds that the cylinder walls are pitted. What should be done when this condition is found?
A. Hone the cylinder and fit oversize pistons
B. Hone the cylinder and install cup expanders
C. Hone the cylinder and install oversize pistons and cup expanders
D. Discard the cylinder

8 When installing wheel cylinders on backing plates made with piston stops, you should
A. remove the piston stops before installing the cylinders
B. hold the wheel cylinder pistons all the way in the cylinders
C. remove the dust boots and reinstall them after the wheel cylinders are installed
D. bleed the cylinders prior to installation

9 When honing a wheel cylinder, the hone should always be lubricated with
A. a petroleum based solvent
B. denatured alcohol
C. brake fluid
D. brake lubricant

10 Two mechanics are discussing the assembly of a wheel cylinder.
Mechanic A says the flat ends of the pistons should face the center of the cylinder.
Mechanic B says the lips or sharp edges of the cups should face the center of the cylinder.
Who is right?
A. A only
B. B only
C. Both A and B
D. Neither A nor B

Chapter 9 Disc Brake Operation and Shoe Replacement

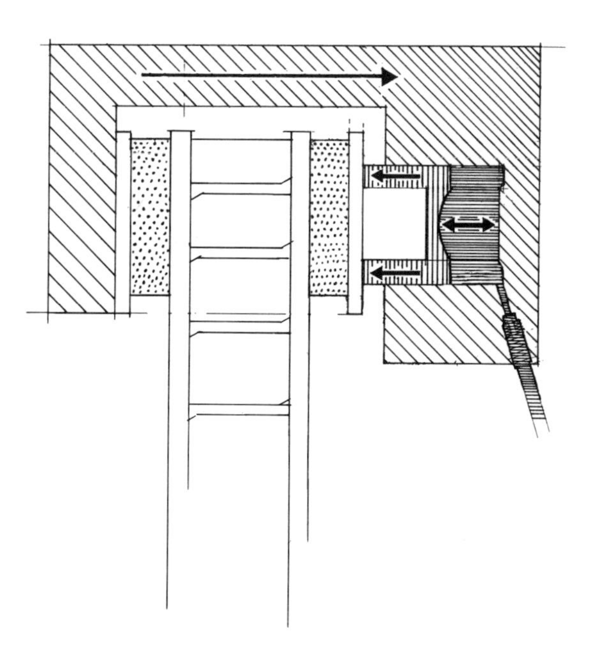

Many cars are fitted with disc brakes. Usually they are part of combination systems in which disc brakes are used at the front wheels and drum brakes are used at the rear wheels. These systems combine the advantages of both types of brakes. The basic principles of braking apply to disc brakes as well as drum brakes. However, disc brake service requires different procedures and techniques from drum brake service. You must learn those procedures and techniques if you are to become a good brake mechanic.

This chapter introduces you to disc brakes. You will learn the principles of operation of the single piston disc brake. You will also learn how to change the brake shoes on the most commonly used disc brake systems. Your objectives are to:

1
Identify the parts in a disc brake assembly.
2
Replace the brake shoes in a Delco-Moraine single piston caliper.
3
Replace the brake shoes in a Ford sliding caliper.
4
Replace the brake shoes in a Kelsey-Hayes sliding caliper.
5
Replace the brake shoes in a Kelsey-Hayes floating caliper.

DISC BRAKE OPERATION Disc brakes differ from drum brakes in that there is a disc instead of a drum. The disc is also called a *rotor.* Figure 9.1 shows the basic principle of disc brake operation. The brake shoes, sometimes called *pads,* are held in a hydraulically operated *caliper,* or clamp. The caliper bridges the rotor like a vise. Applying the brakes causes the caliper to squeeze the shoes together. The shoes grasp the rotor and slow or stop its motion. Figure 9.2 shows the parts of a typical disc brake assembly and their relation to each other.

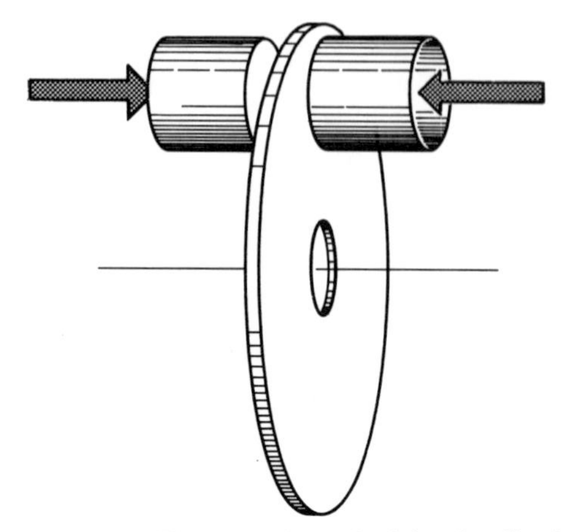

Figure 9.1 The operating principle of a disc brake.

Rotors Disc brakes have many advantages over drum brakes. Most of them are offered by the rotor. A rotor can dissipate heat much faster than a drum. This is because the surfaces of a rotor allow more exposure to the atmosphere. The shape of a rotor allows the use of *splash shields,* or plates, to direct the flow of air over the rotor surfaces. The directed air flow constantly cools the rotor while the car is in motion.

Many small cars use solid rotors similar to the one shown in Figure 9.3. Rotors used on heavy or fast cars are usually ventilated. A ventilated rotor, such as the one shown in Figure 9.4, is cast with cooling fins between the friction surfaces. The internal fins or vanes radiate from the center of the rotor to its outer edge. This design enables the rotor to act as its own cooling fan. As the rotor revolves, the vanes pull air in from the inner edge, force it through the rotor, and discharge it at the outer edge.

Some rotors have internal cooling fins that are shaped like scoops. These rotors pull air in from the outer edge of the rotor and discharge it at the inner edge. The fins are designed to match the direction in which the rotor turns when the car is moving forward. For this reason, the rotors are not interchangeable from side to side. Figure 9.5 shows rotors with curved cooling fins and their correct positions on a car.

Disc brakes do not fade as drum brakes do, even when heated to very high temperatures by repeated use or by panic stops. Brake fade occurs in drum brakes because high temperatures cause the drums to expand outward, increasing the distance between the drum and the lining. But the expansion of a rotor decreases the distance between the rotor and the lining.

Another advantage of disc brakes is that they are self-cleaning. Water, dirt, and particles of worn lining cannot be trapped on a rotor as they are trapped on a drum. The rotation of the rotor throws off water and dirt, and the friction surfaces are wiped dry by the edges of the shoes. This self-cleaning action provides a brake that is relatively unaffected by road and weather conditions.

The advantages of disc brakes are obtained at the expense of some of the advantages offered by drum brakes. The shoes are pushed against the rotor by forces applied perpendicular to the rotor. Therefore, the tendency of the shoes to move with the rotor cannot provide self-energizing braking action. This means that the shoes of a disc brake must be pushed against the rotor with much more force than the shoes of a drum brake must be pushed against the drum. Even though the effort applied by the driver is multiplied by the hydraulic system, it is usually necessary to include a power assist unit in the brake system to boost the driver's effort.

Calipers There are calipers with one, two, or four pistons. However, the type with one piston

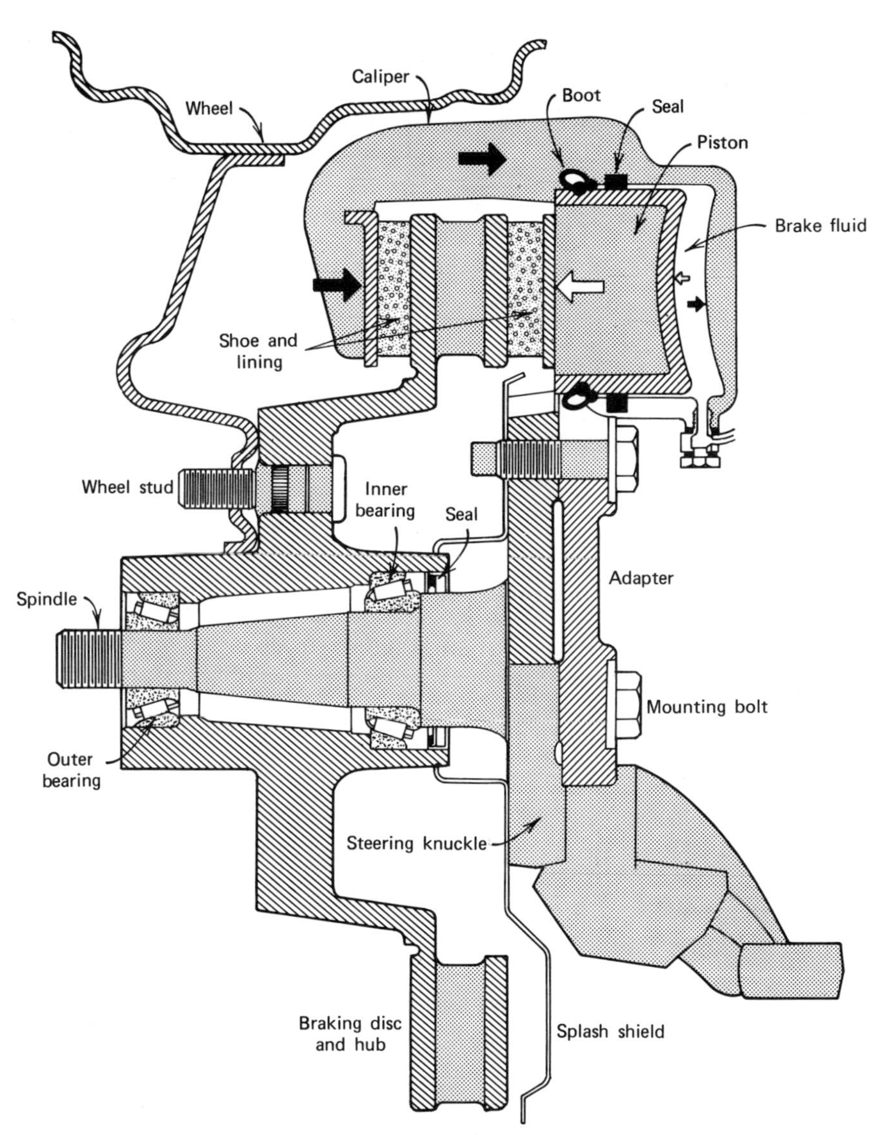

Figure 9.2 A front-sectional view of a disc brake assembly (Chrysler Corporation).

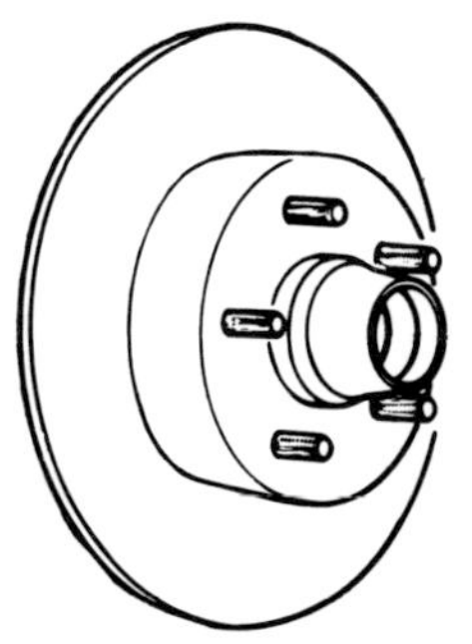

Figure 9.3 A solid rotor.

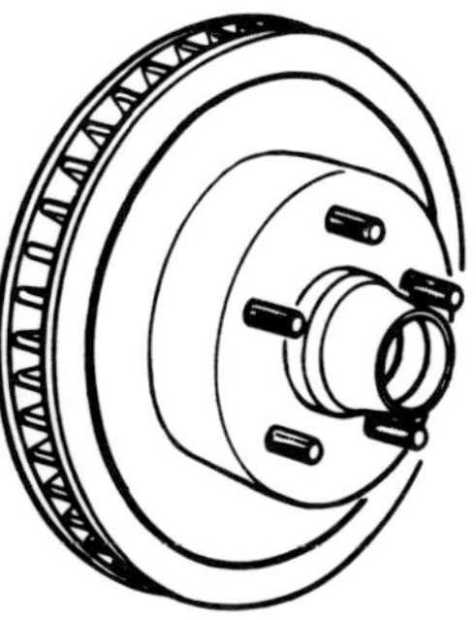

Figure 9.4 A ventilated rotor (courtesy Chevrolet Service Manual, Chevrolet Motor Division).

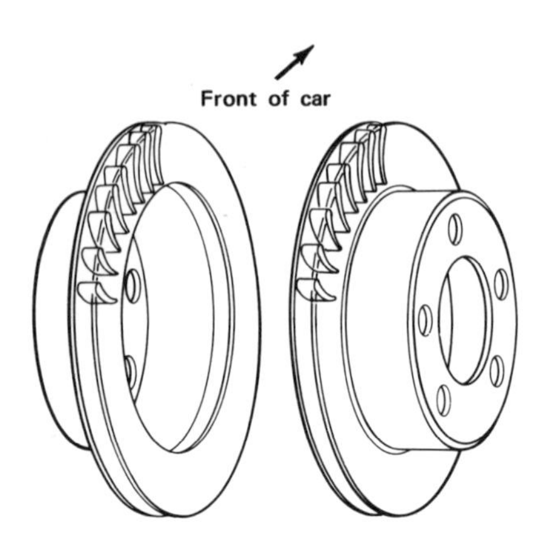

Figure 9.5 Proper positioning of rotors with directional cooling fins (Ford Motor Company).

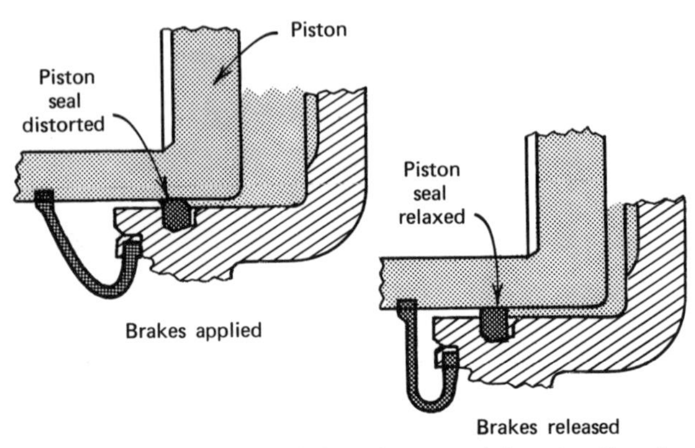

Figure 9.7 The action of the piston seal in retracting the piston. The distance the piston actually moves is exaggerated in this drawing for emphasis (Ford Motor Company).

is most common. The single piston caliper slides or floats in its mount as shown in Figure 9.6.

When the brake pedal is depressed, hydraulic pressure is exerted equally against the bottom of the piston and the bottom of the cylinder bore. The pressure applied to the piston pushes the piston outward and forces the inboard shoe against the inner surface of the rotor. The pressure applied to the bottom of the cylinder causes the caliper to slide inward. This movement forces the outboard shoe against the outer surface of the rotor. Thus, equal force is applied to both sides of the rotor.

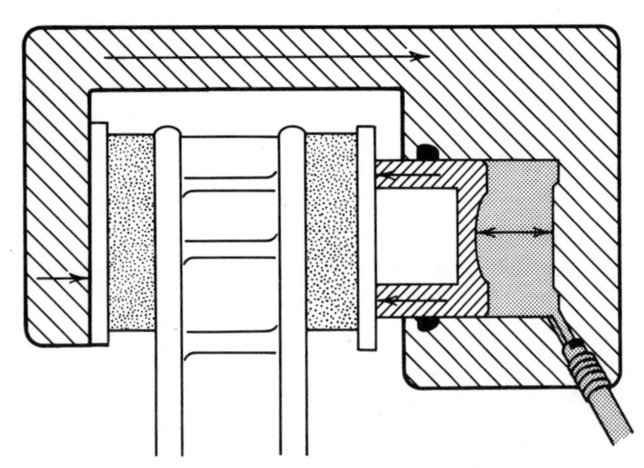

Figure 9.6 Operation of a single piston sliding caliper. Note that since the pressure is equal, the forces pushing the shoes against the rotor are equal (General Motors).

The piston and caliper move very slightly. When the brakes are released, the caliper relaxes. But the shoes do not move away from the rotor. There are no retracting springs to pull the shoes away from the rotor or to pull the piston back into its cylinder bore. Even so, the shoes do not drag on the rotor. If they did, they would quickly wear out.

Two factors act to establish a slight but sufficient clearance between the shoes and the rotor. One is the slight amount of rotor *runout,* or wobble. It pushes the shoes away from the rotor surfaces. The other is the action of the piston seal fitted into the cylinder wall. The outward movement of the piston during brake application distorts the seal. When the brakes are released, the seal returns to its original shape and thus returns the piston to its original position. This function of the piston seal is shown in Figure 9.7.

Both the construction of the floating caliper and the action of the piston seal automatically compensate for lining wear. As the linings wear, the piston moves farther out in its bore on brake applications. It slips through the seal far enough to compensate for the wear. When the brake pedal is released, the piston returns only as far as the seal was distorted. This action

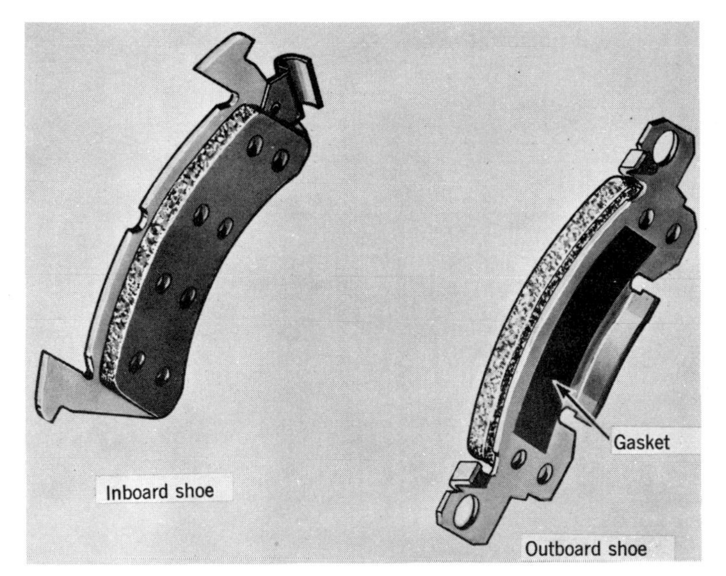

Figure 9.8 Disc brake shoes (General Motors).

makes up for lining wear, maintains lining-to-rotor clearance, and eliminates the need for brake adjustment.

Brake Shoes The brake shoes, or pads, used with disc brakes consist of flat metal plates with pieces of lining bonded or riveted to them as shown in Figure 9.8.

Some shoes have retaining tabs to hold the shoes in position in the caliper. Others have holes or slots for positioning pins or bolts. In some instances the shoes incorporate an extra tab or spring that acts as a wear sensing device. As shown in Figure 9.9, wear sensing tabs touch the rotor when the lining wears to a predetermined thinness. The resulting noise warns the driver that the shoes require replacement.

The shoes in a disc brake system should always be replaced in sets. If one brake assembly requires new shoes, the shoes on the opposite wheel should also be changed to ensure even braking.

Brake Shoe Inspection Usually, the thickness of the lining remaining on the shoes can be observed when the wheel is removed. Most calipers have an inspection port, or hole, so the inboard shoe can be seen. The outboard shoe

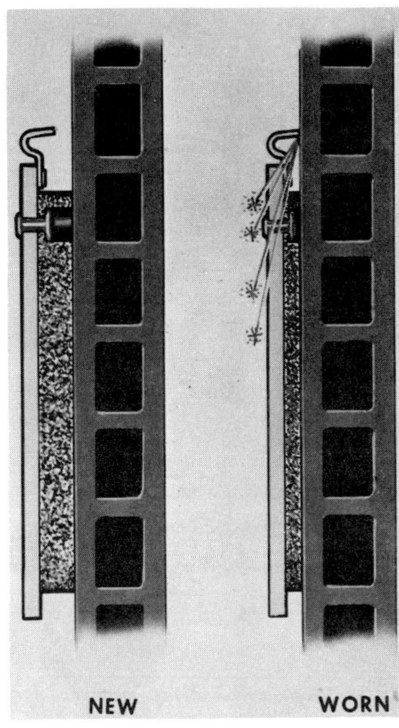

Figure 9.9 Wear indicator operation (General Motors).

should be checked at both its leading and trailing edges. Those are the points where the greatest wear usually occurs. Those points are shown in Figure 9.10. Since the lining is sometimes difficult to see, you should always perform the inspection with the aid of a flashlight or a drop light. Be sure to wear a respirator.

When you check the leading and trailing edges of the shoes, you will often find that the lining is worn unevenly. This is called a tapered wear pattern. In most cases it can be considered normal. You will often find that the lining is worn more on the trailing edge of the shoe than on the leading edge. This is caused by the higher temperatures at the trailing edge of the shoe, as shown in Figure 9.11. Sometimes you will find that the lining is worn more on the leading edge of the inner shoe and the trailing edge of the outer shoe, as shown in Figure 9.12. This wear pattern is caused by the design and mounting of the caliper, which allows the caliper to twist slightly on brake application. Because of these

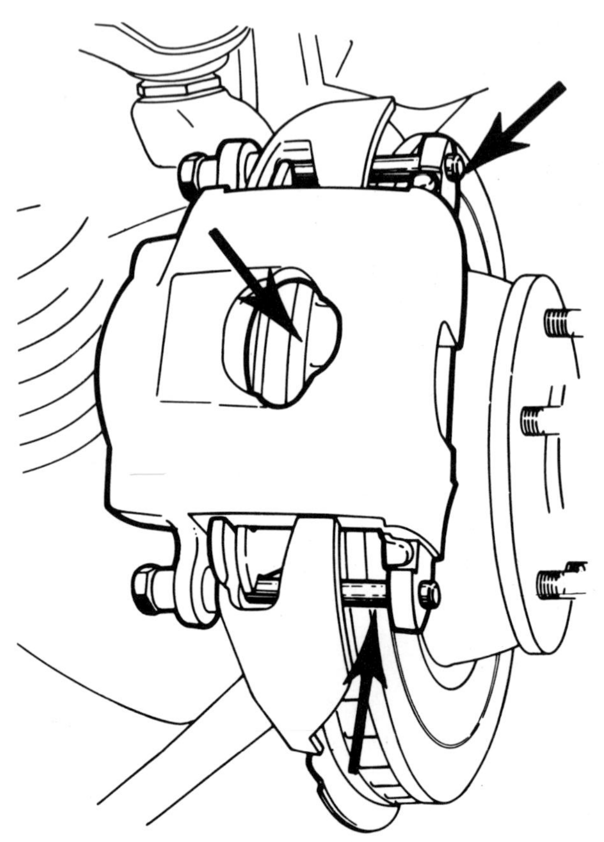

Figure 9.10 Lining inspection areas (General Motors).

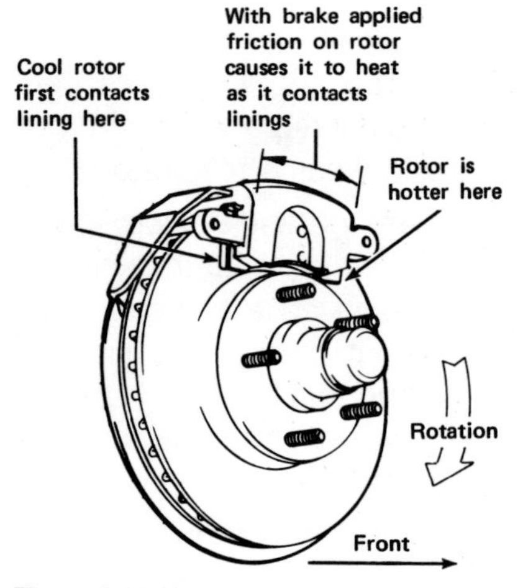

Figure 9.11 Heat pattern in an operating disc brake (General Motors).

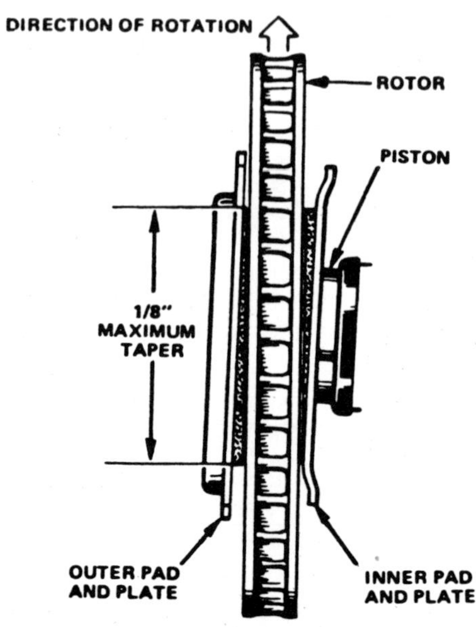

Figure 9.12 Tapered shoe wear caused by caliper twist during braking (Bendix Automotive Aftermarket Operations).

wear patterns, you should always check the leading and trailing edges of both shoes. One edge could appear safe, but the other edge could be dangerously worn.

Not only do brake shoes not always wear evenly, but they do not always wear at the same rate. The type and rate of wear vary considerably with different cars and under different driving conditions. Brake shoes should be replaced when the lining is worn to the minimal thinness allowed by the manufacturer. This specification depends on the type of brakes fitted and on whether bonded or riveted lining is used. Always check an appropriate service manual for this specification. To allow the lining to wear beyond this point could cause rotor damage.

Job 9a

IDENTIFY DISC BRAKE COMPONENTS

SATISFACTORY PERFORMANCE

A satisfactory performance on this job requires that you do the following:

1 Identify the numbered parts on the drawing by placing the number for each part in front of the correct part name.

2 Correctly identify 15 of the 17 parts within 15 minutes.

PERFORMANCE SITUATION

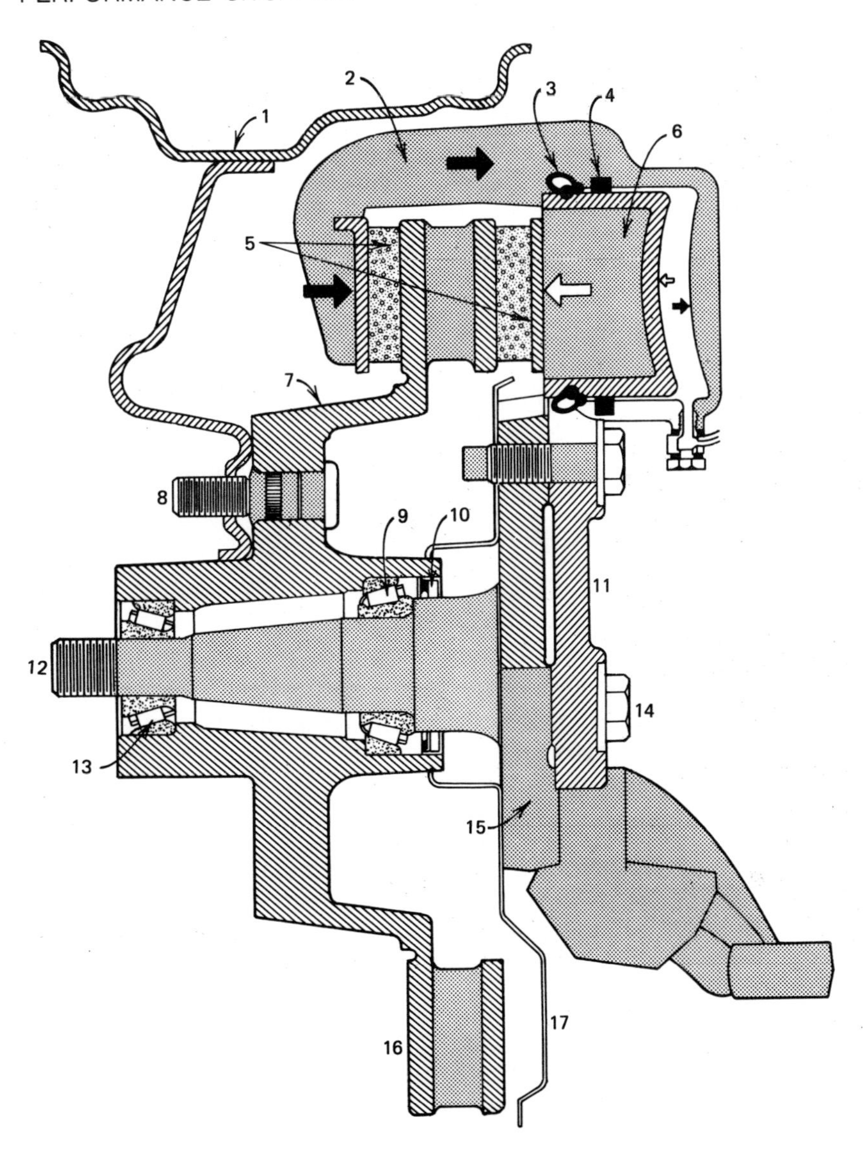

_______ Caliper
_______ Spindle
_______ Steering knuckle
_______ Rotor and hub assembly
_______ Wheel
_______ Inner bearing
_______ Backing plate
_______ Mounting bolt
_______ Brake fluid
_______ Dust boot
_______ Piston seal
_______ Wheel stud
_______ Shoes and lining
_______ Adapter or mount
_______ Outer bearing
_______ Guide plate
_______ Splash shield
_______ Grease retainer
_______ Piston
_______ Reaction arm

SHOE REPLACEMENT When you worked on drum brakes, you found that all car manufacturers use a similar type of brake. Any differences you found were in the types of self-adjusting mechanisms. That was because it took more than 75 years for the drum brake design to evolve, allowing time for the advantages of the self-energizing duo-servo brake to become generally recognized and almost universally used.

Compared to drum brakes, disc brakes are relatively new. Most manufacturers now use single piston calipers, but of different designs. Also, the methods of mounting the calipers differ, even among cars of the same manufacturer. Because of these differences, and because of ongoing changes in design, there are many different procedures for replacing disc brake shoes.

REMOVAL AND INSTALLATION OF BRAKE SHOES IN A DELCO-MORAINE SINGLE PISTON CALIPER The procedure below outlines the steps you must take to replace the brake shoes at the front wheels of a car fitted with Delco-Moraine single piston calipers. There are variations in the design and the mounting of this caliper. Therefore, for the specific procedure you should check the service manual for the car on which you are working. Be sure to wear a respirator.

Removal 1 Remove at least two thirds of the brake fluid from the master cylinder reservoir that supplies the front brake system. This can be done by loosening the front brake line at the master cylinder and carefully pumping the brake pedal until sufficient fluid has leaked out. This method should be used only if you can catch the fluid in a container. If brake fluid comes in contact with painted surfaces, it will destroy the finish. Another method is to remove the fluid from the reservoir with a clean syringe.

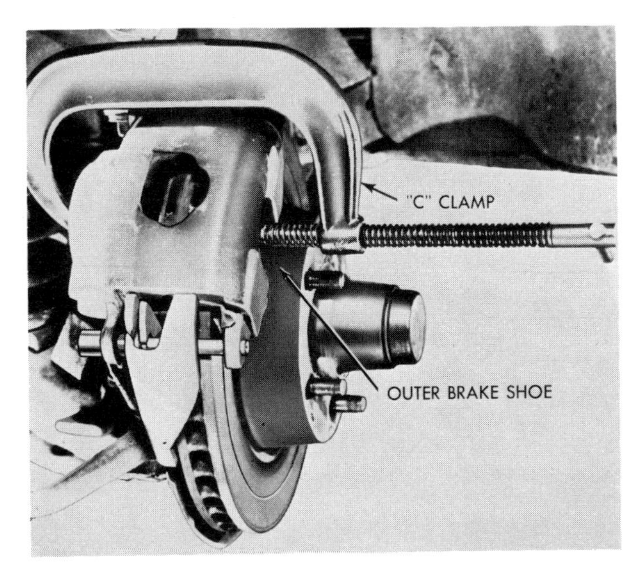

Figure 9.13 Using a C-clamp to push a caliper piston back in its bore (courtesy Chevrolet Service Manual, Chevrolet Motor Division).

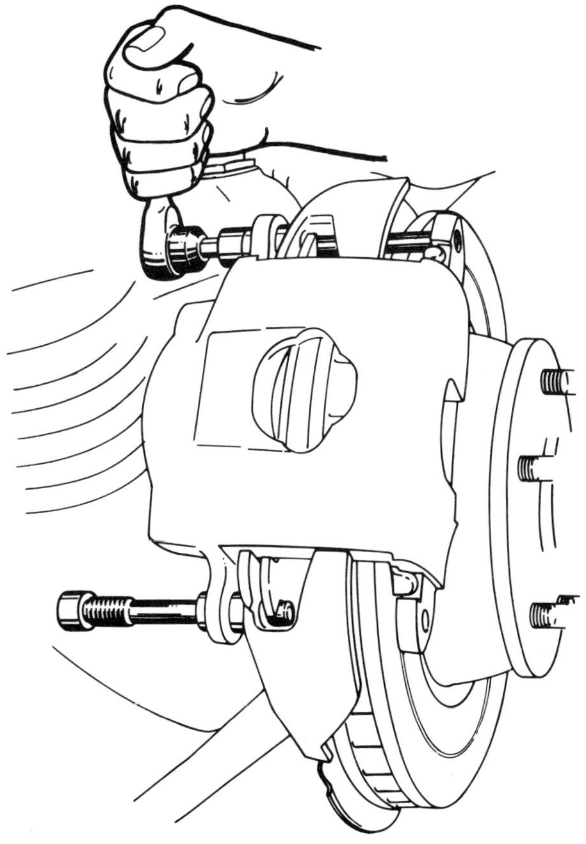

Figure 9.14 Removing caliper mounting bolts (General Motors).

Note. Do not remove all the fluid lest air enter the system. A partial removal of fluid is necessary to prevent the reservoir from overflowing when you later push the caliper piston back into its bore. Discard the drained fluid.

2 Raise and support the car.

3 Remove the wheel.

4 Push the piston back into its bore, using a C-clamp as shown in Figure 9.13. Place the clamp over the caliper with the pad of the clamp against the back of the outboard shoe. Tighten the clamp to pull the caliper assembly outward and to force the piston back into its bore.

5 Using a suitable socket, remove the mounting bolts as shown in Figure 9.14.

6 Lift the caliper up and off the rotor. Support the caliper by placing it securely on part of the suspension system, or tie it up with a piece of wire. Do not allow the caliper to hang from the brake hose. To do so would weaken the hose.

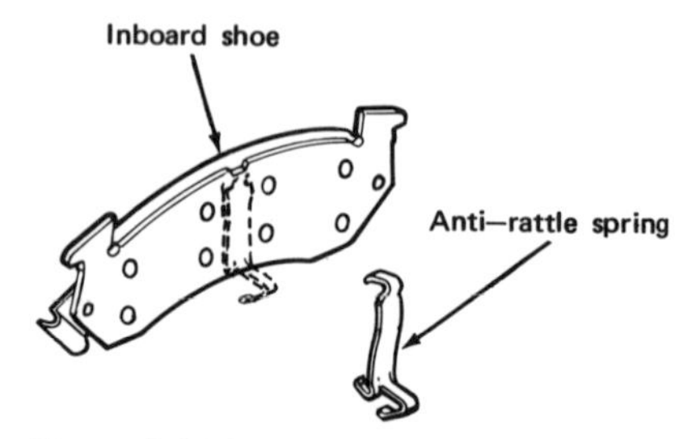

Figure 9.15 An antirattle spring and its position when installed on the inboard shoe (General Motors).

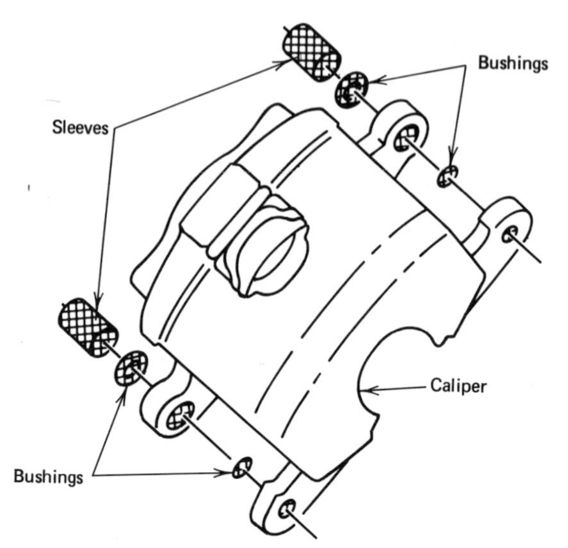

Figure 9.16 Location of the sleeves and bushings in the caliper mounting ears (courtesy Chevrolet Service Manual, Chevrolet Motor Division).

7 Remove the old shoes from the caliper.

8 Remove the antirattle spring from the old inboard shoe, and install it on the new inboard shoe (Figure 9.15).

9 Remove the sleeves and bushings from the mounting ears of the caliper (Figure 9.16). Discard the old sleeves and bushings and replace them with new ones.

Note. **New sleeves and bushings should always be installed whenever the shoes are replaced.**

10 Clean the caliper assembly, giving special attention to the mounting holes and the bushing grooves.

11 Clean the caliper mounting.

12 Check the mounting bolts. If they are damaged, rusted, or corroded, discard them and obtain new bolts.

Note. **The mounting bolts are plated to resist corrosion, and the use of any abrasive to clean them will damage the plating.**

13 Check the caliper for evidence of fluid leakage, and check the dust boot for tears or cracks.

14 If leakage is found or if the dust boot is damaged, remove the caliper and overhaul it or replace it.

Installation

1 Lubricate the new sleeves, the new bushings, and the ends of the mounting bolts as shown in Figure 9.17. Use a silicone lubricant formulated for brake systems.

Note. **The use of petroleum base lubricant can cause eventual damage to the rubber parts.**

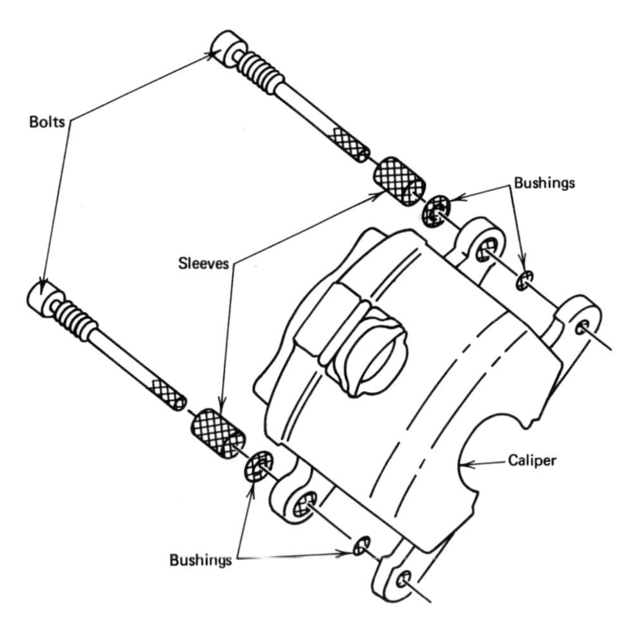

Figure 9.17 Exploded view of a caliper assembly showing where parts should be lubricated (courtesy Chevrolet Service Manual, Chevrolet Motor Division).

2 Install the rubber bushings in the caliper ears.

3 Install the sleeves in the large ears of the caliper. Make sure the sleeve edges are flush with the inner machined surface of the caliper ears.

4 Place the new inboard shoe in position in the caliper. Make sure the antirattle spring is in place (Figure 9.18). Then press the shoe down in place until it touches the piston.

5 Place the new outboard shoe in position as shown in Figure 9.19. Make sure the tab at the bottom of the shoe is lined up with the notch in the caliper.

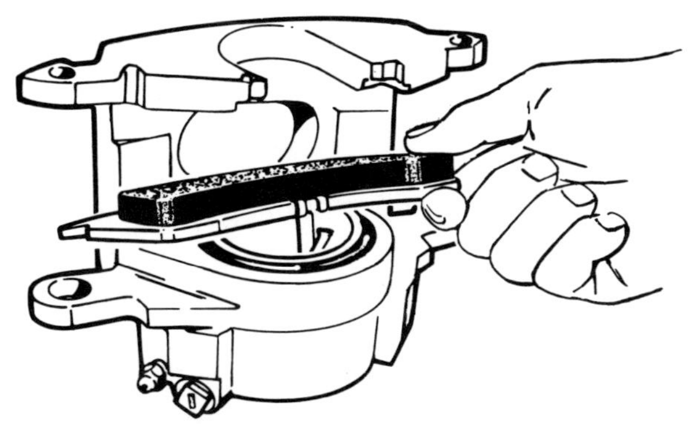

Figure 9.18 Installing the inboard shoe in a caliper (General Motors).

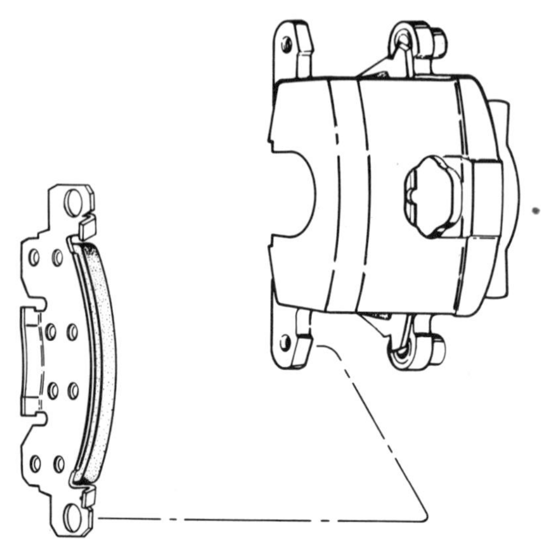

Figure 9.19 The position of the outboard shoe (General Motors).

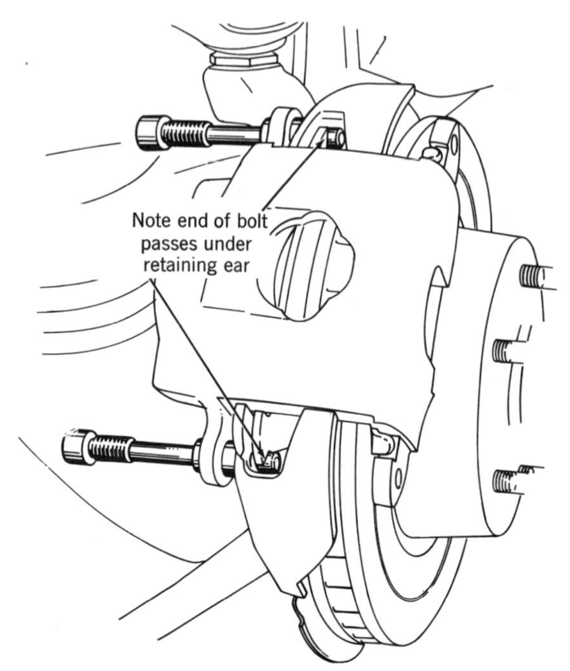

Figure 9.20 How the mounting bolts are installed under the inboard shoe ears (courtesy Chevrolet Service Manual, Chevrolet Motor Division).

6 Hold the shoes in place in the caliper, and lift the caliper. Position the caliper over the rotor, and make sure there is no clearance between the tab at the bottom of the outboard shoe and its notch in the caliper.

7 Slide the caliper down over the rotor, and align the holes in the caliper with the holes in the mounting.

8 Push the mounting bolts through the sleeves in the caliper and through the mount as shown in Figure 9.20. Make sure the bolts pass under the ears on the inboard shoe.

9 Push the bolts through the holes in the outboard shoe and through the holes in the outer caliper ears. Thread the bolts into the mounting bracket, starting the threads by hand to avoid cross-threading.

10 Tighten the mounting bolts to the torque specification of the manufacturer.

11 Using a large pair of arc joint pliers, bend both upper ears of the outboard shoe downward so that no clearance exists between the ears and the caliper housing. During this bending procedure, position the pliers as shown in Figure 9.21.

12 Repeat the removal and installation steps on the opposite wheel.

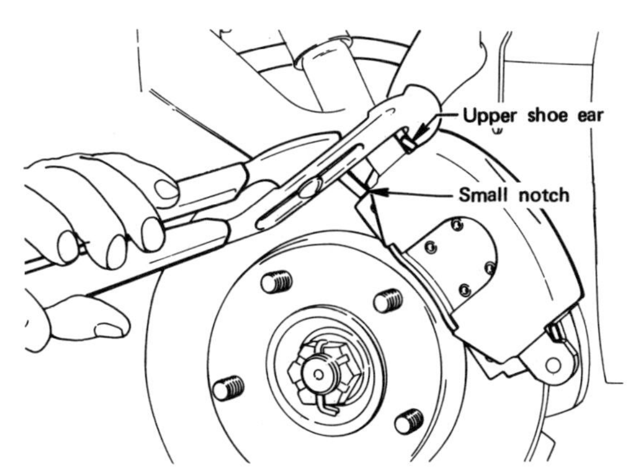

Figure 9.21 Using pliers to bend the ears on the outboard shoe (General Motors).

13 Check the fluid level in the master cylinder and add fluid if needed to bring the level to within $\frac{1}{4}$ in. (about 6 mm) of the top of the casting.

14 Pump the brake pedal several times. This will position the pistons and calipers. Check to see that the pedal is hard and high. Check the fluid level in the master cylinder again, and add to it if necessary.

15 Install the wheels, and lower the car to the floor.

16 Road test the car to check for proper braking action.

Job 9b

REPLACE FRONT DISC BRAKE SHOES: DELCO-MORAINE TYPE

SATISFACTORY PERFORMANCE

A satisfactory performance on this job requires that you do the following;

1 Replace the front brake shoes on a car fitted with Delco-Moraine type calipers.
2 Following the steps in the "Performance Outline" and the manufacturer's procedure and specifications, complete the job within 150 percent of the manufacturer's suggested time.
3 Fill in the blanks under "Information."

PERFORMANCE OUTLINE

1 Partially drain the reservoir.
2 Raise and support the car, and remove the wheel.
3 Push the caliper piston back in its bore.
4 Remove and support the caliper.
5 Remove the shoes.
6 Remove the sleeves and bushings.
7 Clean and inspect all parts.
8 Install new sleeves and bushings.
9 Install the new brake shoes.
10 Install the caliper.
11 Check the pedal feel and fluid level.
12 Repeat operations 1 to 11 at the opposite wheel.
13 Install the wheels, and lower the car to the floor.
14 Road test the car for proper brake operation.

INFORMATION

Vehicle identification ____________________

Reference used ______________ Page(s) ______________

REMOVAL AND INSTALLATION OF BRAKE SHOES IN A FORD SLIDING CALIPER

Here is the procedure for replacing the front brake shoes on a car equipped with Ford sliding calipers. Because of variations in the design and mounting of this caliper, check the appropriate manual for specific procedures.

Removal

1 Remove at least two thirds of the brake fluid from the master cylinder reservoir that supplies the front brake system. Discard the drained fluid.

2 Raise and support the car.

Figure 9.22 Using a screwdriver to pry a caliper piston back into its bore (courtesy American Motors Corporation).

3 Remove the wheel.

4 Push the piston back into its bore. You can do this with a C-clamp, or you can pry back the piston by inserting a screwdriver between the piston and the inboard shoe (Figure 9.22).

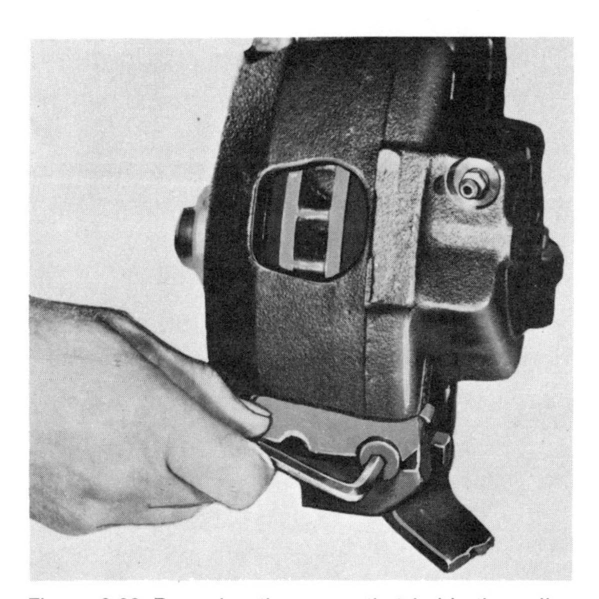

Figure 9.23 Removing the screw that holds the caliper retaining key (Ford Motor Company).

5 Remove the screw holding the caliper retaining key as shown in Figure 9.23.

6 Carefully drive the retaining key and its support spring from the key slot as shown in Figure 9.24.

Figure 9.24 Driving out the caliper retaining key (Ford Motor Company).

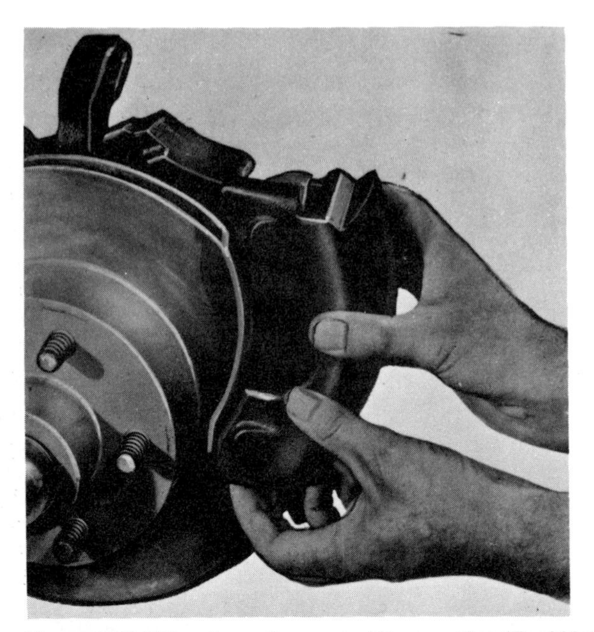

Figure 9.25 Lifting the caliper out of its mounting (Ford Motor Company).

7 Lift the caliper off the rotor and out of its mounting. To do this, push the caliper downward against its anchor plate. Then tip the caliper outward at the top as shown in Figure 9.25.

8 Support the weight of the caliper by hanging it on a piece of wire so that no strain is placed on the brake hose.

9 Remove the old outboard shoe from the caliper. The outboard shoe is tightly fitted as shown in Figure 9.26, but you can dislodge it by tapping it with a hammer.

10 Remove the old inboard shoe from the anchor plate. The antirattle clip, shown in Figure 9.27, must be saved for installation on the new inboard shoe.

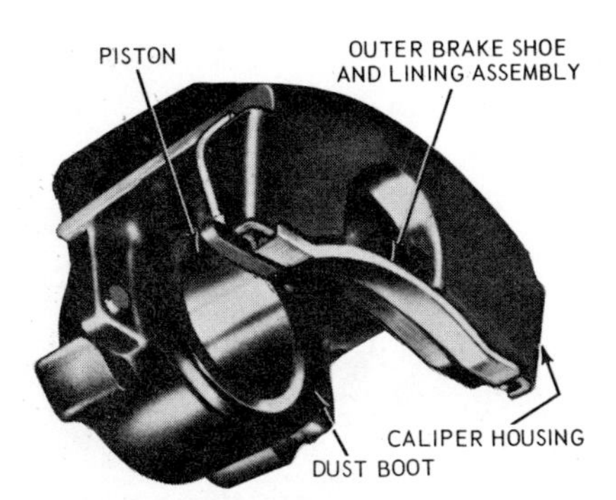

Figure 9.26 Removed caliper with the outboard shoe in place (Ford Motor Company).

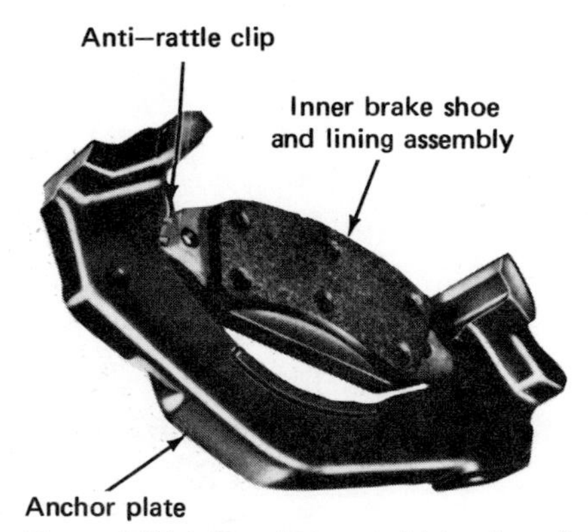

Figure 9.27 Inboard shoe held in place in the anchor plate (Ford Motor Company).

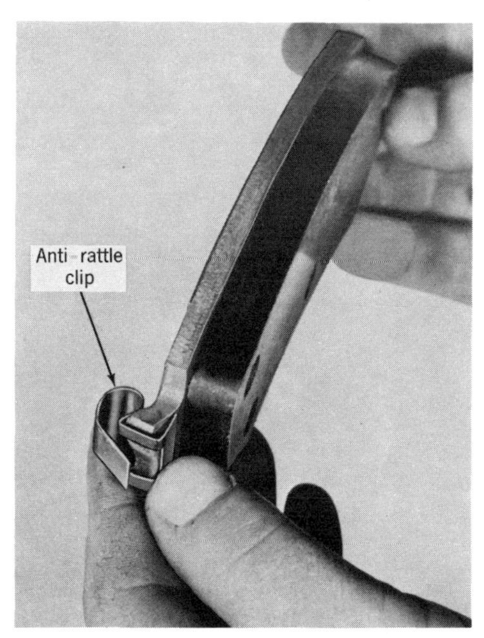

Figure 9.28 Antirattle clip correctly installed on the inboard shoe (courtesy American Motors Corporation).

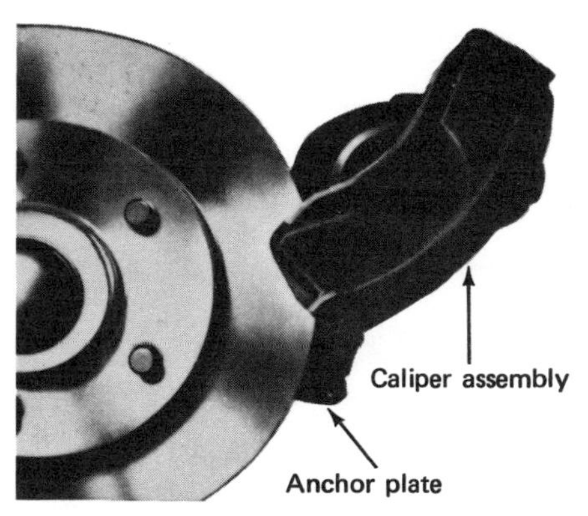

Figure 9.29 First step in caliper installation. Note that the lower mounting groove on the caliper is resting on the anchor plate (Ford Motor Company).

11 Clean the caliper assembly and the anchor plate. Carefully inspect their machined sliding surfaces. Use sandpaper or a wire brush to remove rust or corrosion from those surfaces and from the surfaces that contact the outboard shoe.

12 Check the caliper for evidence of fluid leakage. Check the dust boot for tears or cracks. If you find leakage or a damaged dust boot, remove the caliper for overhaul or replacement.

Installation

1 Check to see that the piston is bottomed in its bore in the caliper. If the piston is not all the way into its bore, place a block of wood over the piston, and use a C-clamp to push the piston all the way in.

2 Lubricate the machined sliding surfaces of the caliper and anchor plate with a lubricant approved by the maufacturer. Be careful not to get the lubricant where it can contaminate the shoes.

3 Install the antirattle clip on the new inboard shoe as shown in Figure 9.28.

4 Install the inboard shoe on the anchor plate. (Refer again to Figure 9.27).

5 Install the new outboard shoe in the caliper. Be sure that the lower flanges on the shoe contact the caliper. Also, be sure that the upper flanges on the shoe go over the shoulders on the caliper legs. (Refer again to Figure 9.26).

6 Position the caliper assembly over the rotor so that the lower mounting groove on the caliper rests on the anchor plate (Figure 9.29).

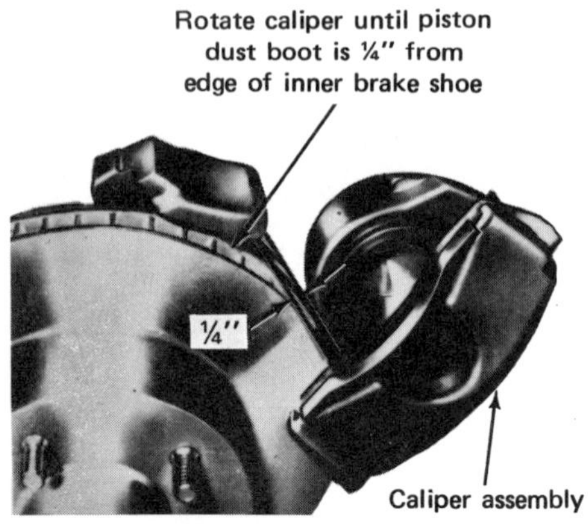

Figure 9.30 Second step in caliper installation (Ford Motor Company).

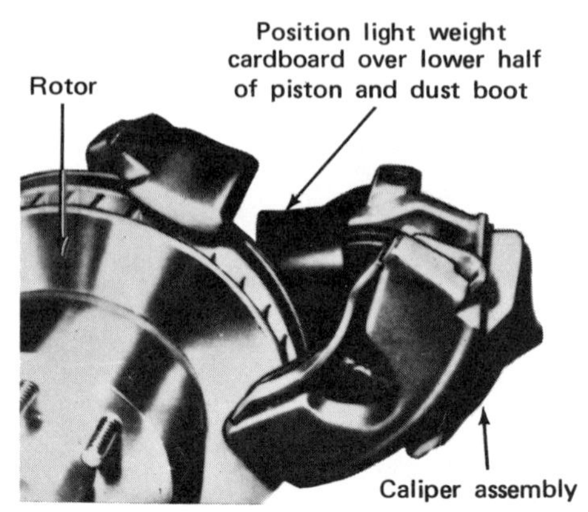

Figure 9.31 Third step in caliper installation. A piece of light cardboard is used to protect the dust boot (Ford Motor Company).

7 Pivot the caliper assembly upward and inward toward the rotor until the edge of the dust boot is $\frac{1}{4}$ in. (about 6 mm) from the edge of the inboard shoe (Figure 9.30).

8 Place a clean piece of lightweight cardboard between the lower half of the dust boot and the shoe, as shown in Figure 9.31. (This prevents the dust boot from getting caught between the piston and the shoe.)

9 Pivot the caliper assembly toward the rotor. When you feel a slight resistance, pull the cardboard down toward the center of the rotor while pushing the caliper over the rotor (Figure 9.32).

10 Remove the cardboard, and push the caliper assembly all the way down over the rotor (Figure 9.33).

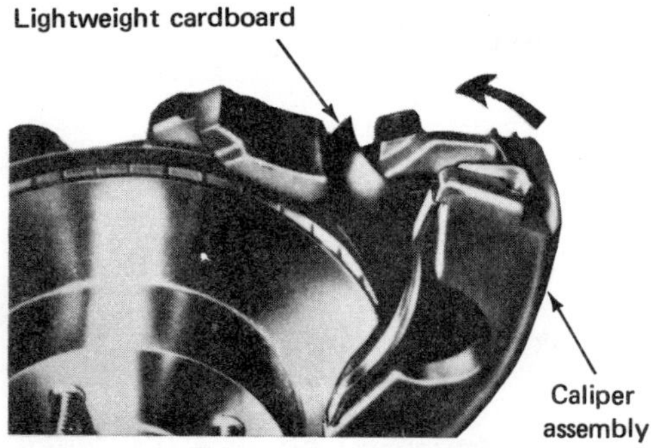

Figure 9.32 Fourth step in caliper installation (Ford Motor Company).

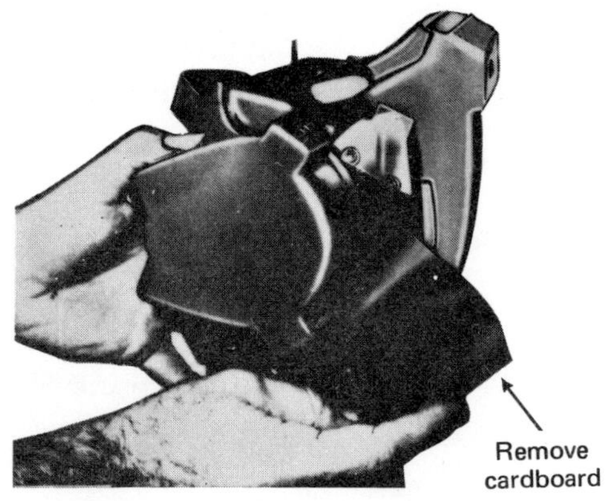

Figure 9.33 Fifth step in caliper installation. The caliper is in place over the rotor and the cardboard is removed (Ford Motor Company).

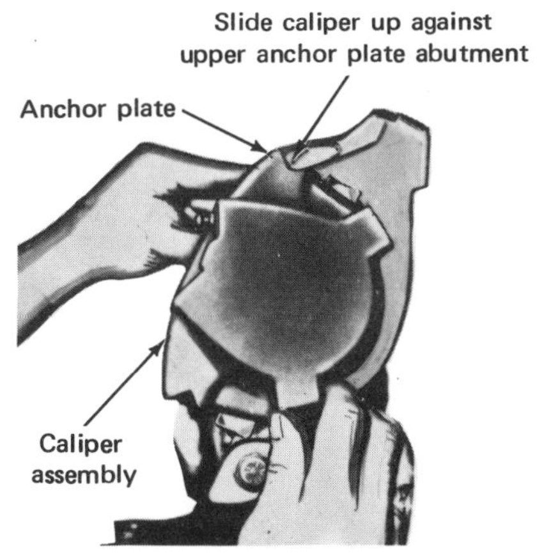

Figure 9.34 Caliper being slid into place against the upper anchor abutment (Ford Motor Company).

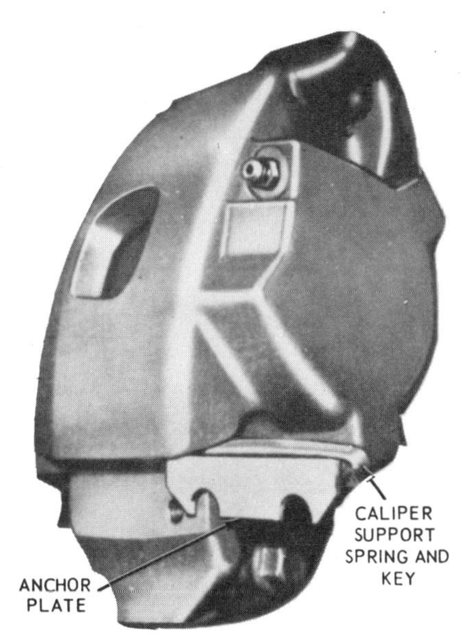

Figure 9.35 Installing the caliper retaining key (Ford Motor Company).

11 Slide the caliper assembly up against the machined surface of the upper anchor plate abutment as shown in Figure 9.34.

12 Position the lower part of the caliper assembly so that the retaining key and its support spring can be inserted into the opening between the caliper and the anchor plate (Figure 9.35).

13 Carefully drive the retaining key and its support spring into the slot until the notch on the key aligns with the threaded hole in the anchor plate.

14 Install the key retaining screw, and tighten to the manufacturer's torque specification.

15 Repeat the removal and installation steps on the opposite wheel.

16 Check the fluid level in the master cylinder. Add fluid if it is needed to bring the level to within $\frac{1}{4}$ in. (about 6 mm) of the top of the casting.

17 Pump the brake pedal several times. This will position the pistons and calipers. Check to see that the pedal is hard and high. Check the fluid level in the master cylinder again, and add to it if necessary.

18 Install the wheels, and lower the car to the floor.

19 Road test the car to check for proper braking action.

Job 9c

REPLACE FRONT DISC BRAKE SHOES: FORD TYPE

SATISFACTORY PERFORMANCE

A satisfactory performance on this job requires that you do the following:

1 Replace the front brake shoes on a car fitted with Ford type calipers.
2 Following the steps in the "Performance Outline" and the manufacturer's procedure and specifications, complete the job within 150 percent of the manufacturer's suggested time.
3 Fill in the blanks under "Information."

PERFORMANCE OUTLINE

1 Partially drain the reservoir.
2 Raise and support the car, and remove the wheel.
3 Bottom the caliper piston in its bore.
4 Remove and support the caliper.
5 Remove the shoes.
6 Clean and inspect all parts.
7 Install the brake shoes.
8 Install the caliper.
9 Check the pedal feel and the fluid level.
10 Repeat operations 1 to 9 at opposite wheel.
11 Install the wheels, and lower the car to floor.
12 Road test the car for proper brake operation.

INFORMATION

Vehicle identification ______________________________

Reference used ____________________ Page(s) ____________________

REMOVAL AND INSTALLATION OF BRAKE SHOES IN A KELSEY-HAYES SLIDING CALIPER

Here is the procedure you must use to replace the front brake shoes on a car equipped with Kelsey-Hayes sliding calipers. There are variations in the design and mounting of this type caliper. Therefore, you should check the manual for the car you are working on for the specific procedure.

Removal

1 Remove at least two thirds of the brake fluid from the master cylinder reservoir that supplies the front brake system. Discard the drained fluid.

2 Raise and support the car.

3 Remove the wheels.

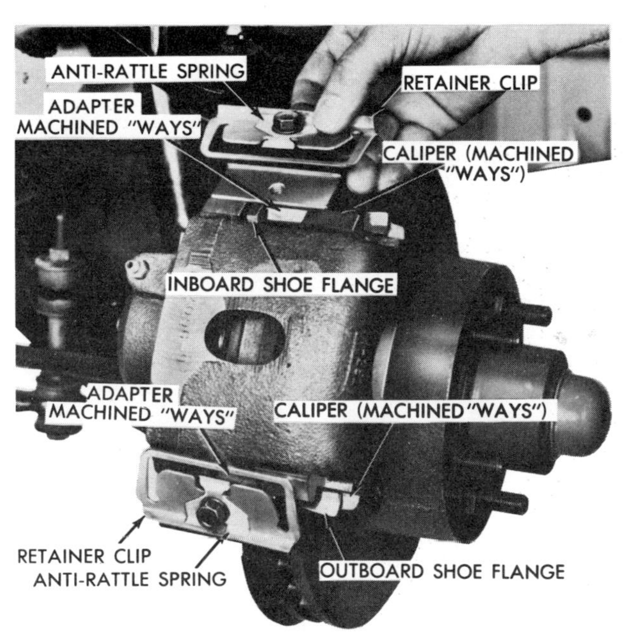

Figure 9.36 Removing caliper retaining clips and springs (Chrysler Corporation).

Figure 9.37 Removing the caliper assembly from the rotor (Chrysler Corporation).

4 Remove the caliper retaining clips and antirattle springs as shown in Figure 9.36.

5 Grasp the caliper assembly and slide it off the rotor as shown in Figure 9.37.

6 Support the weight of the caliper assembly by hanging it on a wire attached to the suspension system. Do not allow it to hang by the brake hose.

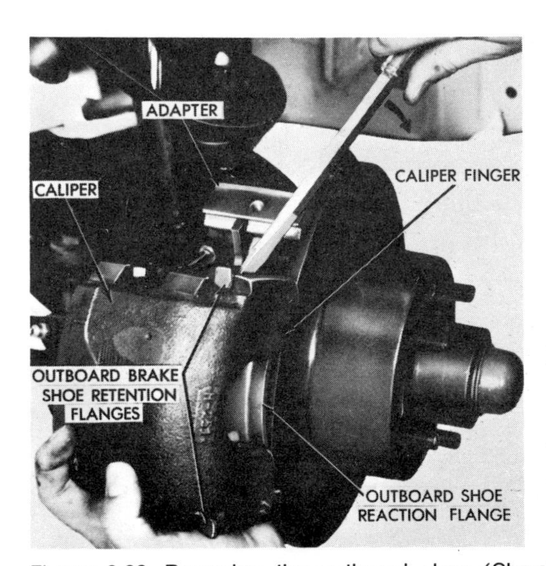

Figure 9.38 Removing the outboard shoe (Chrysler Corporation).

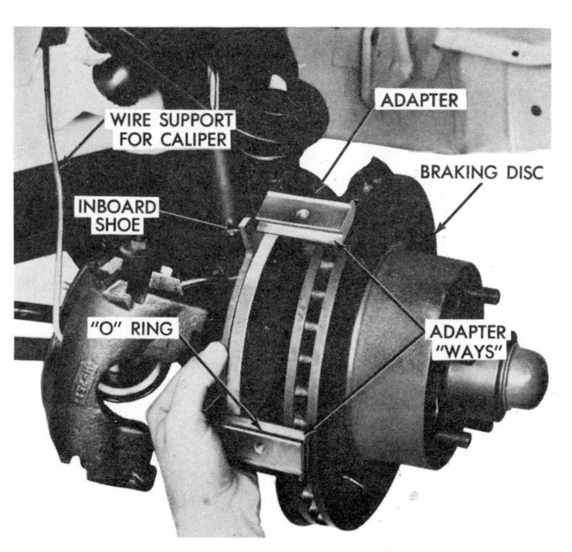

Figure 9-39 Removing the inboard shoe from the adapter (Chrysler Corporation).

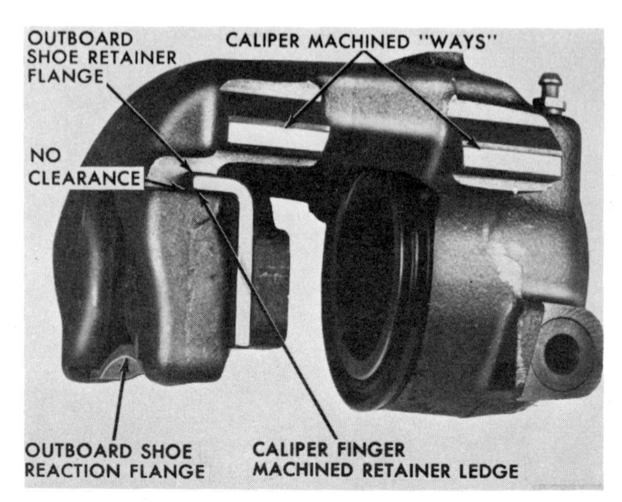

Figure 9.40 Caliper assembly with a properly fitted outboard shoe (Chrysler Corporation).

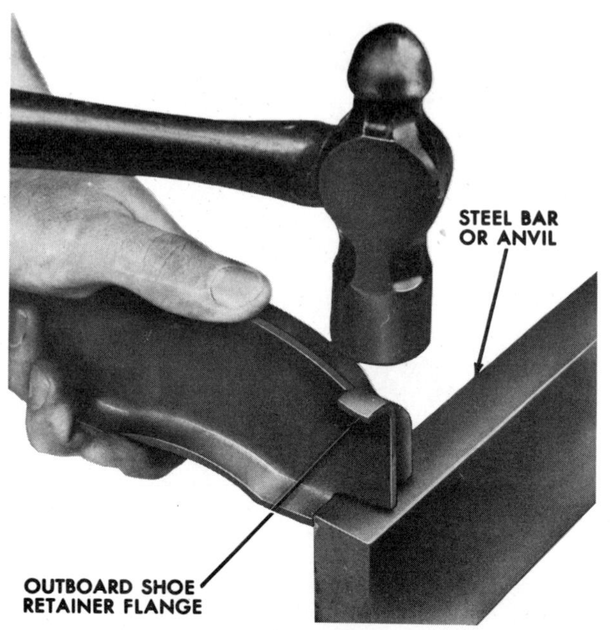

Figure 9.41 Bending the retainer flange on an outboard shoe to achieve a proper fit in the caliper (Chrysler Corporation).

7 Remove the outboard shoe by prying it off with a screwdriver as shown in Figure 9.38.

8 Remove the inboard shoe from the adapter as shown in Figure 9.39.

9 Clean the caliper assembly and the adapter. Carefully check the machined "ways" on the caliper and the adapter. If the ways are rusted or corroded, remove the O-rings from the adapter ways. (See Figure 9.39 again.) Clean the surfaces with a wire brush, and reinstall the O-rings.

10 Check the caliper for evidence of leakage. Check the dust boot for tears or cracks. If you find leakage or dust-boot damage, remove the caliper for overhaul or replacement.

11 Use a C-clamp to bottom the piston in its bore.

Installation

1 Install a new outboard shoe in the caliper. Check the fit of the shoe retainer flanges to the ledges on the caliper (Figure 9.40). There should be no clearance at these points.

Note. **Replacement shoes usually fit loosely against the caliper and must be individually fitted. If the outboard shoes are installed with clearance between the flanges and ledges, they may rattle or squeal in use.**

2 Support the shoe as shown in Figure 9.41, and carefully bend the flanges down by tapping them with a hammer.

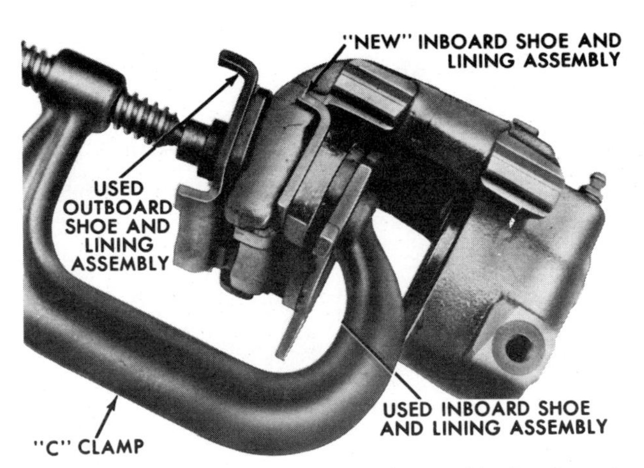

Figure 9.42 Using a C-clamp and the used brake shoes to install a new outboard shoe (Chrysler Corporation).

3 Repeat operations 1 and 2, checking the fit and bending the flanges until the flanges contact the ledges with enough interference to keep the shoe from being pushed into position with finger pressure.

4 Place the old brake shoes against the caliper and the new outboard shoe, as shown in Figure 9.42. Use a C-clamp to seat the new outboard shoe in the caliper.

5 Place the new inboard shoe in position on the adapter. Be sure that the shoe flanges are in place in the adapter ways (Figure 9.39). Slide the shoe in place until it contacts the rotor.

6 Carefully slide the caliper assembly over the rotor (Figure 9.37). Check to see that the dust boot does not get caught between the piston and the inboard shoe.

7 Align the caliper assembly with the machined "ways" of the adapter.

8 Install the retaining springs and antirattle clips (Figure 9.36).

9 Tighten the retaining screws to the manufacturer's specified torque.

10 Repeat the removal and installation steps on the opposite wheel.

11 Check the fluid level in the master cylinder. Add fluid if it is needed to bring the level to within $\frac{1}{4}$ in. (about 6 mm) of the top of the casting.

12 Pump the brake pedal several times to position the pistons and the calipers. Check to see that the pedal is hard and high. Refill the master cylinder if necessary.

13 Install the wheels, and lower the car to the floor.

14 Road test the car to check for proper braking.

Job 9d

REPLACE FRONT DISC BRAKE SHOES: KELSEY-HAYES SLIDING CALIPER TYPE

SATISFACTORY PERFORMANCE

A satisfactory performance on this job requires that you do the following:

1 Replace the front brake shoes on a car fitted with the Kelsey-Hayes type of sliding calipers.
2 Following the steps in the "Performance Outline" and the manufacturer's procedure and specifications, complete the job within 150 percent of the manufacturer's suggested time.
3 Fill in the blanks under "Information."

PERFORMANCE OUTLINE

1 Partially drain the reservoir.
2 Raise and support the car and remove the wheel.
3 Remove and support the caliper.
4 Remove the shoes.
5 Clean and inspect all parts.
6 Bottom the caliper piston in its bore.
7 Install the brake shoes.
8 Install the calipers.
9 Check the pedal feel and the fluid level.
10 Repeat operations 1 to 9 at opposite wheel.
11 Install the wheels, and lower the car to the floor.
12 Road test the car for proper brake operation.

INFORMATION

Vehicle identification ______________________________

Reference used ____________________ Page(s) ____________________

REMOVAL AND INSTALLATION OF BRAKE SHOES IN A KELSEY-HAYES FLOATING CALIPER

Below we give the procedure for replacing the front brake shoes on a car equipped with Kelsey-Hayes floating calipers. Because of variations in the design and mounting of this caliper, you should check the manual of the car you are working on for the specific procedure.

Removal

1 Remove at least two thirds of the brake fluid from the master cylinder reservoir that supplies the front brake system. Discard the fluid.

2 Raise and support the car.

3 Remove the wheels.

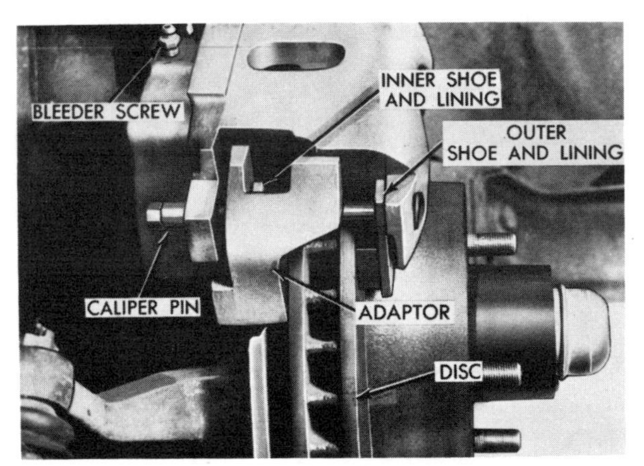

Figure 9.43 A Kelsey-Hayes floating caliper and its related parts (Chrysler Corporation).

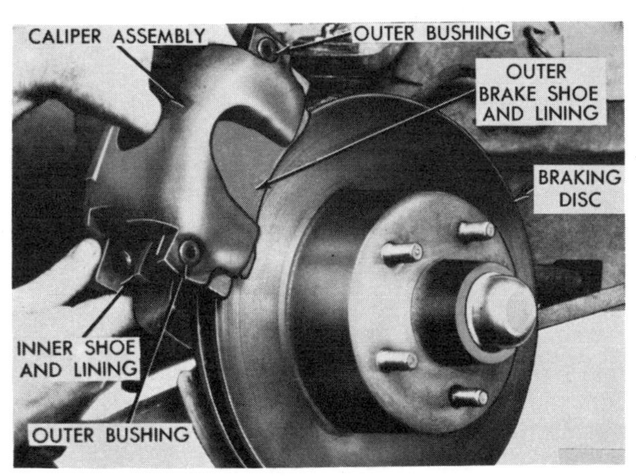

Figure 9.44 Removing the caliper from the rotor (Chrysler Corporation).

4 Use a C-clamp to push the piston back into its bore in the caliper.

5 Remove the threaded caliper guide pins and any shoe positioners and antirattle springs that may be fitted (Figure 9.43).

6 Lift the caliper assembly off the rotor and out of its adapter as shown in Figure 9.44.

7 Support the weight of the caliper by hanging it on a piece of wire. Do not let it hang by the brake hose.

8 Remove the outboard shoe from the caliper as in Figure 9.45.

9 Remove the inboard shoe from the adapter by sliding it upward as shown in Figure 9.46.

10 Remove the rubber bushings from the holes in the inner and outer ears of the caliper. They are easily removed if they are collapsed (Figure 9.47).

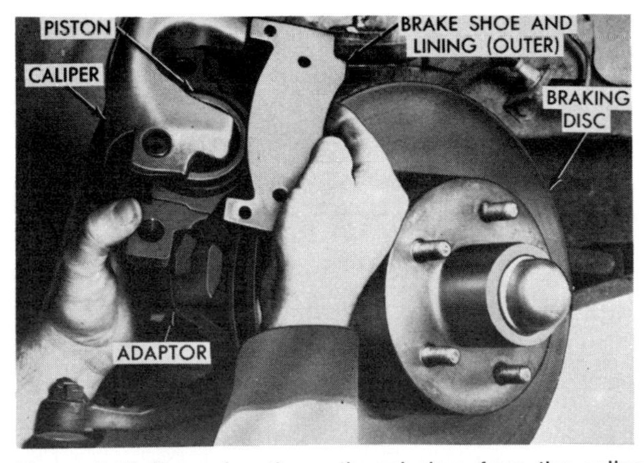

Figure 9.45 Removing the outboard shoe from the caliper (Chrysler Corporation).

Figure 9.46 Removing the inboard shoe from the adapter (Chrysler Corporation).

Figure 9.47 Removing the rubber bushings from the ears of the caliper (Chrysler Corporation).

Figure 9.48 Installing new rubber bushings in the caliper ears (Chrysler Corporation).

11 Clean the caliper assembly and the adapter. If the machined mating surfaces of the caliper or adapter are found to be rusted or corroded, clean them with a wire brush.

12 Check the caliper for evidence of leakage. Check the dust boot for tears or cracks. If you find leakage or dust-boot damage, remove the caliper for overhaul or replacement.

Installation

1 Install new rubber bushings in the holes in the inner and outer ears of the caliper. Compress the bushings with your fingers and twist them into the holes as shown in Figure 9.48.

2 Slide the new inboard shoe down into the adapter (Figure 9.46).

3 Place the outboard shoe in position in the caliper (Figure 9.45).

4 Hold the outboard shoe in position in the caliper, and slide the caliper down in its adapter and over the rotor.

5 Position the caliper assembly so that the holes in the caliper, adapter, and shoes are aligned.

6 Install the caliper guide pins so they pass through the rubber bushings, the adapter, and the shoes.

7 Push the pins in place and thread the pins into the adapter by hand. Use care so that the pins will not be cross-threaded.

8 Tighten the caliper guide pins to the torque specifications of the manufacturer.

9 Repeat the removal and installation steps on the opposite wheel.

10 Check the fluid level in the master cylinder. Add fluid if needed.

11 Pump the brake pedal several times to position the pistons and calipers. Check to see if the pedal is high and hard. Check the master cylinder fluid level and refill if necessary.

12 Install the wheels and lower the car to the floor.

13 Road test the car to check for proper braking action.

Job 9e

REPLACE FRONT DISC BRAKE SHOES: KELSEY-HAYES FLOATING CALIPER TYPE

SATISFACTORY PERFORMANCE

A satisfactory performance on this job requires that you do the following:

1 Replace the front brake shoes on a car fitted with Kelsey-Hayes type floating calipers.
2 Following the steps in the "Performance Outline" and the manufacturer's procedure and specifications, complete the job within 150 percent of the manufacturer's suggested time.
3 Fill in the blanks under "Information."

PERFORMANCE OUTLINE

1 Partially drain the reservoir.
2 Raise and support the car, and remove the wheel.
3 Bottom the caliper piston in its bore.
4 Remove and support the caliper.
5 Remove the shoes.
6 Remove the rubber bushings.
7 Clean and inspect all parts.
8 Install the new rubber bushings.
9 Install the brake shoes.
10 Install the caliper.
11 Check the pedal feel and fluid level.
12 Repeat the operations at the opposite wheel.
13 Install the wheels, and lower the car to the floor.
14 Road test the car for proper brake operation.

INFORMATION

Vehicle identification ______________________

Reference used ______________ Page(s) ______________

SUMMARY

In this chapter you have learned the operating principles of disc brakes. You are aware of both the advantages and disadvantages they offer. You understand the operation of the single piston caliper, and you know the names and functions of all its parts. You are familiar with the different types of single piston calipers used, and you can replace the brake shoes on cars fitted with these units.

SELF-TEST

Each incomplete statement or question in this test is followed by four suggested completions or answers. In each case select the *one* that best completes the statement or answers the question.

1 The caliper mountings used on front wheel disc brakes are attached to the
A. upper control arm

B. steering knuckle
C. splash shield
D. rotor

2 Disc brakes have many advantages, some of which are listed below. Which of the following statements do NOT apply to disc brakes?
A. They are self-cleaning
B. They are self-energizing
C. They dissipate heat faster
D. They resist fading

3 The outboard shoe is pushed against the rotor by the
A. push rod
B. piston
C. caliper
D. caliper mounting

4 Two mechanics are discussing the operation of disc brakes.
Mechanic A says that when the brake pedal is released, the piston moves back in its bore because of the action of the piston seal.
Mechanic B says the piston is pushed back in its bore by rotor runout.
Who is right?
A. A only
B. B only
C. Both A and B
D. Neither A nor B

5 Many manufacturers caution against the use of alcohol for washing brake parts. This is because the alcohol may mix with the brake fluid and
A. lower the boiling point of the fluid
B. attack the rubber parts of the system
C. corrode the metal parts of the system
D. destroy the lubricating qualities of the fluid

6 When replacing brake shoes, you must bottom the caliper piston in its bore. Before you perform this operation you should
A. remove fluid from the master cylinder reservoir
B. loosen the caliper retaining bolts
C. remove the dust boot
D. remove the bleeder valve

7 When replacing shoes in a Delco-Moraine type caliper, you should always replace the
I. sleeves
II. bushings
A. I only
B. II only
C. Both I and II
D. Neither I nor II

8 In a Delco-Moraine type caliper, the antirattle spring on the inboard shoe contacts the
A. caliper
B. caliper mounting
C. rotor
D. piston

9 In a Ford type caliper, the antirattle spring on the inboard shoe contacts the
A. caliper
B. caliper mounting
C. rotor
D. piston

10 Single piston type disc brakes are adjusted by
A. piston movement through the seal
B. star wheel adjusters
C. self-adjusting mechanisms
D. torquing the guide pins

Chapter 10 Caliper and Rotor Service

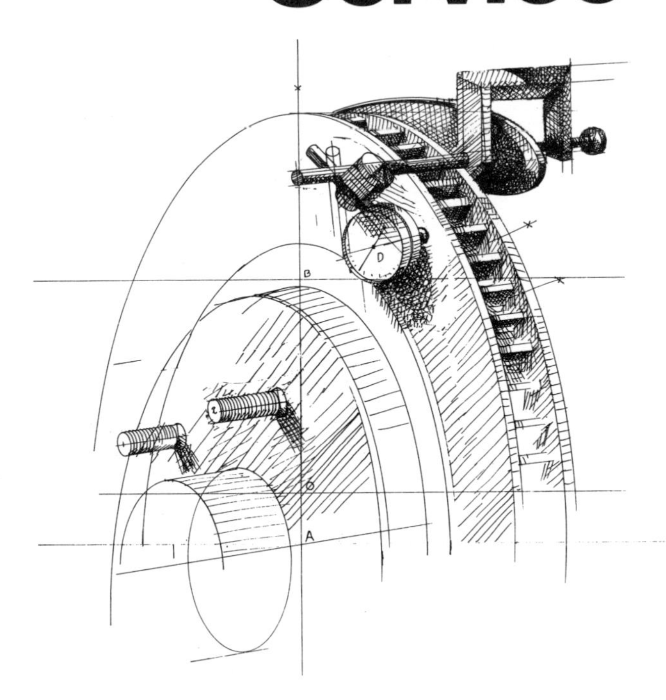

When working on disc brake systems, you will often find calipers and rotors that require service. Caliper and rotor service includes all the operations required to restore those parts to the manufacturer's specifications. Frozen pistons, leaks, and damaged dust boots are all causes for caliper overhaul. Though you know how to overhaul wheel cylinders, the procedures for overhauling calipers are quite different. This chapter covers the procedures you need to know for overhauling the four different calipers with which you worked in the preceding chapter. Though you are familiar with drum defects, you will find that rotor defects are different. You will also find that they require different methods of detection. This chapter shows you how to check rotors and how to restore them to specifications by machining.

Your objectives in this chapter are to:

1
Overhaul Ford and Kelsey-Hayes calipers.
2
Overhaul Delco-Moraine calipers.
3
Determine and measure rotor defects.
4
Machine a rotor on a lathe.

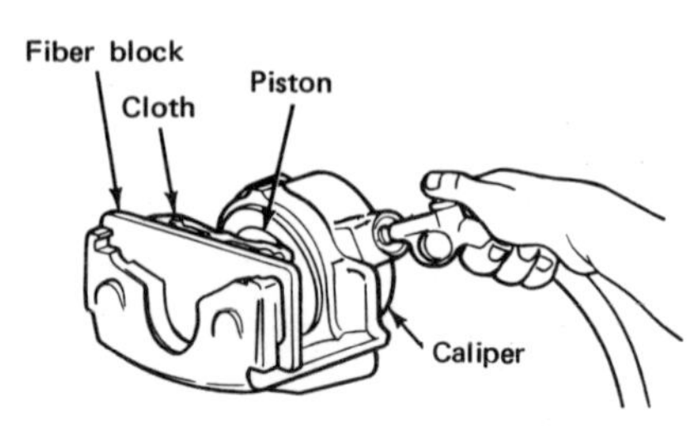

Figure 10.1 Removing a caliper piston with compressed air. Note the placement of the padded block to protect the piston (Ford Motor Company).

CALIPER OVERHAUL Calipers must be removed from the car when they are to be overhauled. When you replaced brake shoes, you removed various types of calipers, but you did not disconnect them from the hydraulic system. In most cases of caliper overhaul, you have to remove only a brake hose or line using the same procedure you followed when you replaced them.

When you overhaul calipers, you should complete your work at one wheel before proceeding to the next. This means that you should remove, overhaul, install, and bleed one caliper before starting work on another. There are two reasons for this:

1 Some calipers fit on either side of a car; therefore, it is easy to get them switched during installation. Although the calipers will fit in place, the position of the bleeder valve will not allow all the air to escape from the cylinder during bleeding.

2 The caliper pistons must be removed from their cylinders. It is easier and much safer to push the pistons out hydraulically while they are still on the car than to blow them out with air pressure on the bench.

Note. **If you have to overhaul a caliper that has already been removed from a car, remove the piston by directing compressed air into the inlet hole of the caliper. A padded piece of cloth supported by a wooden or fiber block should be placed in the caliper opening (Figure 10.1). This will protect the piston when it is blown out. Since the piston can be ejected with considerable force, you should be very careful to keep your fingers out of the way.**

OVERHAULING A DISC BRAKE CALIPER: FORD AND KELSEY-HAYES TYPES

Below is the general procedure for overhauling a typical Ford or Kelsey-Hayes caliper. You should check the car manufacturer's manual for the specific procedure.

Removal

1 Remove the caliper from its mounting.

2 Remove the brake shoes. If the shoes are similar, mark them so you can reinstall them in their original positions.

3 Support the caliper over a drain pan, and carefully pump the brake pedal until the piston is pushed from its bore.

Note. **If the brake pedal is held down with a prop or a weight, the fluid remaining in the master cylinder and line will not run out. This will simplify bleeding.**

4 Disconnect the brake hose from the caliper.

Overhaul

1 Place the caliper between the padded jaws of a vise, and tighten the vise just enough to hold the caliper.

Note. **Clamping a caliper too tightly in a vise may distort the cylinder bore so that the piston will no longer fit.**

Figure 10.2 Removing the dust boot (Chrysler Corporation).

2 Remove the dust boot by peeling it out with your fingers (Figure 10.2).

3 Use a small, pointed, wooden or plastic stick to remove the piston seal (Figure 10.3).

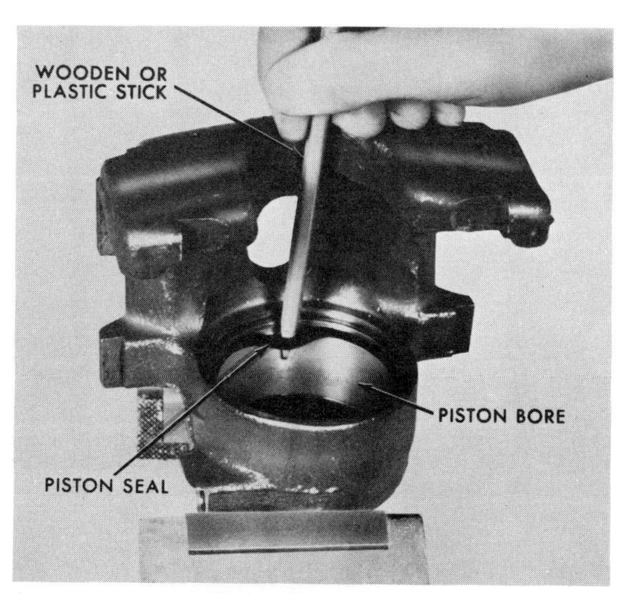

Figure 10.3 Removing the piston seal. Note that a nonmetallic tool is used to avoid scratching the cylinder wall (Chrysler Corporation).

4 Remove the bleeder valve.

5 Clean all parts with brake fluid, and dry them with compressed air. Be sure to blow out the drilled passages.

Note. The use of alcohol or alcohol-based cleaning fluids is not recommended by all manufacturers. Traces of alcohol may remain in the cylinder. Since alcohol has a low boiling point, any of it left in the cylinder might lower the boiling point of the brake fluid.

6 Inspect the piston, and replace it if it is scored or pitted, or if the plating is worn off.

Note. Ignore black stains on the piston. They are caused by the piston seal and are a normal condition.

7 Inspect the cylinder bore for scoring and pitting. Light scratches and corrosion should be removed by polishing the bore with crocus cloth. If the bore is deeply scored or pitted, the caliper should be replaced. Stains on cylinder walls should be ignored. They are normal.

Note. In some cases it may be possible to salvage a scored or pitted caliper by honing. However, most car makers advise against the use of abrasives on the cylinder walls. Therefore, you should consult the manufacturer's manual for recommendations concerning the car on which you are working.

8 Coat the new piston seal with an approved assembly lubricant.

Note. Some manufacturers recommend the use of a specially formulated brake part lubricant. Others recommend the use of brake fluid. Check an appropriate manual to be sure which is recommended for the car you are servicing.

9 Install a new piston seal in the groove in the cylinder bore (Figure 10.4). Gently work the seal in place with your fingers, making sure it is not twisted.

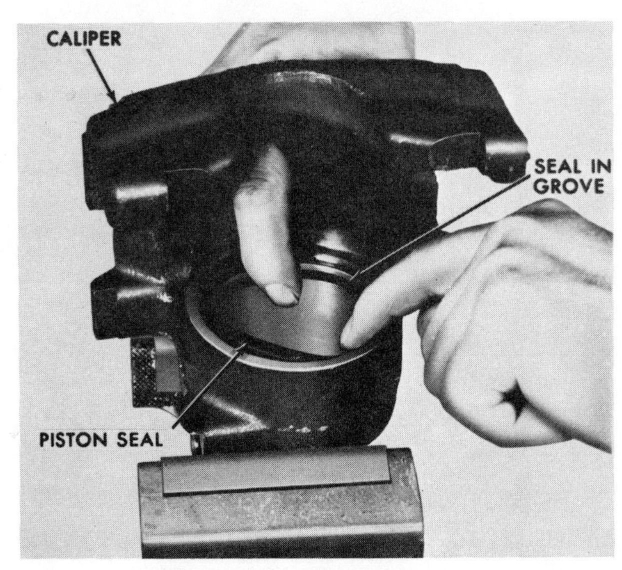

Figure 10.4 Installing the piston seal (Chrysler Corporation).

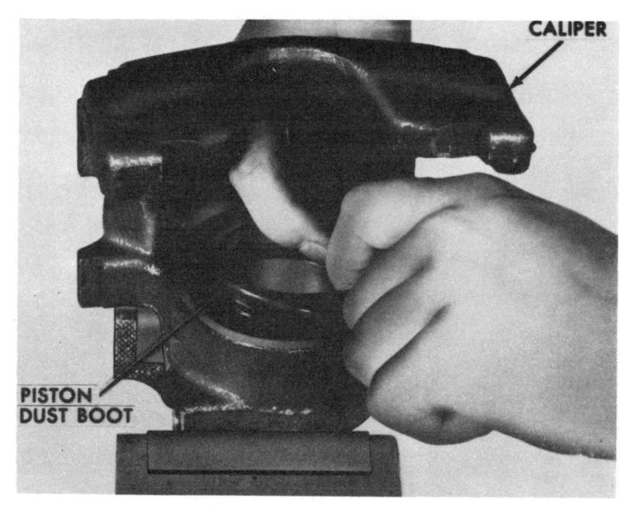

Figure 10.5 Installing the dust boot (Chrysler Corporation).

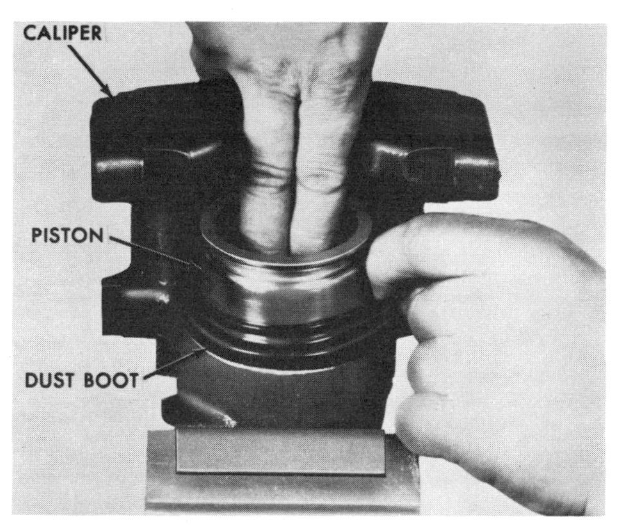

Figure 10.6 Installing the piston (Crysler Corporation).

10 Coat the new dust boot with assembly lubricant. Leave a heavy coating of lubricant inside the boot.

11 Install the boot in the caliper by working it into its groove with your fingers (Figure 10.5).

12 Check the fit of the boot by sliding a finger around the inside of the caliper.

13 Install the bleeder valve.

14 Plug the inlet hole in the caliper.

15 Lubricate the piston with assembly lubricant.

16 Spread the boot open with your fingers, and work the piston down into the boot (Figure 10.6). Push the piston down to start it in its bore.

Note. **The air trapped below the piston will balloon the boot around the piston, causing the boot to slide up into the groove in the piston.**

17 Remove the plug from the inlet hole.

18 Push the piston down to the bottom of its bore. Use care to avoid cocking and jamming the piston.

Note. **In most cases a force of from 50 to 100 pounds, about (23 to 45 kg) is required to bottom a piston.**

Installation

1 Connect the caliper to the brake hose.

2 Install the brake shoes.

3 Install the caliper.

4 Bleed the caliper.

Job 10a

OVERHAUL A FORD OR KELSEY-HAYES CALIPER

SATISFACTORY PERFORMANCE

A satisfactory performance on this job requires that you do the following:

1 Overhaul a Ford or Kelsey-Hayes caliper.
2 Following the steps in the "Performance Outline" and the manufacturer's procedure and specifications, complete the job within 150 percent of the manufacturer's suggested time.
3 Fill in the blanks under "Information."

PERFORMANCE OUTLINE

1 Remove the caliper from the car.
2 Disassemble the caliper.
3 Clean the parts.
4 Inspect the parts.
5 Assemble the caliper.
6 Install the caliper on the car.
7 Bleed the caliper.

INFORMATION

Vehicle identification ____________________

Reference used ____________ Page(s) ____________

OVERHAULING A DISC BRAKE CALIPER: DELCO-MORAINE TYPE The steps that follow give the general procedure for overhauling a typical Delco-Moraine caliper. Check the car manufacturer's manual for the specific procedure.

Removal

1 Remove the caliper from its mounting.

2 Remove the brake shoes.

3 Support the caliper over a drain pan, and carefully pump the brake pedal until the caliper piston is pushed from its bore.

4 Disconnect the brake hose from the caliper.

Overhaul

1 Carefully hold the caliper in a vise.

2 Pry the dust boot out of its retaining groove by using a screwdriver as shown in Figure 10.7.

3 Remove the piston seal from its groove in the cylinder.

4 Remove the bleeder valve.

5 Clean and dry all parts.

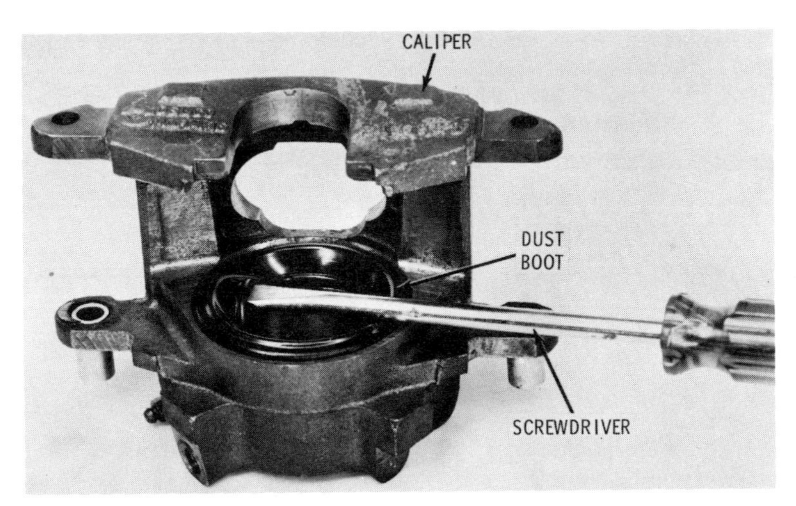

Figure 10.7 Prying out the dust boot with a screwdriver (General Motors.)

6 Inspect the piston.

7 Inspect the cylinder bore.

8 Coat the new piston seal with brake fluid, and install it in its groove in the cylinder.

9 Lubricate the piston with brake fluid, and install the new dust boot as shown in Figure 10.8. Be sure the fold in the boot faces the open (outer) end of the piston.

10 Place the piston carefully in its bore, and push it all the way in until it bottoms.

11 Position the outer edge of the boot in the flanged, or stepped, edge of the cylinder bore.

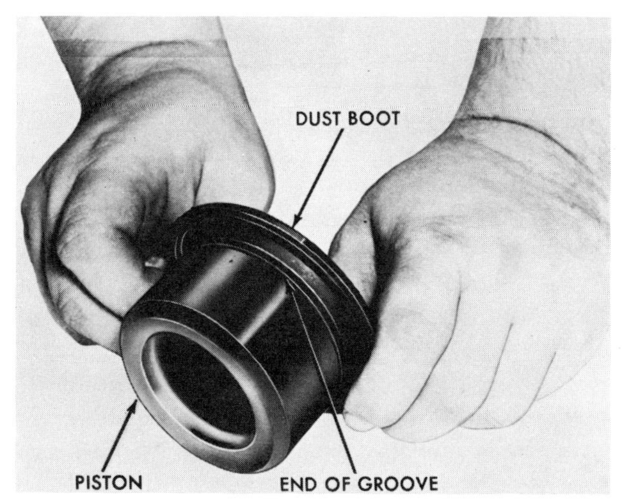

Figure 10.8 Installing the dust boot on the piston (courtesy Chevrolet Service Manual, Chevrolet Motor Division).

12 Using a boot installer as shown in Figure 10.9, seat the boot so that it is below the casting and even all around.

Installation

1 Connect the caliper to the brake hose.

2 Install the brake shoes.

3 Install the caliper.

4 Bleed the caliper.

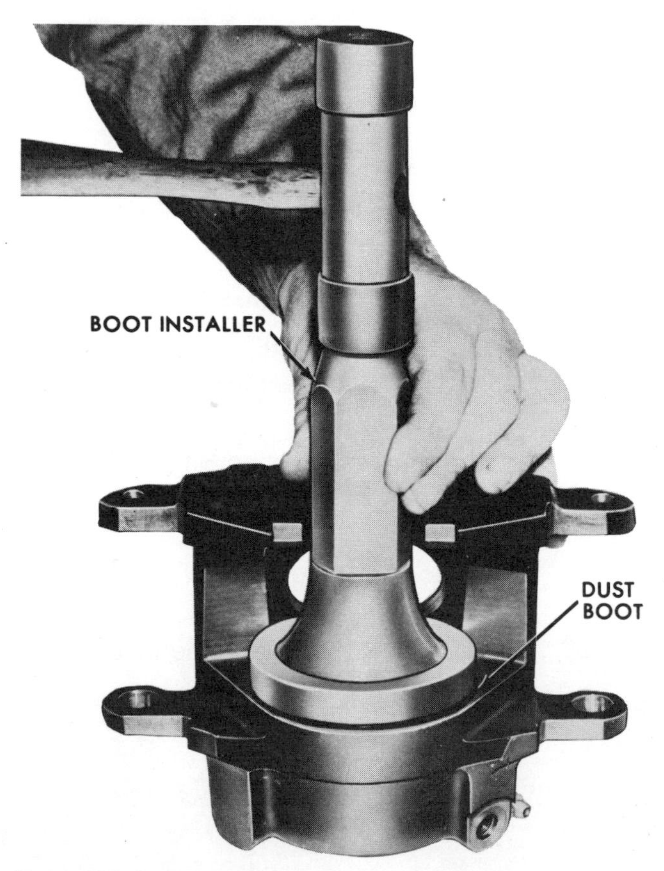

Figure 10.9 Seating the dust boot with a boot installer (courtesy Chevrolet Service Manual, Chevrolet Motor Division).

Job 10b

OVERHAUL A
DELCO-MORAINE
CALIPER

SATISFACTORY PERFORMANCE

A satisfactory performance on this job requires that you do the following:

1 Overhaul a Delco-Moraine caliper.
2 Following the steps in the "Performance Outline" and the manufacturer's procedure and specifications, complete the job within 150 percent of the manufacturer's suggested time.
3 Fill in the blanks under "Information."

PERFORMANCE OUTLINE

1 Remove the caliper from the car.
2 Disassemble the caliper.
3 Clean the parts.
4 Inspect the parts.
5 Assemble the caliper.
6 Install the caliper on the car.
7 Bleed the caliper.

INFORMATION

Vehicle identification ______________________________

Reference used ______________ Page(s) ______________

ROTOR SERVICE Rotors are usually made of cast iron. In most instances, those used for front wheel brakes are integral, or in one piece, with their hubs. Rotors used for rear wheels do not have hubs. They are attached to flanges on the rear axles much the same as floating brake drums are attached. To obtain the advantages of a self-energizing brake in the parking brake system, some car makers incorporate a small brake drum into the rear-wheel rotors (Figure 10.10). Two small brake shoes expand inside the drum when the parking brake is applied. Thus, the parking brake is completely independent of the service brakes.

Like brake drums, rotors must be inspected for wear and damage. Visual inspection may disclose some of the more obvious defects. But a thorough inspection requires careful use of precision measuring instruments.

Visual Inspection You can usually perform a quick, visual inspection without removing the rotor. When the wheel is removed, the outer surface of the rotor is exposed. The inner sur-

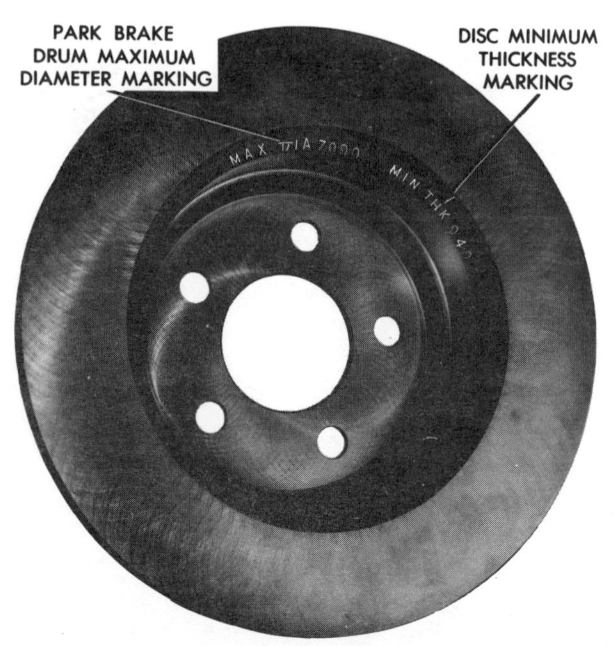

Figure 10.10 A parking brake drum built into a rear wheel rotor. Note that the specs for maximum drum diameter and minimum rotor thickness are both given (Chrysler Corporation).

face may be concealed by the splash shield, but you can usually look between the rotor and its shield with the aid of a drop light or a flashlight.

In making a visual inspection you should look for two types of defects—cracks and scoring.

Cracks Any rotor exhibiting a crack should be replaced. Otherwise, it could fail and cause a serious accident.

Scoring Scoring is the most obvious type of wear. The presence of light scoring on a rotor is no problem. Most manufacturers consider light scoring as a normal, acceptable wear pattern. Machining is required only when the scoring has caused a rough surface or when the grooves exceed 0.015 in. (0.39 mm) in depth. When you find a scored rotor, you should check the manufacturer's manual for recommendations.

Determination and Measurement of Rotor Defects To provide effective, safe braking, rotors must turn without excessive wobble. The surfaces on both sides of the rotor must be parallel, and those surfaces must be located so as to fall within the operating range of the caliper. Also, the thickness of the rotor must not be reduced too much by wear or machining. All manufacturers provide specifications covering these factors. In some instances they require that special measuring tools be used.

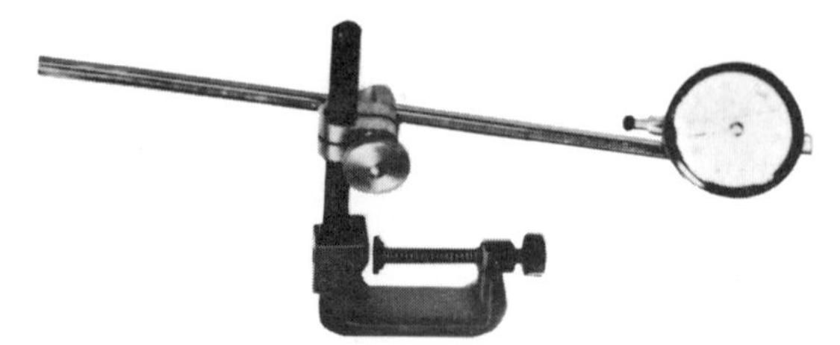

Figure 10.11 A dial indicator (Chrysler Corporation).

Rotor Runout Excessive rotor runout is a common defect. Runout is a lateral, or side-to-side, movement of the rotor as it turns. This movement, or wobble, is caused by a warping and twisting of the rotor by the heat and pressures of braking. Rotor runout can be compared to brake drum out-of-round in that the rotor surface moves alternately toward and away from the lining. If the runout is excessive, it will push the shoes away from the rotor. This will increase the lining-to-rotor clearance, causing the caliper piston to move back in its bore. When the brakes are applied, the piston must then move outward farther than normal. This, in turn, increases pedal travel. At times it also causes the pedal to pulsate, or bounce, as the shoes follow the wavy rotor surfaces.

Checking Rotor Runout Before you can check for rotor runout, you must remove the wheel from the hub and all play from the wheel bearings. If the bearings are not adjusted to remove all play, the rotor can wobble on the spindle, and the actual runout cannot be accurately measured.

Rotor runout is measured with a *dial indicator.*

A dial indicator is an instrument that measures movement in thousandths of an inch (0.001 in.). The amount of movement is indicated by a pointer that moves on a dial. A typical dial indicator is shown in Figure 10.11. Various means of anchoring the dial indicator are provided to allow the instrument to be positioned where desired.

MEASURING ROTOR RUNOUT

Here is the procedure you should use to measure rotor runout.

1 Raise and support the car.

2 Remove the wheel.

3 Adjust the wheel bearings to remove all play.

4 Mount the dial indicator on a solid surface such as the caliper, steering knuckle, or splash shield, as shown in Figure 10.12. Position the plunger so that it contacts the rotor surface about 1 in. (about 2.5 cm) from the edge.

Figure 10.12 A dial indicator mounted to measure rotor runout (General Motors).

5 Turn the rotor one complete turn. Any runout will move the plunger and cause the pointer to move across the dial.

6 Record the total runout.

Note. Any runout that exceeds 0.005 in. (about 0.127 mm) is considered excessive. Since different manufacturers have different specifications, you should always check an appropriate manual for this specification. A rotor with excessive runout should be machined or replaced.

7 Remove the dial indicator.
8 Adjust the wheel bearings, and install a new cotter pin.
9 Install the wheel and lower the car to the floor.

Job 10c

MEASURE ROTOR RUNOUT

SATISFACTORY PERFORMANCE

A satisfactory performance on this job requires that you do the following:

1 Measure the runout of a front wheel disc brake rotor.
2 Following the steps in the "Performance Outline," complete the job within 20 minutes.
3 Fill in the blanks under "Information."

PERFORMANCE OUTLINE

1 Raise and support the car.
2 Remove the wheel.
3 Eliminate play from the wheel bearings.
4 Mount the dial indicator.
5 Check for runout.
6 Remove the dial indicator.
7 Adjust the wheel bearings.
8 Install the wheel and lower the car to the floor.

INFORMATION

Vehicle identification ______________________

Rotor position ______________ Total runout ______________

Maximum allowable runout ______________________

Reference used ______________ Page(s) ______________

Comments ______________________

Rotor Parallelism
The inner and outer surfaces of a rotor must be parallel to each other. Lack of parallelism is similar to runout in that it causes the pistons to be pushed back too far into their caliper bores. This defect is a common cause of vibration and pedal pulsation, and it should always be considered when those conditions exist.

Checking Rotor Parallelism Rotor parallelism can be checked by several methods. One method is a continuation of the check for runout. It involves indicating the high and low areas on the

Figure 10.13 A rotor marked with its minimum thickness dimension (General Motors).

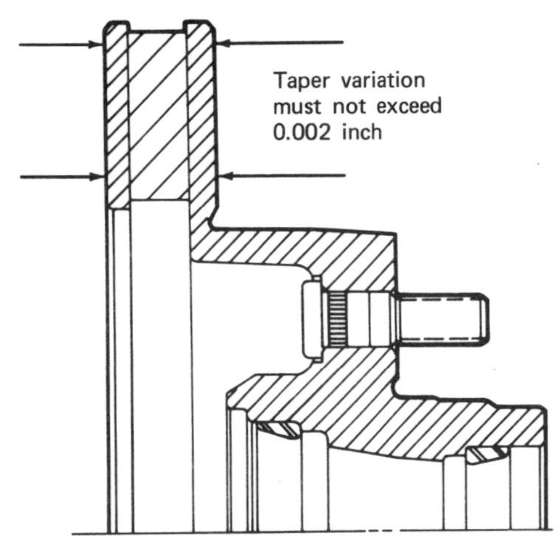

Figure 10.14 Sectional view of a rotor showing where measurements for taper variation should be taken (courtesy American Motors Corporation).

rotor with chalk marks while checking runout. Position the dial indicator to measure the runout of the inner rotor surface. If the inner surface duplicates the outer surface runout in reverse, the rotor surfaces are parallel. Any variation in measured runout means a rotor of varying thickness and thus a lack of parallelism.

Another method of checking rotor parallelism requires that you use a micrometer. Mark off the rotor in six to eight segments, and then measure the rotor thickness in each segment. If the rotor surfaces are parallel, all the measurements will be equal. Specifications for rotor parallelism are rather critical. Some manufacturers state that parallelism should be maintained to 0.0005 in. (about 0.0127 mm). Here too, you should consult the manufacturer's shop manual for the recommended procedure and specifications. Rotors that lack parallelism must be machined or replaced.

Rotor thickness

Every rotor has a minimum thickness specification assigned by its manufacturer. Many rotors are marked with this dimension, as shown in Figure 10.13. This dimension is sometimes referred to as a *discard dimension.* Rotors that do not meet this specification should be discarded. This dimension is a wear dimension only. Under no circumstances should you ever machine a rotor to this dimension.

Checking Rotor Thickness Rotor thickness, like rotor parallelism, is measured with a micrometer. Therefore, many mechanics combine the two operations. If the minimum thickness is not marked on the rotor, you will have to refer to the manufacturer's manual for the specification.

Taper Variation

Both surfaces of a rotor should be perfectly flat. Any taper wear exceeding the manufacturer's specifications causes the shoes to wear on an angle.

Checking Taper Variation Taper variation is another condition you can check with a micrometer. You can combine this check with the checks for parallelism and thickness. Most manufacturers allow a taper variation of 0.002 in. (about 0.091 mm). Even so, you should always consult the manual for the car you are working on. Figure 10.14 shows the spots you should measure. Measurements should be taken at several places around the rotor.

Job 10d

MEASURE ROTOR DEFECTS

SATISFACTORY PERFORMANCE

A satisfactory performance on this job requires that you do the following:

1 Measure the thickness, parallelism, and taper variation of the rotor provided.
2 Complete the job within 15 minutes and record your readings below.
3 Fill in the blanks under "Information."

PERFORMANCE SITUATION

	Specification	*Actual Measurement*
Minimum thickness dimension	______________	______________
Taper variation	______________	______________
Parallelism	______________	______________

INFORMATION

Rotor identification ______________________________

Reference used ______________ Page(s) ______________

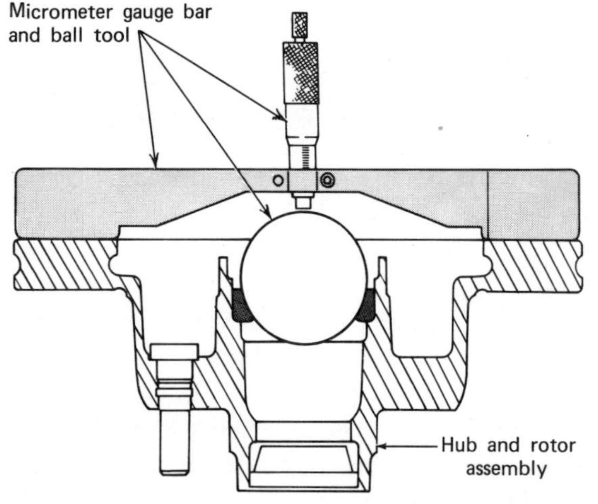

Figure 10.15 Measuring for excessive stock removal. Note the position of the special micrometer gauge bar and ball tool (Ford Motor Company).

Inner Surface Stock Removal Limits

Since a rotor has two frictional surfaces, normal wear and any required machining may remove more *stock,* or material, from one side than from the other. Uneven wear is considered normal, and it is allowed within certain limits. Machining one side of a rotor more than the other side is also allowed, again within certain limits. Some manufacturers specify a limit on how much stock may be removed from the inner surface of their rotors. This limit is called the *maximum allowable stock removal.*

Checking for Excessive Stock Removal

Since this limit is established by each manufacturer for each different rotor design, special instruments are required for measuring it. Figure 10.15 shows one type of gauge set up to measure

Figure 10.16 Resurfacing a rotor. The grinder head is fitted with a sandpaper disc that spins rapidly while the rotor is turning (Chrysler Corporation).

for excessive stock removal on a front hub and rotor. The reading, obtained on a special micrometer, is checked against a chart provided with the instrument. Any rotor not meeting specifications should be replaced.

Machining Rotors

Two methods may be used to restore a rotor to specifications: (1) resurfacing and (2) reconditioning. The method you choose will depend on the condition of the rotor.

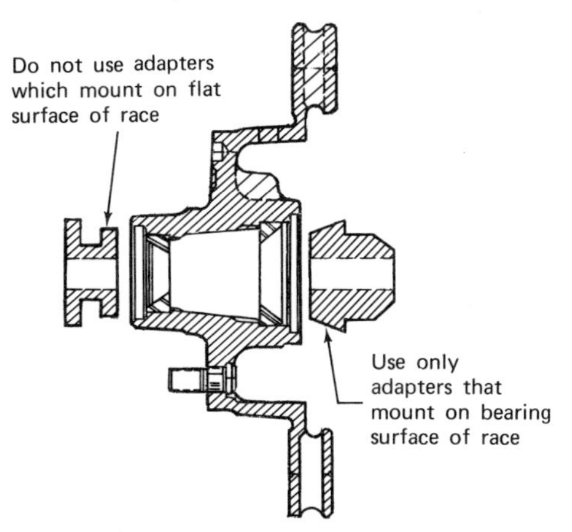

Figure 10.18 Selection of adaptors for mounting a rotor on a lathe arbor (General Motors).

Figure 10.17 Reconditioning a rotor. The twin cutting tools remove material from both sides of the rotor at the same time (Chrysler Corporation).

Resurfacing A rotor can be resurfaced when its surfaces are rusted, contaminated with lining deposits, or lightly scored. Resurfacing consists of grinding the rotor surfaces with a sanding disc while the rotor is turning on a lathe. This operation is shown in Figure 10.16. Resurfacing removes very little material from the rotor, but it smoothes out minor imperfections.

Reconditioning A rotor that lacks parallelism or has excessive runout, excessive taper variation, or deep scoring should be reconditioned. The procedure for reconditioning a rotor is similar to the procedure for reconditioning a brake drum. Cutting tools, or bits, are used for cutting away the surfaces of the rotor to eliminate defects. This operation is shown in Figure 10.17.

The specifications for rotors, however, are much more critical than the specifications for drums. Therefore, you must use extreme care in mounting a rotor on the lathe arbor. Rotors are mounted with the aid of adapters and cones in the same way that brake drums are mounted. When mounting a hubbed rotor, be sure the adapters contact the bearing cups as shown in Figure 10.18.

Like drums, rotors will vibrate while you are machining them. Their vibration can prevent you from obtaining the desired finish on the rotor surfaces. For this reason, these vibrations must be dampened out. Always install a thin rubber dampening belt on a rotor before you machine it. Refer to Figures 10.16 and 10.17 for the correct way to install the belt.

The finish on the rotor surfaces must fall within a certain range of smoothness and must be nondirectional. This requires that any rotor that is reconditioned also be resurfaced. Most manufacturers specify a finish of between 20 and 60 microinches. This finish cannot be measured with the instruments available in the average shop. It can be obtained, however, by the proper selection of the sanding disc used on the grinder. You should check the instruction manual for the lathe you are using to determine the proper grade of sandpaper. Grinding the disc after it has been reconditioned will provide a nondirectional finish, as shown in Figure 10.19.

Figure 10.19 A rotor after resurfacing. Note the desired nondirectional pattern on the surface (courtesy American Motors Corporation).

The operating procedures for rotor lathes vary with the lathe design. Before using the lathe in your shop you should study the instruction manual furnished with it.

Job 10e

MACHINE A DISC BRAKE ROTOR

SATISFACTORY PERFORMANCE

A satisfactory performance on this job requires that you do the following:

1 Machine a rotor on a rotor lathe so that it meets the specifications of its manufacturer.
2 Following the steps in the "Performance Outline" and the procedure required for the lathe used, complete the job within 60 minutes.
3 Fill in the blanks under "Information."

PERFORMANCE OUTLINE

1 Select the correct adapters.
2 Mount the rotor on the lathe arbor.
3 Adjust the rotor position to obtain minimum runout.
4 Cut both sides of the rotor to eliminate defects.
5 Grind the rotor surfaces to obtain a smooth nondirectional pattern.
6 Check the minimum thickness dimension.

INFORMATION

Rotor identification ______________________

Minimum thickness dimension ______________________

Actual thickness ______________________

SUMMARY

In studying this chapter and performing the assigned jobs, you have learned how to overhaul the various single piston calipers used in disc brake systems. You have developed diagnostic skills in detecting and measuring defects in rotors, and you have learned how to restore them to specifications.

SELF-TEST

Each incomplete statement or question in this test is followed by four suggested completions or answers. In each case select the *one* that best completes the statement or answers the question.

1 Although some calipers will fit either side of a car, they should not be interchanged because the
 A. shoes must be reversed
 B. dust boots may not seal properly
 C. pistons move in opposite directions
 D. bleeder valve will be in the wrong position

2 If hydraulic pressure cannot be used to remove a caliper piston, the piston can be
 A. pushed out with a C-clamp
 B. driven out with a soft punch
 C. pried out with a screwdriver
 D. blown out with compressed air

3 Two mechanics are discussing disc brake calipers.

Mechanic A says that the piston seal is fitted into a groove in the piston.
Mechanic B says the piston seal should be installed with the aid of a ring compressor.
Who is right?
A. A only
B. B only
C. Both A and B
D. Neither A nor B

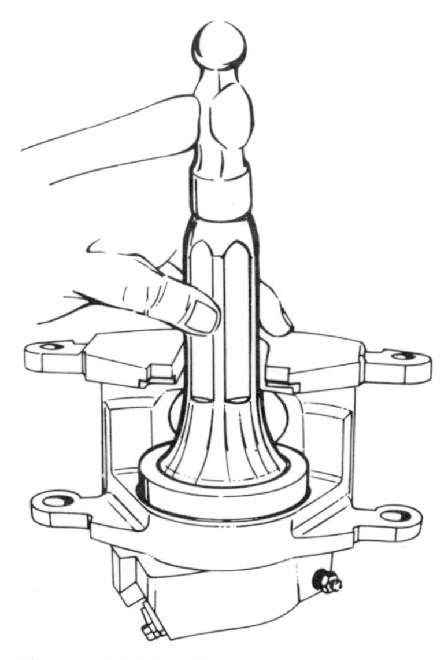
(General Motors).

4 The drawing above illustrates an operation performed in the overhaul of a caliper. What operation is shown?
A. Bottoming a piston
B. Seating a dust boot
C. Installing a lock ring
D. Expanding a piston seal

5 Which of the following is NOT a factor to consider when checking rotors?
A. Runout
B. Parallelism
C. Out-of-round
D. Taper variation

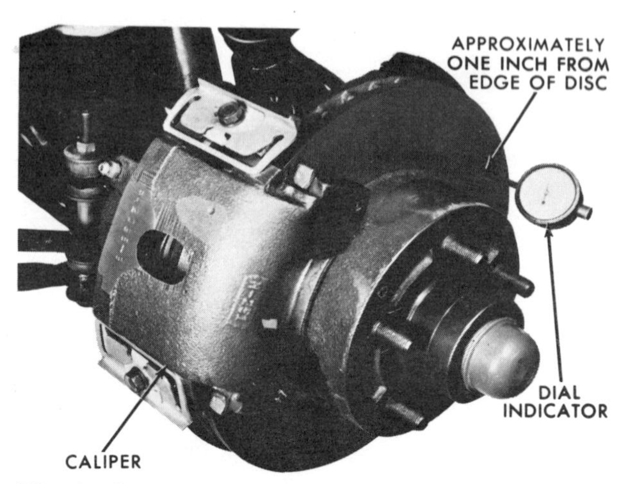

(Chrysler Corporation).

6 The photograph above shows a dial indicator set up to detect and measure a particular rotor defect. What defect can be found with this particular set up?
A. Runout
B. Scoring
C. Out-of-round
D. Taper variation

7 Two mechanics are discussing disc brake rotors.
Mechanic A says that when a rotor is machined, an equal amount of material should be removed from both surfaces.
Mechanic B says that when a rotor is worn to below its minimum thickness dimension, it should be discarded.
Who is right?
A. A only
B. B only
C. Both A and B
D. Neither A nor B

8 A car owner states that the brake pedal pulsates, or bounces, during braking. Which of the following conditions would NOT cause this to occur?
A. Lack of parallelism

B. Loose wheel bearings
C. Excessive rotor runout
D. Excessive taper variation

9 Which of the following rotor defects CANNOT be detected with a micrometer?
A. Excessive runout
B. Lack of parallelism
C. Excessive taper variation
D. Insufficient rotor thickness

10 Two mechanics are discussing disc brake rotor service.
Mechanic A says that the bearing cups should be removed from the rotor before it is mounted on the lathe arbor.
Mechanic B says that a nondirectional finish on the rotor surfaces can be obtained by grinding.
Who is right?
A. A only
B. B only
C. Both A and B
D. Neither A nor B

Chapter 11 Master Cylinders and Valves

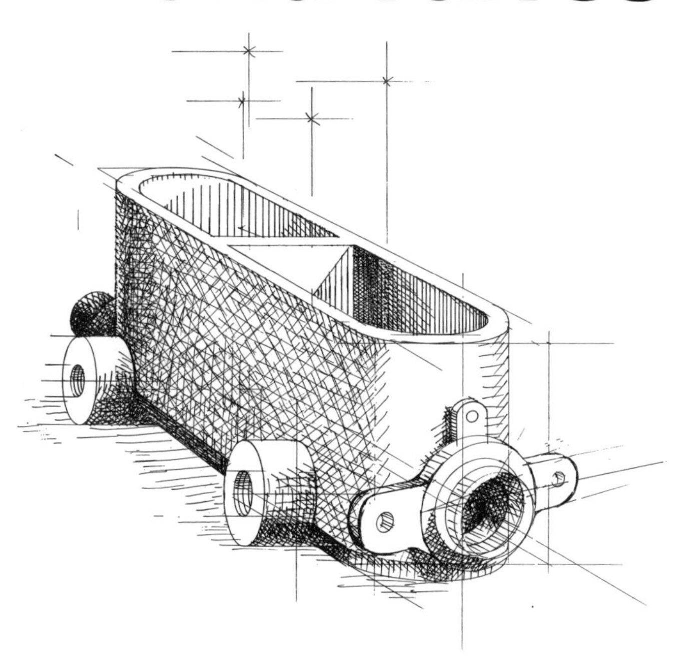

The force a driver applies to the brake pedal is converted to hydraulic pressure in the master cylinder. As mentioned before, pressure in a hydraulic system is equal throughout and, therefore, a failure of one part of the system affects the entire system. For this reason, all cars now produced are equipped with dual braking systems. One system brakes the front wheels, and the other brakes the rear wheels. If one of the systems fails, the one that remains in operation can stop the car.

Dual braking systems require dual master cylinders. In turn, dual master cylinders and combination (disc and drum) brake systems require certain valves to ensure effective braking under all conditions.

Your objectives in this chapter are to:

1
Identify the parts of dual master cylinders and their operation.

2
Replace dual master cylinders.

3
Overhaul dual master cylinders.

4
Identify the function and location of brake system valves.

5
Replace brake system valves.

MASTER CYLINDERS A dual master cylinder is a combination of two single master cylinders in a single casting body. The pistons share the same cylinder, but each system is actuated independently. This situation provides two separate systems, and leakage or failure in one system will not impair the operation of the other system. Common practice is to operate the front brakes with one section and the rear brakes with the other. Figure 11.1 shows a sectioned view of a typical dual master cylinder and labels the parts.

Note. Car makers have yet to agree on common names for master cylinder parts. Therefore the names used in this chapter are not universal. However, those names are used consistently in this chapter.

Dual master cylinders are used with drum brake, disc brake, and combination systems. When used with drum brakes, both sections of the dual master cylinder have residual pressure check valves in their outlet ports. When used with disc brakes or combination systems, the section that serves the disc brakes does not have a residual pressure check valve; a disc brake system must have no residual pressure.

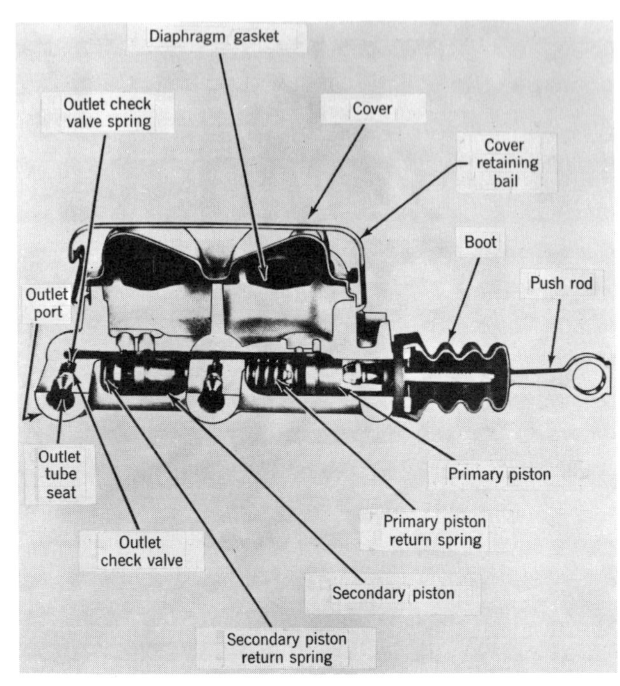

Figure 11.1 A dual master cylinder (Ford Motor Company).

Residual pressure in a disc brake system would hold the shoes in contact with the rotor. This contact would cause dragging brakes, overheating, and unnecessary wear.

Job 11a

IDENTIFY MASTER CYLINDER PARTS

SATISFACTORY PERFORMANCE

A satisfactory performance on this job requires that you do the following:

1 Identify the numbered parts on the drawing by placing the number for each part in front of the correct part name below the drawing.
2 Correctly identify all the numbered parts within 15 minutes.

PERFORMANCE SITUATION

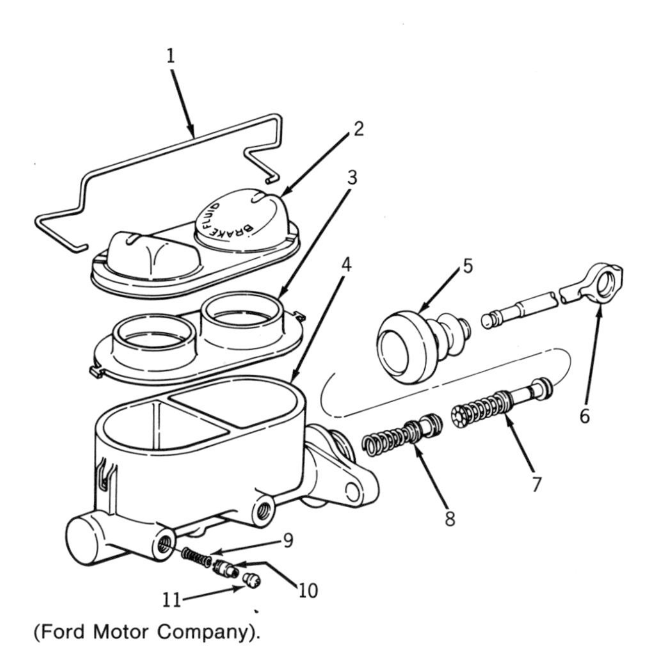

(Ford Motor Company).

______ Diaphragm gasket	______ Primary piston
______ Outlet check valve	______ Inlet valve
______ Cover	______ Boot
______ Outlet tube seat	______ Push rod
______ Secondary piston	______ Cylinder body
______ Check valve spring	______ Cover retaining bail

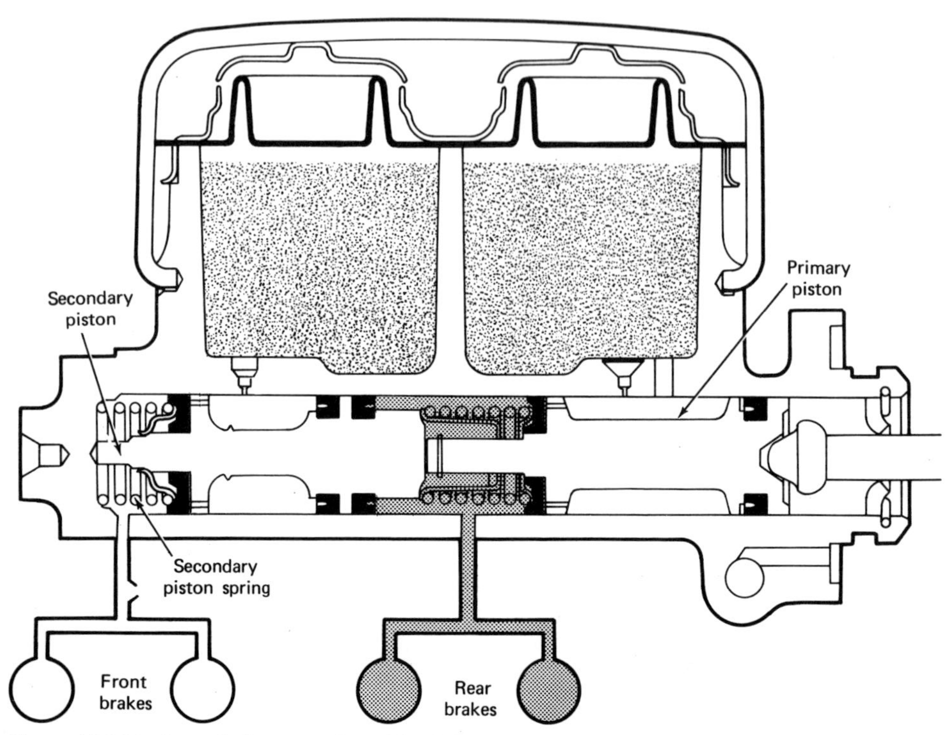

Figure 11.2 Master cylinder operation with a failure in the front brake system (courtesy Chevrolet Service Manual, Chevrolet Motor Division).

Operation of the Dual Master Cylinder When force is applied to a brake pedal, the force, multiplied by the pedal leverage, is applied to the master cylinder primary piston through the push rod. The primary piston is pushed forward. The secondary piston is also pushed forward by the primary piston spring and the fluid between the primary and secondary pistons. As a result, hydraulic pressure is built up equally in both systems. When the pedal is released, the return springs push back the pistons, and the hydraulic pressure is released.

Though one of the dual braking systems can stop the car if the other system fails, the stopping distance is increased. Consequently, there is an increase in the possibility of an accident. There are two indications that one of the systems has failed. First, the driver notices increased pedal travel and a need to apply greater effort to the pedal. Second, a warning light in the instrument panel turns on.

Front Brake System Failure If there is a failure in the hydraulic system for the front brakes, both pistons move forward when the brake pedal is depressed. However, the failure in the front system does not allow a pressure buildup ahead of the secondary piston. The primary piston continues to move forward until the extended nose on the secondary piston bottoms in the cylinder bore. Continued force applied by the driver then causes pressure to build up between the primary piston and the secondary piston. That pressure operates the rear brakes (Figure 11.2).

Rear Brake System Failure If there is a failure in the hydraulic system for the rear brakes, the primary piston moves forward when the brake pedal is depressed. However, no pressure is built up between the primary and secondary pistons; the primary piston continues to move forward until the piston extension contacts the

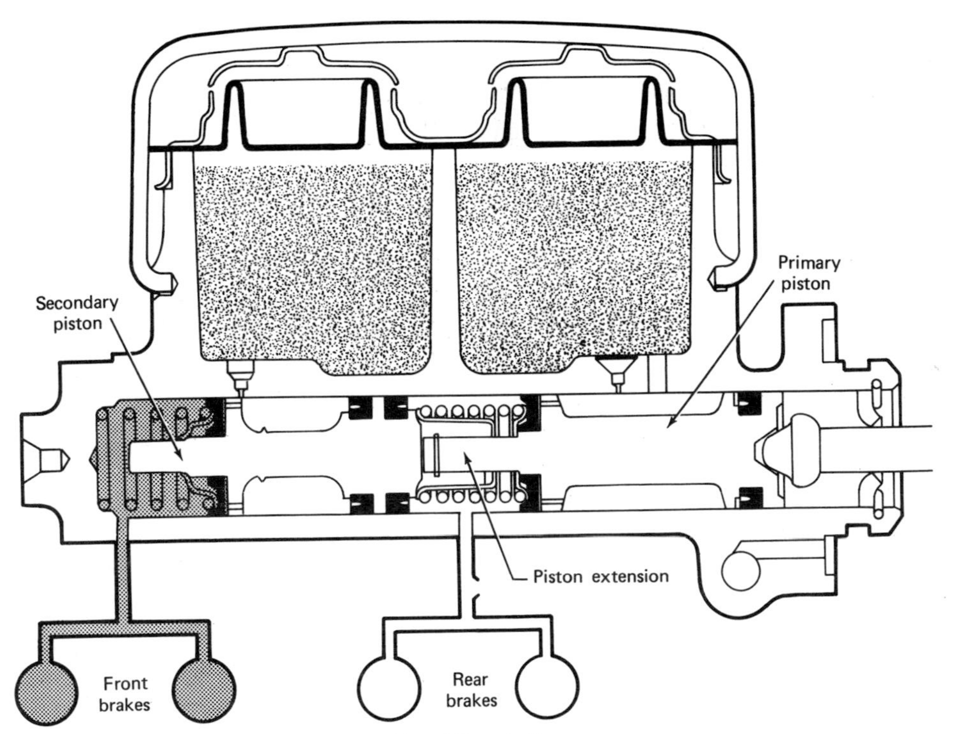

Figure 11.3 Master cylinder operation with a failure in the rear brake system (courtesy Chevrolet Service Manual, Chevrolet Motor Division).

back of the secondary piston. The force of the push rod is then directed against the secondary piston, which moves forward and applies pressure to the front brake system (Figure 11.3).

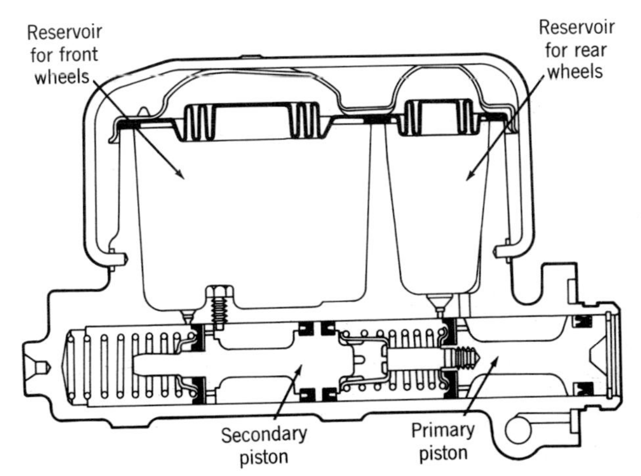

Figure 11.4 Master cylinder used with a combination brake system (courtesy Chevrolet Service Manual, Chevrolet Motor Division).

Construction of Dual Master Cylinders Though all dual master cylinders operate the same way, they include many different designs. Their two reservoirs, for instance, are often different in capacity and shape. Master cylinders for cars that have only drum brake systems have reservoirs that are about equal in capacity. The master cylinders shown in Figures 11.1, 11.2, and 11.3 are of this type. Master cylinders for cars that have combination (disc and drum) brake systems have reservoirs that differ from each other in capacity. The reservoir for the disc brake system is always larger (Figure 11.4).

The disc brake system requires a larger reservoir because the cylinder bores in calipers are much larger than the bores in wheel cylinders. As the caliper pistons move out to compensate for wear on the brake lining, fluid from the reservoir must fill up the additional space without leaving the reservoir empty.

Though in Figure 11.4 the reservoir closest to

the front of the car supplies the front brake system, that is not always the case. Many a master cylinder uses the primary piston to actuate the front brakes. Therefore the rear reservoir in such a master cylinder supplies the front brake system, as shown in Figure 11.5.

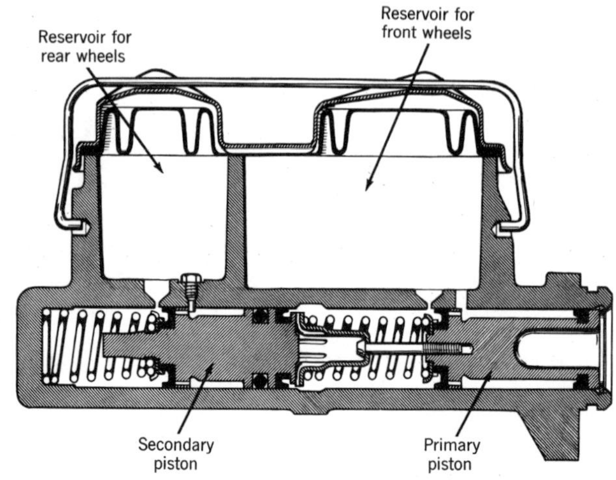

Figure 11.5 Master cylinder used with a combination brake system. Note the change in position of the reservoirs (courtesy Chevrolet Service Manual, Chevrolet Motor Division).

REMOVAL AND INSTALLATION OF A MASTER CYLINDER

For removing and installing a master cylinder, you should consult the manufacturer's manual for the car. However, the general procedures for removing and installing a typical master cylinder are given below:

Removal

1 Remove the bolt or pin that connects the master cylinder push rod to the brake pedal (Figure 11.6).

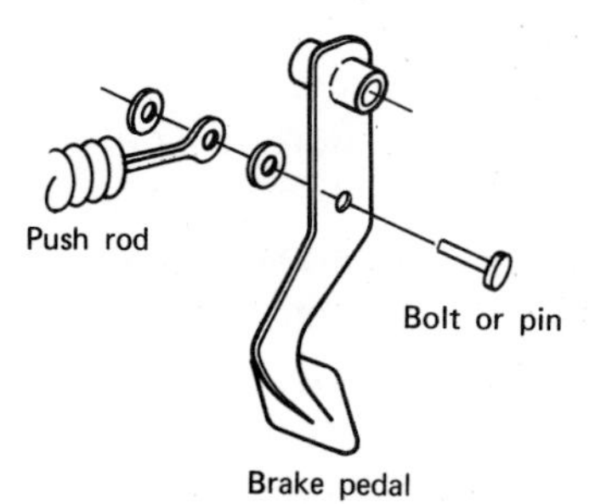

Figure 11.6 Method of connecting the push rod to the brake pedal.

Note. **This step is unnecessary if the master cylinder is attached to a power assist unit.**

2 Using a tubing wrench, disconnect the lines from the master cylinder

3 Plug the master cylinder ports with rubber plugs or dummy fittings. This prevents fluid leakage and the damage that such leakage causes to painted finishes.

4 To prevent dirt from entering the system, seal the open ends of the brake lines with tape.

5 Remove the nuts or bolts that secure the master cylinder to the firewall or power assist unit.

6 Remove the master cylinder by pulling it forward.

Bench Bleeding. Before any master cylinder is installed, it should be bench bled to force the air out of it. Bench bleeding minimizes the chance of pushing air into the brake lines on installation. In many cases, it makes bleeding the entire system unnecessary. The steps involved in bench bleeding follow.

1 Place the master cylinder between the jaws of a vise, and tighten the vise gently.

2 Install bleeder tubes in the outlet ports. Position the tubes so they will return the fluid to the reservoirs as shown in Figure 11.7.

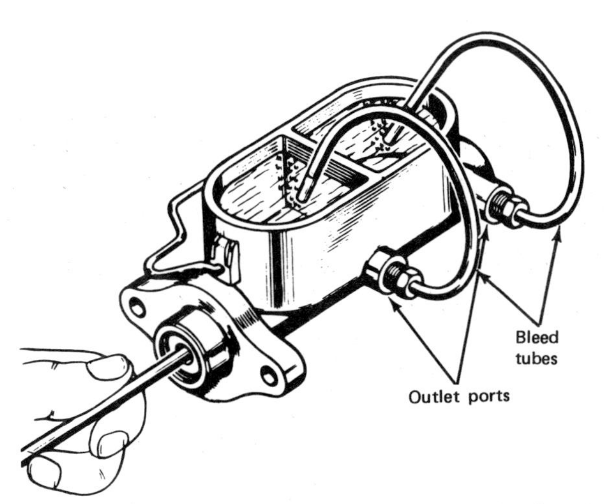

Figure 11.7 Bench bleeding a master cylinder (courtesy American Motors Corporation).

3 Fill the reservoirs with fluid.

4 Slowly push in on the push rod to move the pistons through their bores. Allow the pistons to return.

Note. If the master cylinder is not fitted with a push rod, use a wooden dowel to push the pistons.

5 Pump the pistons several times while watching the ends of the bleeder tubes for bubbles. Continue pumping until there are no bubbles.

6 Remove the bleeder tubes, and plug the ports.

7 Install the reservoir gasket and the cover.

Installation This is the general procedure for installing a typical master cylinder.

1 Place the master cylinder in position. Take care that the end of the push rod (if there is one) is aligned with the pedal linkage.

2 Install all attaching bolts or nuts, but do not tighten them.

3 Remove the tape from the lines and the plugs from the ports. Thread the tubing nuts into the fittings at the outlet ports using only finger pressure so as to avoid stripping the threads.

4 Tighten the attaching bolts or nuts.

5 Install the bolt or pin that connects the master cylinder push rod to the brake pedal.

Note. **This step is unnecessary if the master cylinder is attached to a power assist unit.**

6 Check for free play at the pedal. Adjust the pedal play if necessary.

Note. **Adjustment procedure and specifications for free play should be obtained from the manufacturer's manual.**

7 Using a tubing wrench, tighten the lines at the master cylinder.

8 Loosen the tubing nuts one-half turn. Wrap the fittings with a wiper or a rag to absorb any leakage.

9 Slowly depress and release the brake pedal several times to bleed any air that may have been trapped at the fittings.

10 Using a tubing wrench, tighten the lines, and then clean up any spilled fluid.

11 Check the pedal feel to decide whether additional bleeding is needed.

12 Bleed the brake system if bleeding is needed.

13 Refill the master cylinder reservoirs.

JOB 11b

REPLACE A MASTER CYLINDER

SATISFACTORY PERFORMANCE

A satisfactory performance on this job requires that you do the following:

1 Replace the master cylinder on the car assigned.
2 Following the steps in the "Performance Outline" and the manufacturer's procedure and specifications, complete the job within 150 percent of the manufacturer's suggested time.
3 Fill in the blanks under "Information."

PERFORMANCE OUTLINE

1 Remove the master cylinder.
2 Bench bleed the replacement master cylinder.
3 Install the replacement master cylinder.
4 Adjust pedal free play if necessary.
5 Bleed the cylinder at fittings.
6 Bleed the system if necessary.

INFORMATION

Vehicle identification ______________________________

Type of master cylinder replaced ______________________________

Reference used ____________________ Page(s) ____________________

MASTER CYLINDER OVERHAUL

There are many design and construction differences in the master cylinders built by various manufacturers. Because those differences require different procedures for assembly and disassembly, you should always consult the manual about the master cylinder you are overhauling. However, the steps below give the procedure for overhauling a typical master cylinder. Figure 11.8 shows the parts of a typical dual master cylinder and labels them with the names that will be used for them in the rest of this chapter.

Disassembly

1 Remove the cover and the seal.

2 Drain and discard all fluid.

3 Remove the piston stop.

4 If there are a push rod and a boot, remove them.

5 Using snap-ring pliers, remove the snap ring from its groove in the open end of the cylinder (Figure 11.9).

6 If there are retainers, remove them.

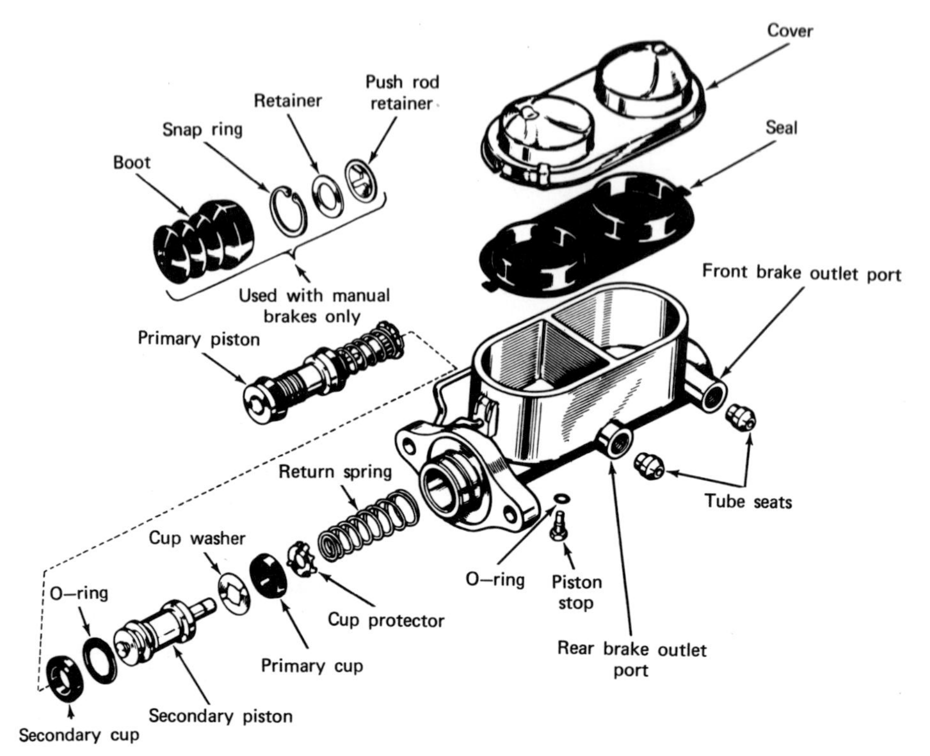

Figure 11.8 Exploded view of a typical master cylinder (courtesy American Motors Corporation).

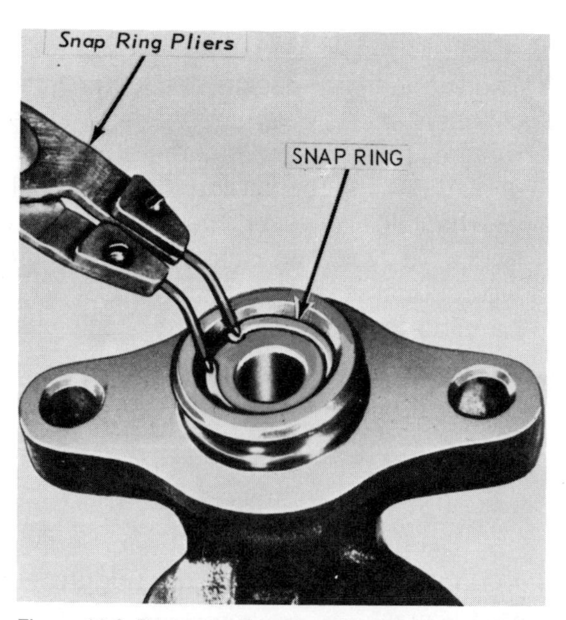

Figure 11.9 Removing the snap ring (Ford Motor Company).

7 Remove the primary piston assembly.

8 Remove the secondary piston assembly, the cup protector, and the return spring.

Note. The secondary piston can be easily removed by air pressure applied through the piston stop hole.

9 Clean all parts with brake fluid or a recommended brake-part cleaner. Be sure that all recesses and passages are clean and clear. Air pressure can be used to remove dirt from those places.

10 Inspect all parts for wear and pitting.

Note. If the cylinder bore shows deep pitting, the master cylinder should be replaced. Light pitting or etching can be removed with a hone.

Tube Seat Replacement. The tube seats should not be replaced unless they are damaged. Some manufacturers do not consider tube seat replacement to be a service operation. They recommend that the master cylinder be replaced if a tube seat is damaged. The following procedure can be used if tube seat replacement is recommended by the manufacturer.

1 Thread a 6-32 $\times \frac{5}{8}$ in. self-tapping screw into the defective tube seat.

2 Place the ends of two screwdrivers under the head of the screw as shown in Figure 11.10. Then pry the screw upward, pulling the tube seat from its bore.

3 Clean the tube seat bore in the outlet hole.

4 Position the new tube seat in the outlet hole, making sure it is not cocked.

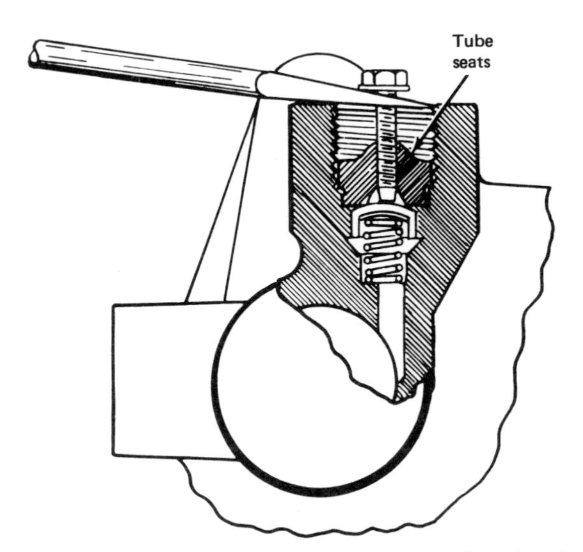

Figure 11.10 Prying out tube seats (courtesy American Motors Corporation).

5 By using a brake line tubing nut as a press, turn the nut down into the outlet hole until the tube seat bottoms in its bore.

6 Remove the tubing nut, and inspect the tube seat for any burrs that may have been formed.

Assembly The parts you replace during master cylinder assembly will of course depend on the parts that are supplied in the repair kit for the master cylinder. In the assembly procedure it has been assumed that:

1 All rubber parts will be replaced.

2 The original secondary piston will be reused.

3 The primary piston will be replaced as an assembly.

The following steps are recommended for assembling the parts of a master cylinder:

1 Coat all parts except the cylinder body with clean brake fluid.

2 Install the O-ring and the secondary cup on the secondary piston.

3 Install the cup washer, primary cup, cup protector, and return spring on the secondary piston.

Note. The lips of the cups must face outward.

4 Coat the cylinder bore with clean brake fluid.

5 Carefully insert the secondary piston assembly, spring-end first, into the cylinder bore.

6 Carefully insert the primary piston assembly, spring-end first, into the cylinder bore.

7 Press the end of the primary piston slightly into the bore.

8 Install the retainers if retainers are provided.

9 Using snap-ring pliers, install the snap ring (Figure 11.9).

10 Push in on the primary piston. (A number-2 Phillips screwdriver can be used as a substitute push rod.) Hold the piston in, and install the piston stop. Do not forget to use a new O-ring.

11 Tighten the piston stop to the torque specifications of the manufacturer.

12 Install the push rod and boot if they are provided.

13 Install the seal and the cover.

Job 11c

OVERHAUL A MASTER CYLINDER

SATISFACTORY PERFORMANCE

A satisfactory performance on this job requires that you do the following:

1 Overhaul a master cylinder.
2 Following the steps in the "Performance Outline" and the manufacturer's procedure and specifications, complete the job within 150 percent of the manufacturer's suggested time.
3 Fill in the blanks under "Information."

PERFORMANCE OUTLINE

1 Disassemble the master cylinder.
2 Clean all parts.
3 Inspect all parts.
4 Replace the tube seats if necessary.
5 Lubricate the parts.
6 Assemble the master cylinder.

INFORMATION

Cylinder identification ______________________________

Reference used ____________________ Page(s) ______________

VALVES Most hydraulic brake systems are fitted with automatic valves. One type of valve turns on a warning light to notify the driver of a system failure. Another type ensures balance in braking between the front and rear wheels. Still another valve delays the application of the front brakes until a certain pressure is built up in the system.

These valves are found as individual units on some cars and are combined in a single assembly on others. The types of valves used on any car are determined by the manufacturer. They are chosen to provide the best braking action for the particular brake system used on the car. Proper diagnosis of braking problems requires that you understand the function of these valves and the problems caused by their failure.

Pressure Differential Valves All autos equipped with a dual master cylinder have a warning light on the instrument panel. That light notifies the driver if one of the dual braking systems fails. The light is turned on by a switch incorporated into a *pressure differential valve* similar to the one shown in Figure 11.11.

The pressure differential valve is connected to the two brake lines that connect the master cylinder with the other parts of the two braking systems. It detects differences in pressure between the two systems. The pressures in dual braking systems are equal when the systems are functioning properly and, therefore, a difference in pressure indicates an abnormal condition.

The most common type of pressure differential valve consists of a spool-shaped piston floating in a passage that connects both systems. As

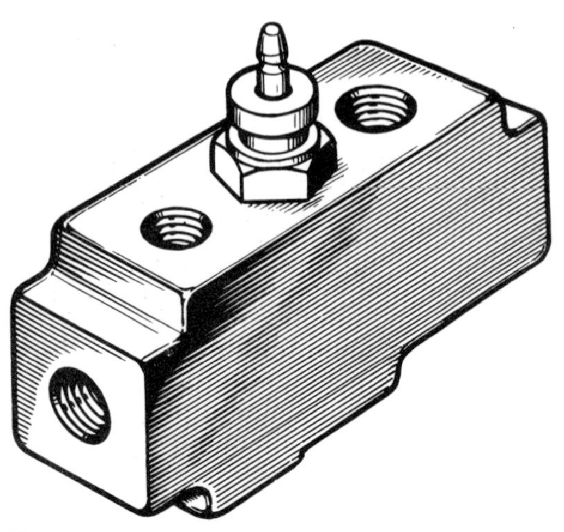

Figure 11.11 A pressure differential valve (courtesy Chevrolet Service Manual, Chevrolet Motor Division).

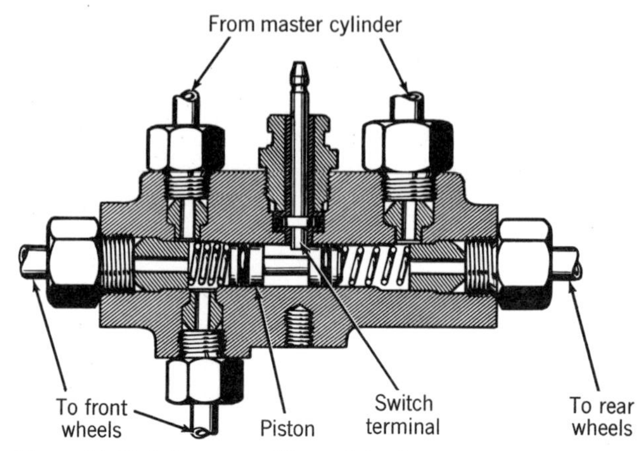

Figure 11.13 A Pressure differential valve with the piston in an off-center position. The position of the valve indicates a failure in the front brake system (courtesy Chevrolet Service Manual, Chevrolet Motor Division).

long as the pressure in both systems is equal, the piston will remain in the center of the passage (Figure 11.12). But if the pressure in one system drops, the higher pressure in the other system moves the piston off center (Figure 11.13).

When the piston is moved off center, one of its inner edges comes into contact with the switch terminal. This contact completes the ground circuit for the warning light, and the light goes on.

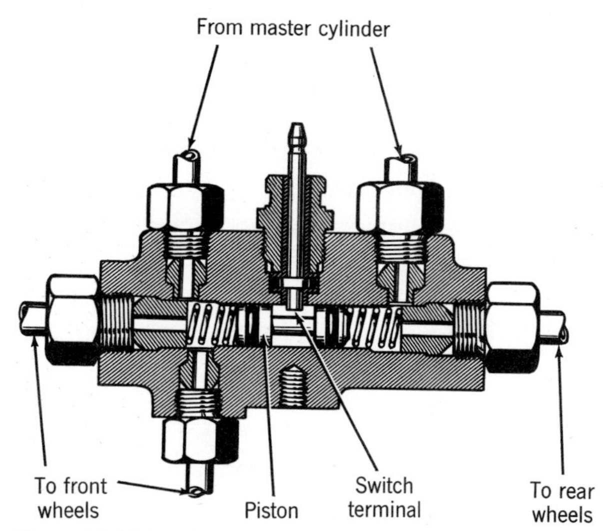

Figure 11.12 A pressure differential valve with the piston in the centered position (courtesy Chevrolet Service Manual, Chevrolet Motor Division).

There are pressure differential valves of different designs, but basically there are only two types—(1) the self-centering type and (2) the noncentering type.

The self-centering type uses springs on both sides of the piston to recenter the piston when the brake pedal is released. With this type of valve the warning light operates only while the brake pedal is depressed. When the brake pedal is released, the light goes out. The valve shown in Figure 11.12 is of the self-centering type.

The noncentering type of pressure differential valve does not use springs. When the piston is moved to either side, it does not return when the brake pedal is released. With this type of valve, the warning light remains on until the system has been repaired and the switch has been hydraulically recentered.

In most instances, normal brake bleeding procedures cause the piston in either type of valve to shift, causing the light to go on. If the valve is of the self-centering type, it returns to the center position when the bleeding operations have been completed. Noncentering valves usually become recentered after a few hard brake applications. If you encounter a valve that will not center itself, you may be able to move it in the following manner:

1 Check to see that there are no leaks in the system, that the system is free of air, and that the master cylinder reservoirs are full.
2 Loosen one of the lines or open one bleeder valve.
3 Slowly press down on the brake pedal.
4 If the light goes out, slowly release the pedal and tighten the loose fitting or valve. If the light does not go out, tighten the fitting or valve you have loosened, and loosen one in the other system.
5 Slowly press down on the brake pedal again. The light should go out. Tighten the loose fitting or valve.

Pressure differential valves are not adjustable and parts are not available for their repair. When a pressure differential valve is defective, you must replace it.

Proportioning Valves *Proportioning valves* are used to improve the balance of braking between the front and rear wheels. Stops from high speeds using high system pressures can cause the rear wheels to lock and skid. This problem is due to the weight shift to the front wheels that takes place during braking. Cars equipped with combination brake systems are especially prone to rear wheel locking because the drum brakes used on the rear are self-energizing whereas the disc brakes used on the front are not.

Many cars use a proportioning valve in the rear brake system to eliminate this tendency to skid. The valve reduces the pressure applied to the rear wheel cylinders in relation to the pressure

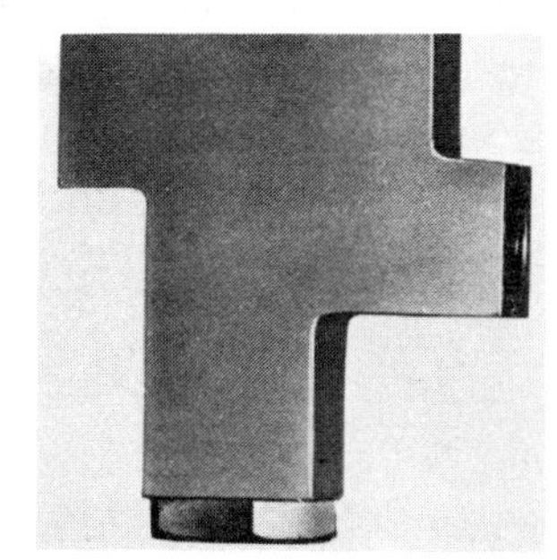

Figure 11.14 A proportioning valve (Chrysler Corporation).

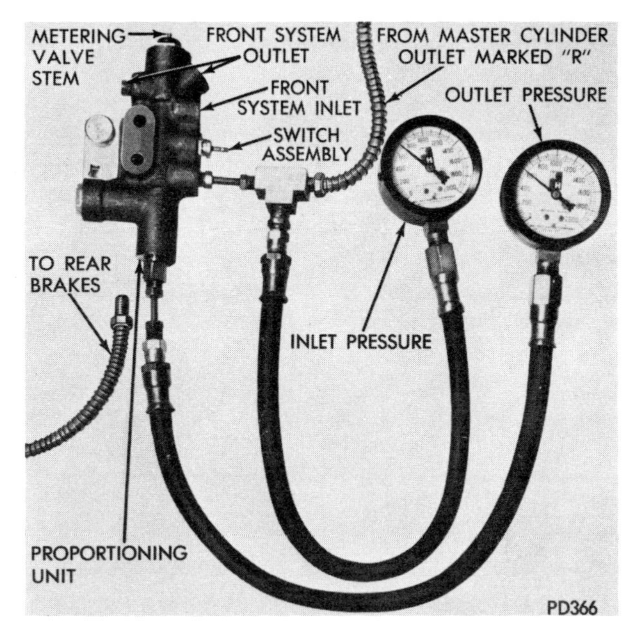

Figure 11.15 Testing a proportioning valve. Note that the outlet pressure is lower than the inlet pressure (Chrysler Corporation).

applied to the front calipers. The actual amount of pressure reduction depends on the amount needed for each type of car, and it is determined by the manufacturer. Proportioning valves may be in separate units as shown in Figure 11.14, or they may be incorporated into a combination valve.

A defective proportioning valve is indicated when the rear wheels tend to lock during sudden stops. Proportioning valve operation can be tested only by using gauges as recommended in the maufacturer's service manual. Two high-pressure gauges are needed for the test. One gauge is connected to the line from the rear brake section of the master cylinder by means of a T-fitting. The second is connected to the outlet port of the proportioning valve. When the brake pedal is depressed, the first gauge indicates the output pressure of the master cylinder, and the second indicates the reduced pressure. Figure 11.15 shows two gauges being used to test a proportioning valve. Proportioning valves are neither adjustable nor repairable. When defective they must be replaced.

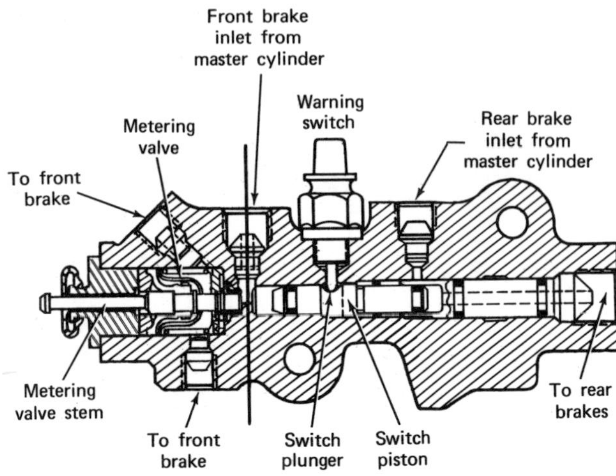

Figure 11.16 A metering valve combined with a pressure differential valve (courtesy American Motors Corporation).

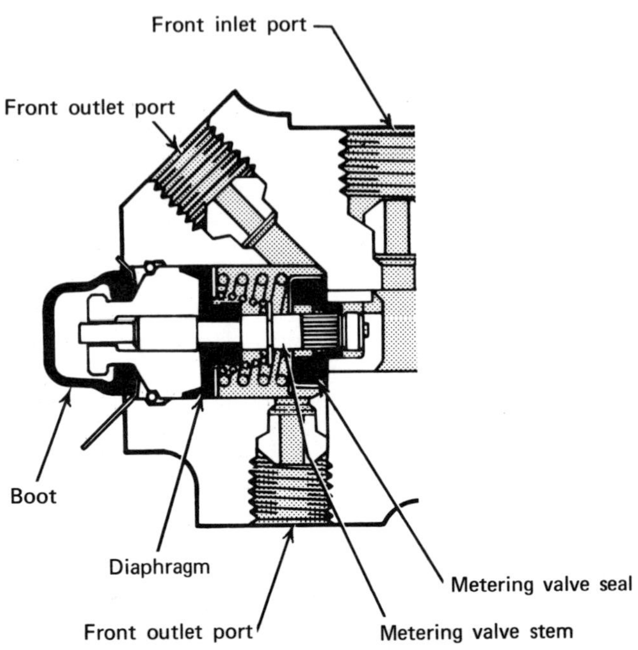

Figure 11.18 Sectional view of a metering valve. Note the boot covering the valve stem (courtesy Chevrolet Service Manual, Chevrolet Motor Division).

Metering Valves *Metering valves* are used on many cars with combination brake systems. These valves delay the flow of brake fluid to the front calipers until the pressure in the system rises to a certain point. The delay permits the system pressure (1) to overcome the tension of the rear brake-shoe retracting springs, (2) to expand the shoes, and (3) to apply the rear brakes before the front brakes are applied. Metering valves are used to prevent the front brakes from locking during light braking on icy or slippery roads.

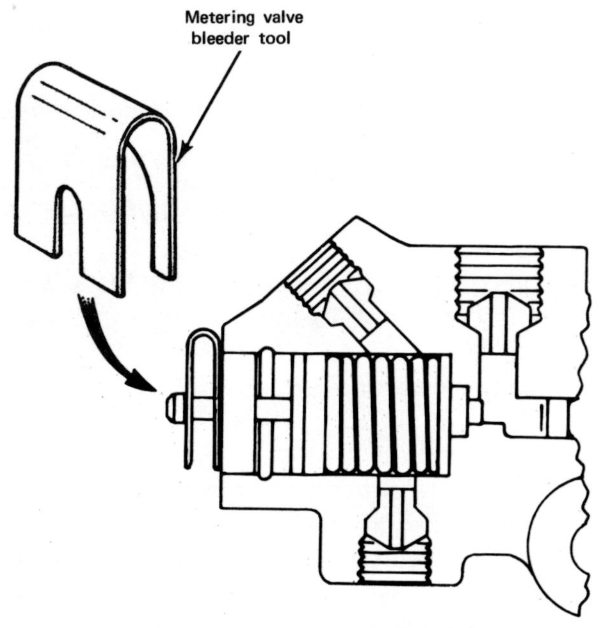

Figure 11.17 Stem-type metering valve held open by a tool (Ford Motor Company).

Metering valves are separate from other valves on some cars. On most cars, however, they are combined with another valve in a common housing (Figure 11.16).

Metering valves are designed to remain open when there is no pressure in the brake system. This operation allows the fluid in the system to expand and contract with temperature changes. As soon as the brakes are applied, the valve closes (usually in the 3 to 5 psi range), and pressure is applied only to the rear brake system. As braking pressure increases, the valve opens (usually in the 30 to 140 psi range), and the front brake system is pressurized. Since the metering valve functions only in a relatively low pressure range, it has no effect on braking during stops from high speeds.

A metering valve will close under the pressure required for bleeding operations. Therefore, when you bleed a system that incorporates a metering valve, you must hold the valve open. All metering valves have an external stem or button to use for holding the valve open.

Stem Type To hold open metering valves that have an exposed stem, such as the valve in Figure 11.16, pull the stem outward. You can pull it with your fingers, but the simple tool shown in Figure 11.17 makes the job easier. When you have to bleed a system that has a stem type of metering valve, compress the tool between your fingers and place it over the stem. When the tool is released, its spring action pulls the stem outward, holding the valve open.

Button Type. The button type of metering valve has an exposed button. As shown in Figure 11.18, the button is actually a rubber boot that covers the valve stem. To hold this type of valve open, you can push the button inward as shown in Figure 11.19. You can also hold the button type of valve open with a simple tool (Figure 11.20). The slotted end of this tool is slipped under one of the valve mounting bolts. If you hold the button in and tighten the bolt, the valve will remain open for bleeding operations.

Whenever you use a tool to hold a metering valve open while you bleed brakes, do not forget to remove the tool after you have completed the bleeding operations.

Metering valves can be checked by two methods:

1 Depress the brake pedal slowly. Usually you can feel a slight bump after the pedal moves about one inch. The bump indicates that the valve is functioning.
2 Have someone depress the brake pedal slowly while you watch or feel the stem or button. The movement of the stem or button, which can usually be seen or felt, indicates that the valve is functioning.

Faulty metering valves cannot be adjusted or repaired. If a faulty metering valve is combined with another valve or valves in a common housing, the entire assembly must be replaced.

Combination Valves A *combination valve* is, as its name implies, the combining of all of the valves in a particular system. Such a valve can contain a pressure differential valve, a proportioning valve, and a metering valve (Figure 11.21). Combination valves are usually

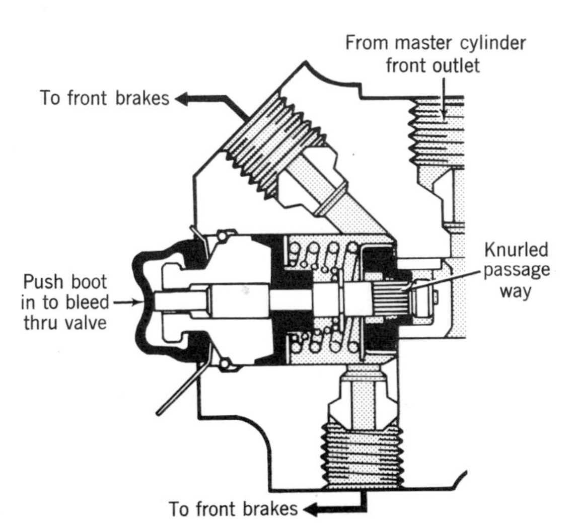

Figure 11.19 Sectional view of a metering valve held in the open position (courtesy Chevrolet Service Manual, Chevrolet Motor Division).

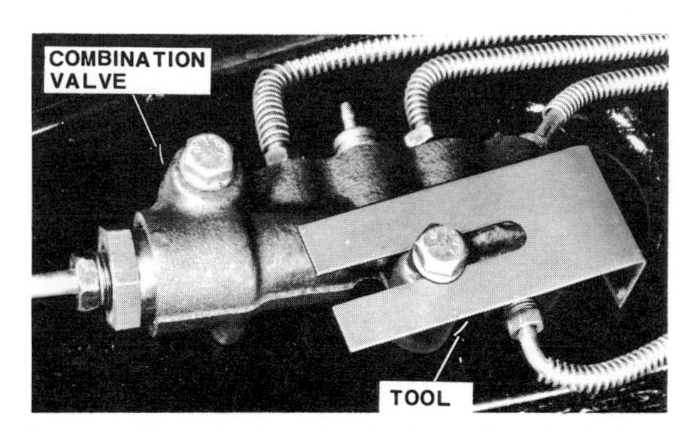

Figure 11.20 A metering valve bleeder tool holding the valve open (courtesy American Motors Corporation).

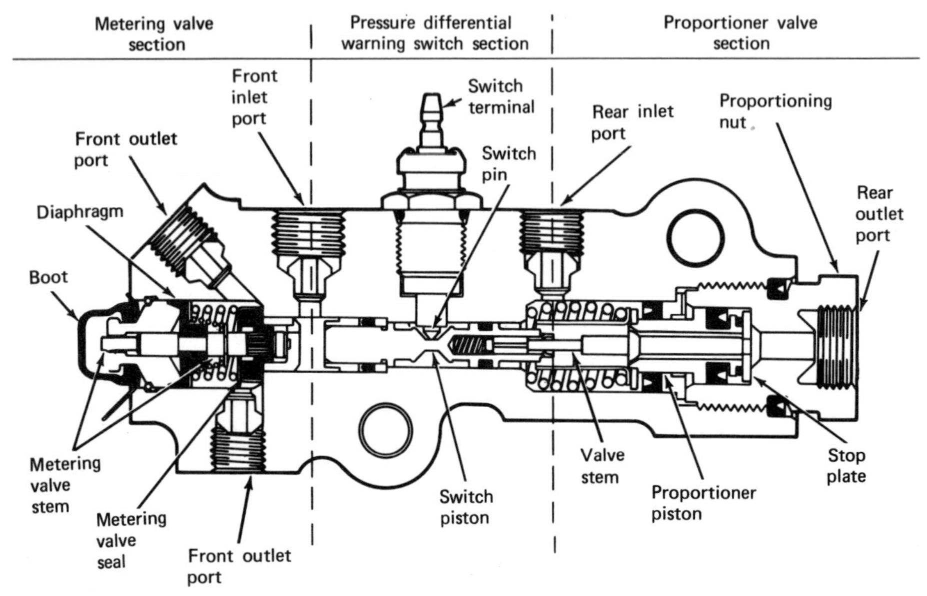

Figure 11.21 A three-function combination valve (courtesy American Motors Corporation).

mounted on the frame, the firewall, or a side panel near the master cylinder. The position of such valves, their appearance, and their service procedure varies, even among cars made by the same manufacturer.

Faulty combination valves must be replaced as a unit, even if only one of the combined valves is defective.

Job 11d

IDENTIFY VALVE FUNCTION AND LOCATION

SATISFACTORY PERFORMANCE

A satisfactory performance on this job requires that you do the following:

1 Place the number of each valve listed below in front of the items below the list that best relate to that valve. Not all the statements relate to the valves listed.

2 Place the correct valve number in front of 8 items within 10 minutes.

PERFORMANCE SITUATION

1 Pressure differential valves
2 Proportioning valves
3 Metering valves

_______ delay the application of pressure to the front brakes.

_______ may be located in the line to the rear brakes.

_______ operate only at low pressures.

_______ are a part of both the front and rear systems.

_______ maintain a residual pressure in the system.

_______ operate switches to turn on warning lights.

_______ decrease the pressure applied to the rear brakes.

_______ must be considered when bleeding the front calipers.

_______ operate only at high pressures.

_______ may be located in the line to the front brakes.

REMOVAL AND INSTALLATION OF HYDRAULIC SYSTEM VALVES

The types and locations of hydraulic system valves vary with each make or model of car. For this reason proper removal and installation procedures can be found only in the manufacturers' manuals. Certain precautions, however, must be taken in replacing any hydraulic system valves. Those precautions are listed below.

Removal Precautions

1 Always use tubing wrenches when you remove the lines from the valves.

2 Always support the valve with an open-end wrench when you loosen the fittings. (Pliers are needed in some instances.)

3 If the valve is supported at the frame or body, precaution number 2 may not be necessary, but loosen all fittings before you loosen the supporting bolts or clamps.

4 Take care not to bend the tubing when you disengage the flared ends. This precaution minimizes alignment problems during installation.

Installation Precautions

1 Position all lines and start all threads by hand. Do not use a wrench on any fitting until you are sure the threads are properly aligned and started. Do not tighten any fitting until all the fittings have been started.

2 After all fittings are hand-threaded in place, install any attaching bolts or clamps.

3 Use a tubing wrench to tighten all lines, supporting the valve if necessary.

4 Thoroughly clean the valve, the fittings, and the lines near the valve to ensure the detection of leaks.

5 Bleed the system in the sequence specified by the manufacturer.

6 Check for leaks.

Job 11e

REPLACE A HYDRAULIC VALVE

SATISFACTORY PERFORMANCE

A satisfactory performance on this job requires that you do the following:

1 Replace the assigned hydraulic system valve on the car provided.
2 Following the steps in the "Performance Outline" and the manufacturer's procedure and specifications, complete the job within 150 percent of the manufacturer's suggested time.
3 Fill in the blanks under "Information."

PERFORMANCE OUTLINE

1 Loosen all fittings.
2 Remove retaining bolts or clamps.
3 Remove the valve.
4 Position the valve and start all threads by hand.
5 Install retaining bolts or clamps.
6 Tighten all fittings.
7 Bleed the system.
8 Check for leaks.

INFORMATION

Vehicle identification ____________________

Valve replaced ____________ Location ____________

Reference used ____________ Page(s) ____________

SUMMARY

Having completed this chapter, you now understand the construction and operation of the dual master cylinder. You can remove such a master cylinder from a car, overhaul it, and install it. You have also learned of the various valves that may be used in braking systems. You understand their importance and function in the systems. You can diagnose problems related to the malfunction of these valves, and you can replace the valves if they are defective.

SELF-TEST

Each incomplete statement or question in this test is followed by four suggested completions or answers. In each case select the *one* that best completes the statement or answers the question.

1 Two mechanics are discussing master cylinders.
Mechanic A says that disc brakes require larger reservoirs than do drum brakes.

Mechanic B says that disc brakes require residual pressure (check) valves.
Who is right?
A. A only
B. B only
C. Both A and B
D. Neither A nor B

2 Tube seats are usually used in
A. filler ports
B. outlet ports
C. bleeder ports
D. compensating ports

3 Two mechanics are discussing the operation of a dual master cylinder.
Mechanic A says that the push rod is connected to the primary piston.
Mechanic B says that the secondary piston is pushed forward by the fluid trapped ahead of the primary position.
Who is right?
A. A only
B. B only
C. Both A and B
D. Neither A nor B

4 The brake warning light on the instrument panel lights up when the brake pedal is depressed, but the light goes out when the pedal is released. The most probable reason for this occurrence is that
I. one of the brake systems has a malfunction
II. the car is equipped with a self-centering pressure differential valve
A. I only
B. II only
C. Either I or II
D. Both I and II

5 While checking the fluid level in a master cylinder, a mechanic notices that the larger reservoir is very low but the smaller reservoir is full. The most probable reason for this finding is that
A. the lining on the front brake shoes is badly worn
B. the master cylinder bypass valve is leaking
C. the compensating port is plugged
D. the pressure differential valve is not centered

6 A car owner states that brake fluid is leaking from behind the brake pedal and dripping on the carpet. The most probable cause of this problem is a defective
A. primary cup on the primary piston
B. secondary cup on the primary piston
C. primary cup on the secondary piston
D. secondary cup on the secondary piston

7 A proportioning valve is used
I. on cars with combination brake systems
II. to reduce the hydraulic pressure at the rear brakes
A. I only
B. II only
C. Both I and II
D. Neither I nor II

8 The application of hydraulic pressure to the front calipers on some cars is delayed by a
A. metering valve
B. proportioning valve
C. residual pressure valve
D. pressure differential valve

9 A car owner states that the front wheels of a car lock up on icy roads, even when braking with light pedal pressure. The most probable cause of this problem is a defective
A. metering valve
B. proportioning valve
C. residual pressure valve
D. pressure differential valve

10 A car owner states that the rear wheels lock and slide when attempting to bring the car to a fast stop from high speeds. The most probable cause of this problem is a defective
A. metering valve
B. proportioning valve
C. residual pressure valve
D. pressure differential valve

Chapter 12 Parking Brakes

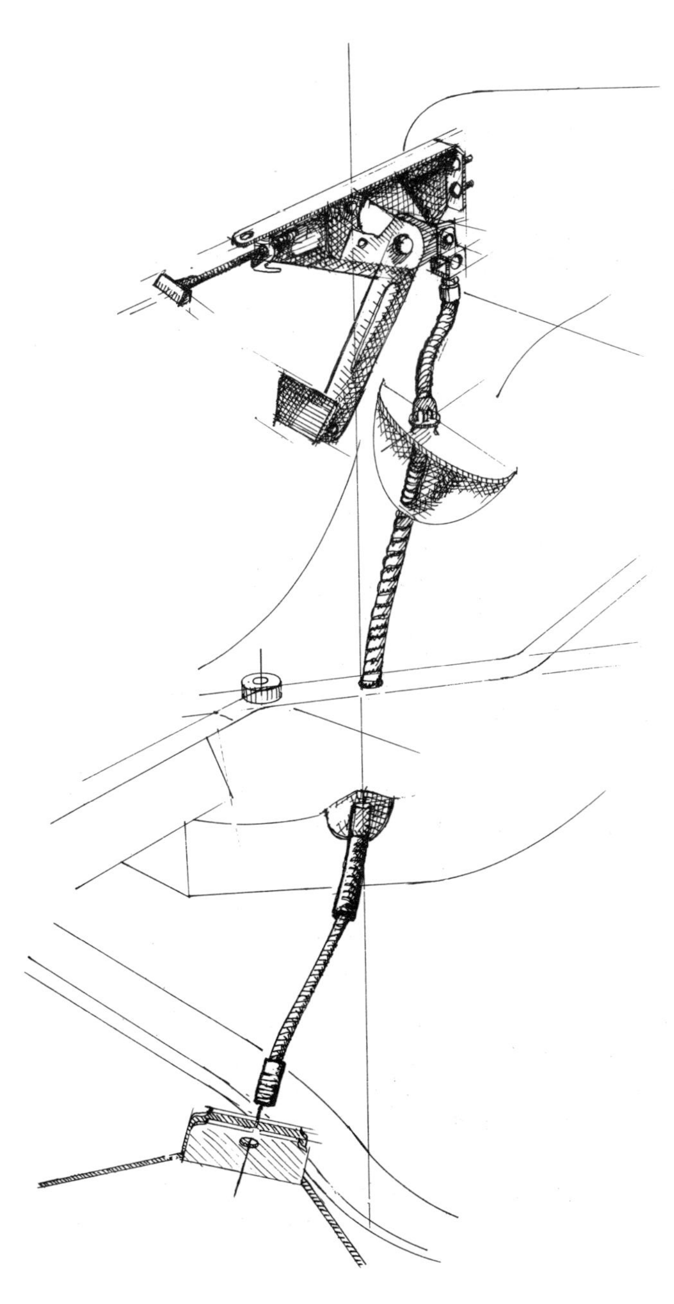

Parking brake systems provide a means of applying the rear brakes and holding them in the applied position. They are not designed to overcome the kinetic energy of a moving automobile. They are meant only to keep an already stopped auto from moving. Because of their relative weakness, parking brake systems should always be kept in proper adjustment and in good condition. This chapter covers different types of parking brake systems and the procedures you need to know for servicing them.

Your objectives in this chapter are to:

1
Identify the parts of the parking brake system.

2
Identify the function of the parts of the parking brake system.

3
Adjust parking brakes.

4
Replace rear parking brake cables.

5
Replace front parking brake cables.

6
Replace parking brake shoes in a rear disc brake.

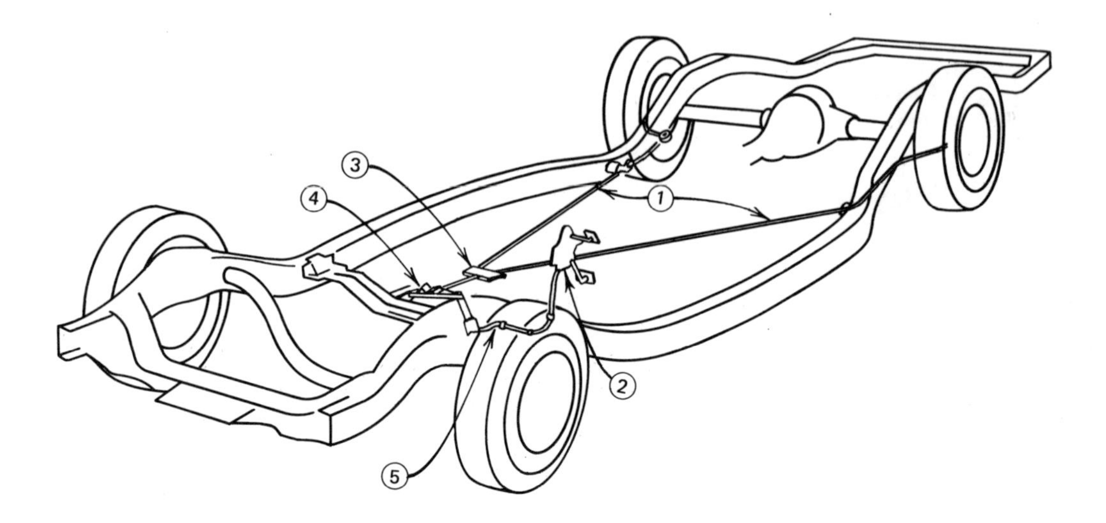

Figure 12.1 A typical parking brake cable-and-lever system (Ford Motor Company).

PARKING BRAKES IN DRUM BRAKE SYSTEMS The parking brake system used by most manufacturers is a mechanical system that expands the rear brake shoes inside their drums. When the driver of a car applies the parking brakes, the effort with which the brake lever is moved is transmitted to the rear shoes by cables. Levers in the system multiply the physical effort of the driver enough to force the rear brake shoes into tight contact with the drums. A typical parking brake cable and lever system is shown in Figure 12.1.

Compared to service brake systems, parking brake systems are relatively inefficient. The force with which they are applied depends on the strength of the driver. You should always remember this fact when diagnosing problems that involve slipping parking brakes. A parking brake system in good operating condition may fail to hold a parked car if the brakes are improperly applied. The service brakes should be applied before the parking brakes are applied. This action allows the more efficient hydraulic system to force the shoes into contact with the drum. The less efficient mechanical system, then, merely holds them in place.

Job 12a

IDENTIFY PARKING BRAKE SYSTEM PARTS

SATISFACTORY PERFORMANCE

A satisfactory performance on this job requires that you do the following:

1 Identify the parts of a cable-and-lever system by matching the numbered parts on the drawing with a list of part names.

2 Correctly identify all the numbered parts within 10 minutes.

PERFORMANCE SITUATION

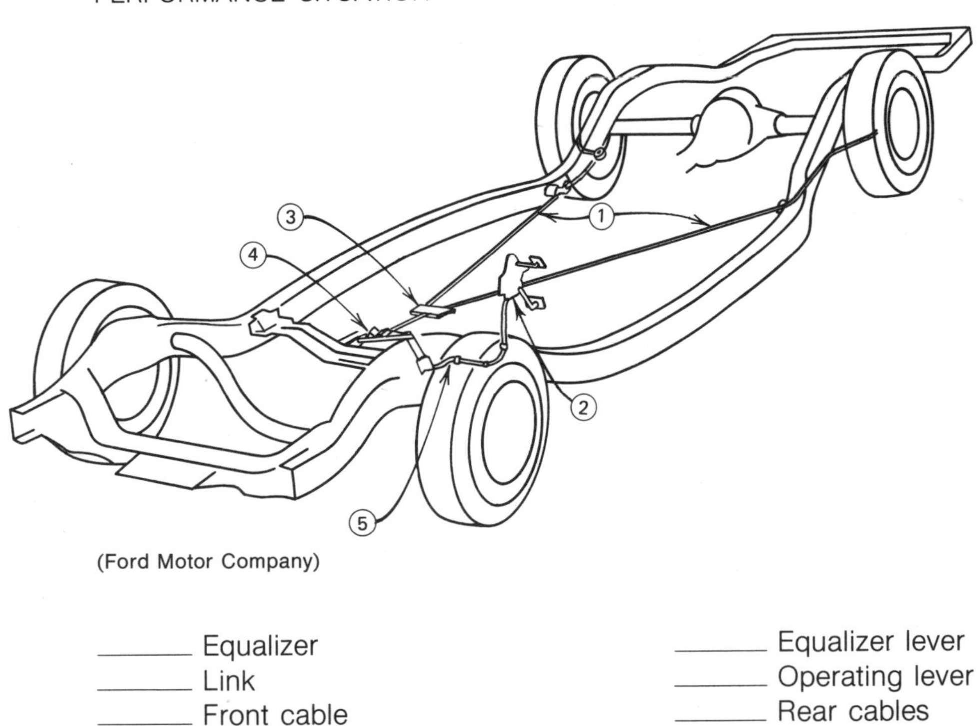

(Ford Motor Company)

_______ Equalizer
_______ Link
_______ Front cable
_______ Equalizer lever
_______ Operating lever
_______ Rear cables

Parking Brake Operation Most parking brake systems use three levers to multiply the physical effort of the driver. The first lever is the operating lever (Figure 12.1). When the operating lever is moved, the driver's effort is multiplied and used to pull the front cable. The front cable, in turn, pulls the equalizer lever.

The equalizer lever multiplies the effort of the operating lever and pulls the rear cables. This pulling effort passes through an equalizer, which ensures equal pull on both rear cables. The equalizer functions by allowing the rear brake cables to slip slightly so as to balance out small differences in cable length or adjustment. The rear cables, in turn, pull the *parking brake levers.*

The parking brake levers are attached to the secondary shoes in the rear brakes. A parking brake lever and its related parts are shown in Figure 12.2. When the parking brake lever is pulled forward, it forces the link against the link spring, compressing the spring. The link continues to move, forcing the primary shoe against the brake drum. When the primary shoe contacts the drum, the motion of the link is stopped. The parking brake lever then pivots on the end of the link, and the top of the lever forces the secondary shoe against the drum. The action of the parking brake lever again multiplies the driver's effort.

Because of the placement of the *fulcrums,* or pivot points, on the parking brake lever, drum-type parking brakes are self-energizing. They also provide servo action, but only when the car is moving or trying to move forward, as it does when facing downhill. For this reason, drum-type parking brakes are not as efficient when the car is facing uphill as they are when it is facing downhill.

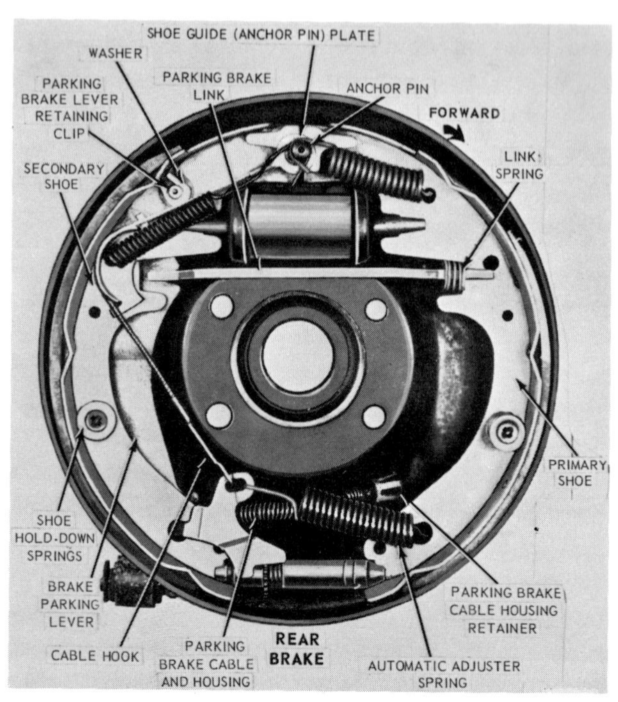

Figure 12.2 A rear brake assembly showing the parking brake lever, link, and spring (Ford Motor Company).

Job 12b

IDENTIFY THE FUNCTION OF PARKING BRAKE PARTS

SATISFACTORY PERFORMANCE

A satisfactory performance on this job requires that you do the following:

1 Identify the function of the listed brake parts by inserting the number of each part in the space provided in front of the list of part functions.
2 Identify the function of all the parts within 15 minutes.

PERFORMANCE SITUATION

1 Operating lever	5 Secondary shoe
2 Equalizer	6 Front cable
3 Parking brake lever	7 Rear cables
4 Link	8 Equalizer lever

______ provides an even pull on both rear cables.

______ pulls the parking brake levers.

______ pulls the front cable.

______ transmits the movement of the parking brake lever to the primary shoe.

______ pivots on the secondary shoe.

______ pulls the rear cables.

______ pulls the equalizer lever.

______ provides a mounting point for the parking brake lever.

PARKING BRAKE ADJUSTMENT

A parking brake is considered properly adjusted when it meets the following criteria:

1 The brakes are fully applied and holding after the pedal or lever has been moved through less than half its possible travel.

2 The brakes are fully released when the pedal or lever is in the released position.

Since the parking brakes actuate the rear brake shoes, the service brakes should have the proper lining-to-drum clearance. Therefore, before attempting to adjust a parking brake, you should inspect the lining, drums, and related parts. You should check the operation of the star adjuster and should adjust the brakes to obtain the proper lining-to-drum clearance.

In most instances parking brake adjustment consists of shortening the length of one or more of the cables to remove unnecessary slack. The adjustment is usually made by means of an adjusting nut at the equalizer. (Figure 12.3). This nut is easy to reach once the car has been raised.

Parking brake adjustment, however, involves more than just turning the adjusting nut until you obtain the desired braking action. The operation of self-adjusting mechanisms can be affected by an improperly adjusted parking brake cable. Most manufacturers have established specific procedures for adjusting the parking brakes on their various models. Therefore, you should always follow the procedures given in the appropriate shop manual.

Here is a typical procedure for adjusting parking brakes in drum brake systems:

1 Set the transmission shift lever in the neutral position.

2 Set the parking brake operating lever in the released position.

3 Raise the car and support it with jack stands placed under the suspension.

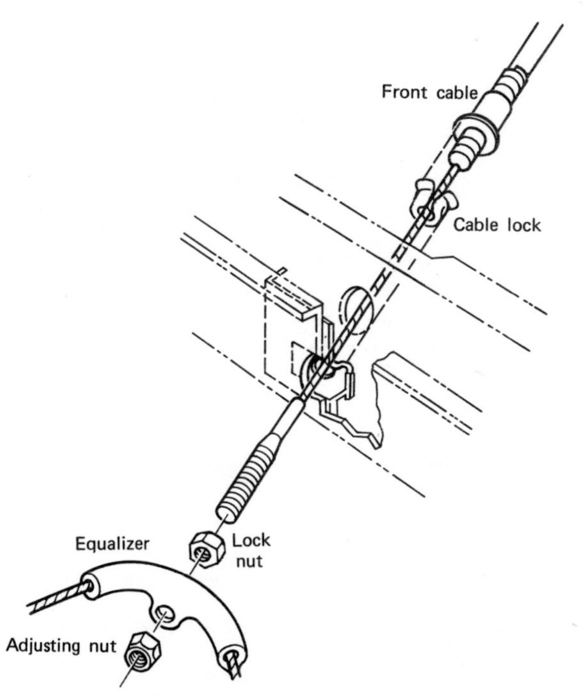

Figure 12.3 Typical provision for brake adjustment at the equalizer (Chrysler Corporation).

4 Loosen the lock nut (Figure 12.3).

5 Tighten the adjusting nut against the equalizer until the rear brakes barely start to drag.

6 Loosen the adjusting nut until the brakes are fully released.

7 Tighten the lock nut.

Note. This step is necessary for retaining the adjustment.

8 Check the operation of the parking brake.

9 Lower the car.

Job 12c

ADJUST A PARKING BRAKE

SATISFACTORY PERFORMANCE

A satisfactory performance on this job requires that you do the following:

1 Adjust the parking brake on the car assigned.
2 Following the steps in the "Performance Outline" and the manufacturer's procedure and specifications, complete the job within 150 percent of the manufacturer's suggested time.
3 Fill in the blanks under "Information."

PERFORMANCE OUTLINE

1 Set the transmission shift lever and the parking brake lever in their proper positions.
2 Raise the car and support it.
3 Adjust the parking brake.
4 Check the brake operation.
5 Lower the car.

INFORMATION

Vehicle identification ______________________

Reference used ______________ Page(s) ______________

Location of adjustment ______________________

Wrench size of adjusting nut ______________________

How was the condition of the lining checked? ______________

What was the approximate thickness of the lining? ______________

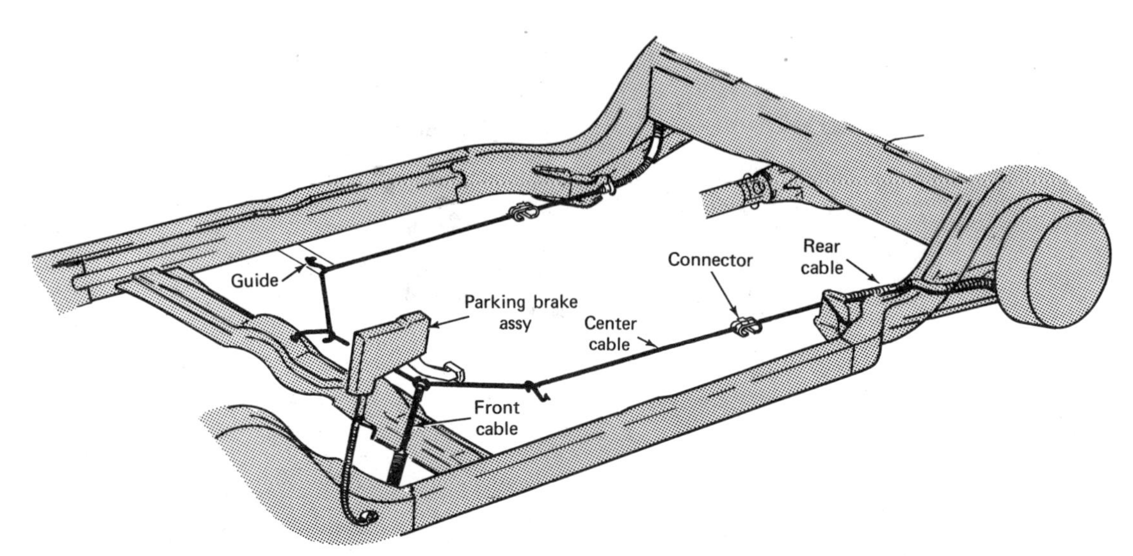

Figure 12.4 Typical parking brake cable routing (General Motors).

Parking Brake Cables Parking brake cables are usually mounted under the floor of the car. There they are exposed to road splash, which causes them to corrode. Water trapped between a cable and its housing causes rust, which may restrict the movement of the cable. At times cables become so badly rusted that they will not move. Frozen cables and cables that have been kinked or otherwise damaged should be replaced. Commonly, a separate rear cable is used for each rear wheel, and both rear cables are connected to the equalizer by a center, or equalizer, cable as shown in Figure 12.4.

The center cable is easy to replace. You merely have to loosen the adjustment nut until the cable has enough slack to allow you to unhook it from the connectors (Figure 12.5). To complete the installation, place the new cable in position and tighten the adjusting nut.

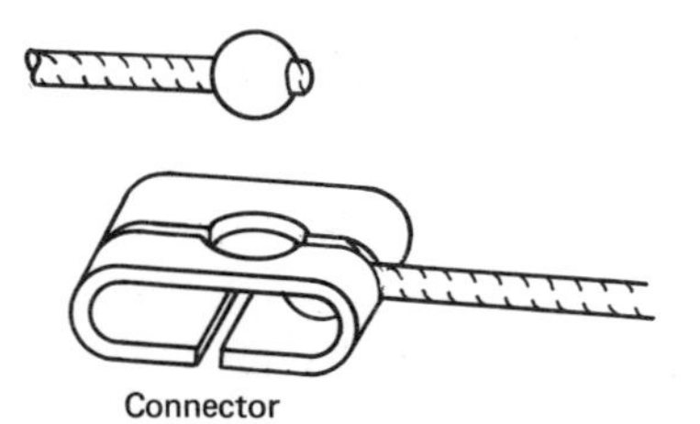

Figure 12.5 A typical cable connector. Note the ball end on the cable (Chrysler Corporation).

REAR PARKING BRAKE CABLE REPLACEMENT

For the correct procedure for replacing a rear parking brake cable, you should check the appropriate car manual. However, the following procedure is typical:

Removal

1 Raise the car and support it by the suspension.

2 Release the parking brake.

3 Remove the wheel.

4 Disconnect the cable from the equalizer.

5 Remove the horseshoe clip that secures the cable housing to its bracket on the frame (Figure 12.6).

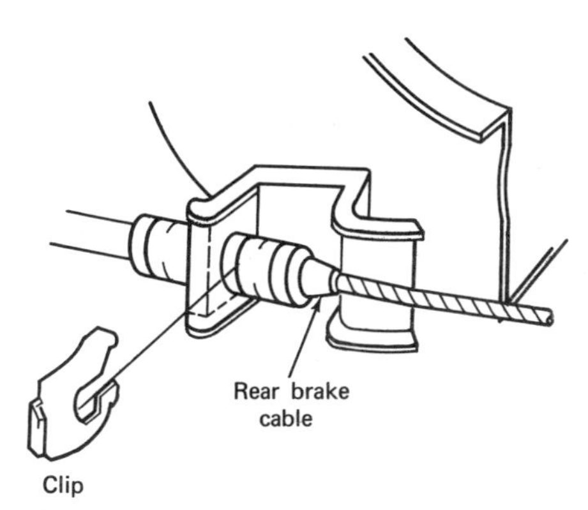

Figure 12.6 Horseshoe clip and cable housing bracket (Chrysler Corporation).

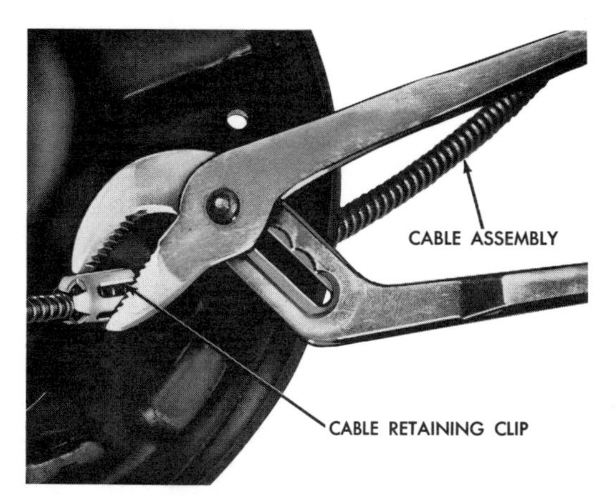

Figure 12.7 Removing the cable assembly from the backing plate (Chrysler Corporation).

6 Remove the brake drum.

7 Remove the brake shoes and attaching hardware to gain access to the cable.

8 Using a pair of pliers as shown in Figure 12.7, compress the fingers, or prongs, on the cable retaining clip, and pull the cable assembly from its hole in the backing plate.

Installation

1 Push the new cable assembly through the hole in the backing plate. Then make sure that the retaining fingers are expanded and that the cable assembly is locked in place.

2 Position the forward end of the cable assembly at the mounting bracket, and install the horseshoe clip (Figure 12.6).

3 Install the brake shoes and all attaching hardware. Make sure the cable is attached to the lever on the secondary shoe.

4 Install the drum.

5 Connect the end of the cable to the equalizer.

6 Install the wheel.

7 Adjust the parking brakes.

8 Lower the car.

9 Check the parking brake operation.

Job 12d

REPLACE A REAR PARKING BRAKE CABLE

SATISFACTORY PERFORMANCE

A satisfactory performance on this job requires that you do the following:

1 Replace a rear parking brake cable on the car assigned.
2 Following the steps in the "Performance Outline" and the manufacturer's procedure and specifications, complete the job within 150 percent of the manufacturer's suggested time.
3 Fill in the blanks under "Information."

PERFORMANCE OUTLINE

1 Raise the car and support it.
2 Remove the wheel and drum.
3 Disconnect the cable from the equalizer and from the frame.
4 Remove the brake shoes.
5 Remove the cable from the backing plate.
6 Install the cable, and secure it at the backing plate and at the frame.
7 Install the brake shoes, drum, and wheel.
8 Connect the cable to the equalizer.
9 Adjust the parking brakes.
10 Lower the car.
11 Check the operation of the brake.

INFORMATION

Vehicle identification ______________________

Reference used ______________ Page(s) ______________

Condition of lining ______________________

FRONT PARKING BRAKE CABLE REPLACEMENT On most cars the parking brake control lever assembly is located under the instrument panel. You will have to work under the instrument panel to disconnect the forward end of the front cable. Therefore, you should disconnect one of the battery cables before starting the job. This step eliminates the possibility of your causing a short circuit while you are working. Figure 12.8 shows the parts of a typical front cable. Note that there are several clips used to support the cable. They hold the cable

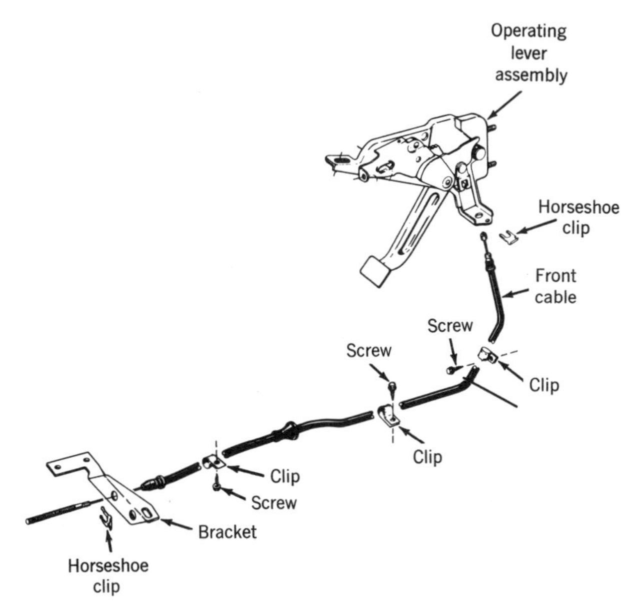

Figure 12.8 Typical installation of a front parking brake cable (courtesy American Motors Corporation).

against the floor or frame, and they must be removed from the old cable and transfered to the new.

Different procedures for the removal and installation of a front parking brake cable are required for different cars. You should always consult an appropriate manual before starting the job. The following steps, however, outline a typical procedure:

Removal

1 Set the parking brake lever in the released position.

2 Disconnect one of the battery cables.

3 Raise and support the car.

4 Remove the adjusting nut, equalizer, and lock nut from the lower end of the cable (Figure 12.3).

5 Remove the horseshoe clip from the lower end of the cable.

6 Remove the screws that hold any cable support clips to frame.

7 Working under the dash, remove the horseshoe clip that holds the cable to the operating lever assembly.

8 Remove the cable by pulling it down through its hole in the floor.

Installation

1 Transfer to the new cable any clips that were used on the old.

2 Push the new cable up through its hole in the floor.

3 Connect the cable to the operating lever assembly.

4 Install the horseshoe clip.

5 Position the lower end of the cable in its bracket.

6 Install the horseshoe clip.

7 Align and secure the cable support clips.

8 Install the locknut, equalizer, and adjusting nut on the end of the cable.

9 Adjust the parking brakes.

10 Lower the car.

11 Connect the battery cable.

12 Check the operation of the parking brake.

Job 12e

REPLACE A FRONT PARKING BRAKE CABLE

SATISFACTORY PERFORMANCE

A satisfactory performance on this job requires that you do the following:

1 Replace the front parking brake cable on the car assigned.
2 Following the steps in the "Performance Outline" and the manufacturer's procedure and specifications, complete the job within 200 percent of the manufacturer's suggested time.
3 Fill in the blanks under "Information."

PERFORMANCE OUTLINE

1 Disconnect the battery cable.
2 Raise and support the car.
3 Remove attaching parts at the rear end of the cable.
4 Remove any cable support clips.
5 Disconnect the cable from the operating lever assembly.
6 Connect the replacement cable to the operating lever assembly.
7 Install the attaching parts at the rear of cable.
8 Install any cable support clips.
9 Adjust the parking brakes.
10 Lower the car.
11 Connect the battery cable.
12 Check the operation of the parking brake.

INFORMATION

Vehicle identification ______________________________

Reference used ________________ Page(s) ________________

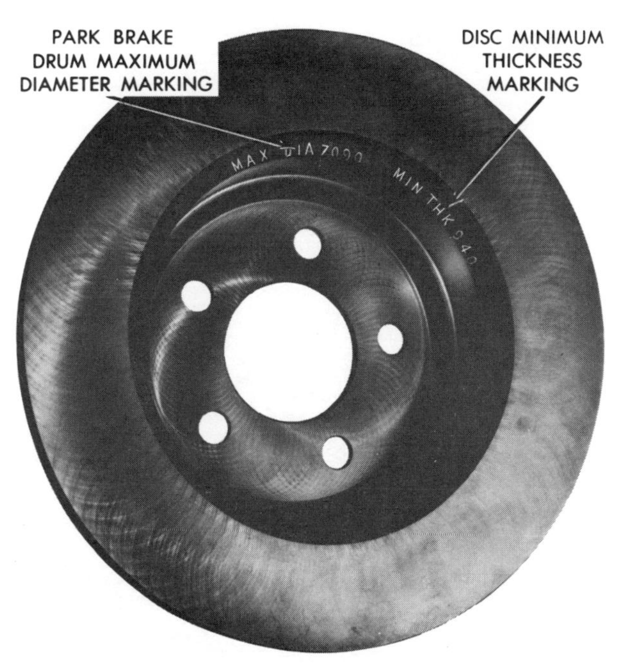

Figure 12.9 A parking brake drum built into the center of a rear rotor. Note that the specs for maximum drum diameter and minimum rotor thickness are both given (Chrysler Corporation).

PARKING BRAKES IN DISC BRAKE SYSTEMS Some cars have disc brakes at all four wheels. Such cars have parking brake systems that are different from those of cars that have drum or combination brakes. One type of parking brake used with disc brake systems has brake drums built into the center of the rear rotors. Small brake shoes expand inside the drums when the parking brake is applied (Figure 12.9). Another type of parking brake mechanically actuates the rear calipers so that the disc brake shoes grip the rotors when the parking brake is applied.

Parking Brakes That Use a Drum in the Center of the Rotor This type of parking brake provides some of the advantages of the self-energizing duo-servo brake. The brake shoes are mounted in a way that is similar to those in a conventional drum brake. The shoes are connected at the bottom by a star wheel adjuster. At the top the shoes contact a cam instead of an

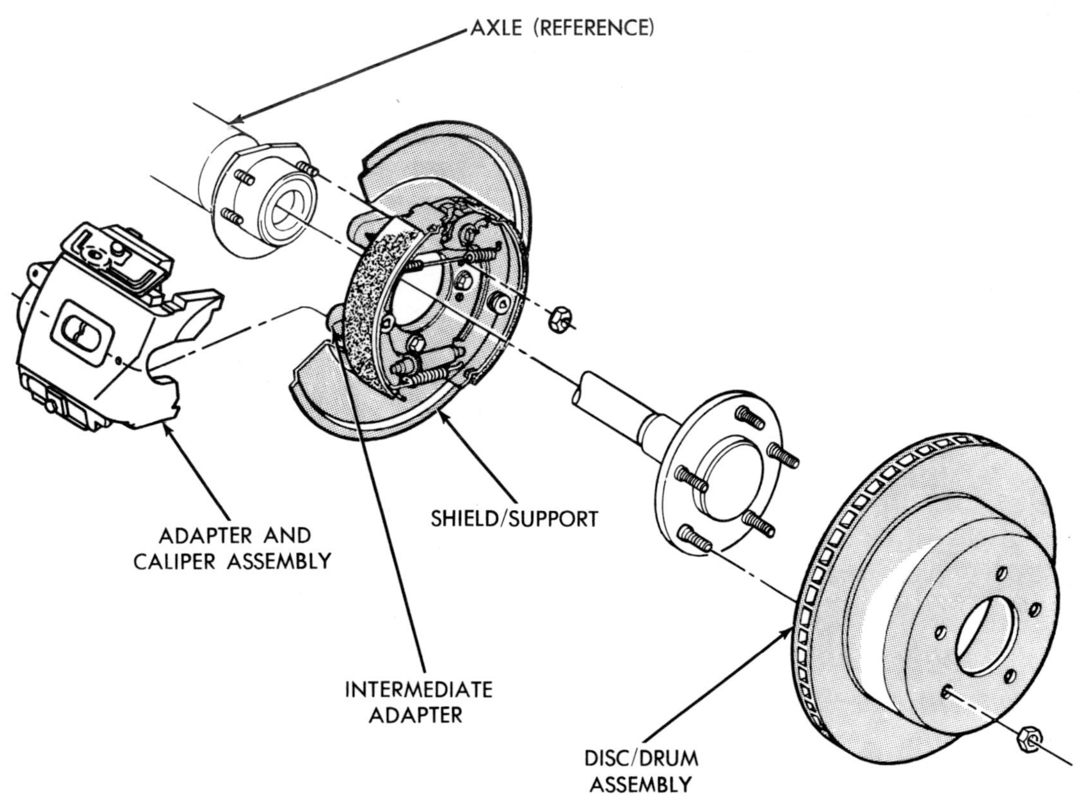

Figure 12.10 A typical parking brake that uses a drum built into the center of a rotor (Chrysler Corporation).

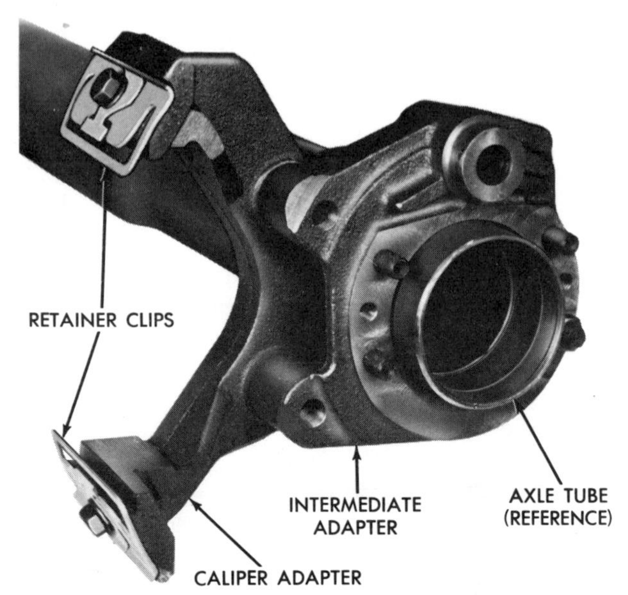

Figure 12.11 View showing how the intermediate adapter is mounted on the axle flange. Note that the caliper adapter is bolted to the intermediate adapter (Chrysler Corporation).

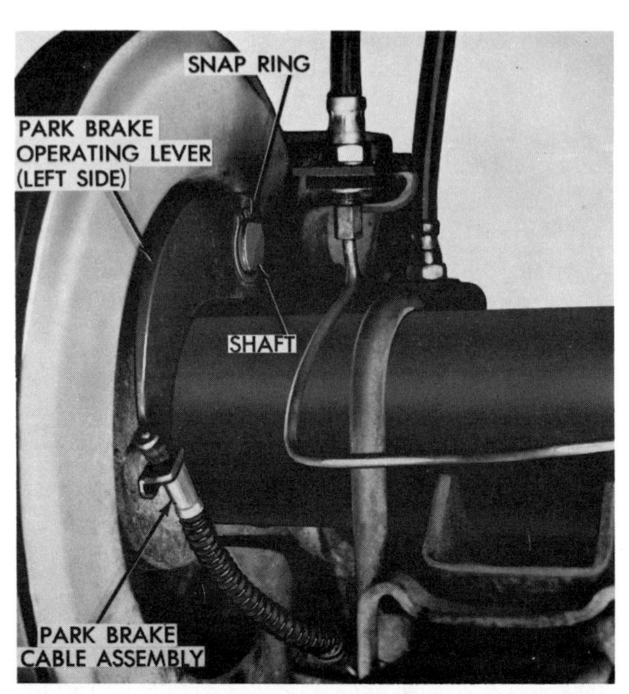

Figure 12.12 View showing how the parking brake lever and cable are mounted behind the splash shield (Chrysler Corporation).

anchor pin. When the parking brake is applied, the cam turns and spreads the shoes apart, forcing them into contact with the drum. The shoes are positioned on the disc brake splash shield, which acts as a backing plate. They are held in place with spring-and-pin type hold-downs. Figure 12.10 shows an exploded view of a typical brake of this type.

The combined splash shield and backing plate is mounted on the rear axle flange by means of the *intermediate adapter.* As shown in Figure 12.11, the intermediate adapter also provides the mounting point for the caliper adapter. The intermediate adapter has a large hole for the parking brake shaft. The shaft is rotated by the parking brake lever, and by means of an actuator, it moves the cam that expands the shoes. The parking brake lever is mounted externally as shown in Figure 12.12. It is pulled by a cable passing under the axle housing. The cables and equalizer are routed and mounted in a manner similar to those on drum brake systems.

Adjustment Self-adjusters are not used on these parking brakes. Therefore, you must adjust them by turning the star wheel adjuster with a brake spoon. An access slot is provided in the splash shield. Because of the small size of the drum and interference by the splash shield, a specially curved brake spoon is usually needed. Figure 12.13 illustrates how a special brake spoon is used to loosen and tighten the adjustment. A variant design provides an access hole in the drum. The hole is exposed when the wheel is removed. A special brake spoon is not needed with this design. A screwdriver can be used as shown in Figure 12.14.

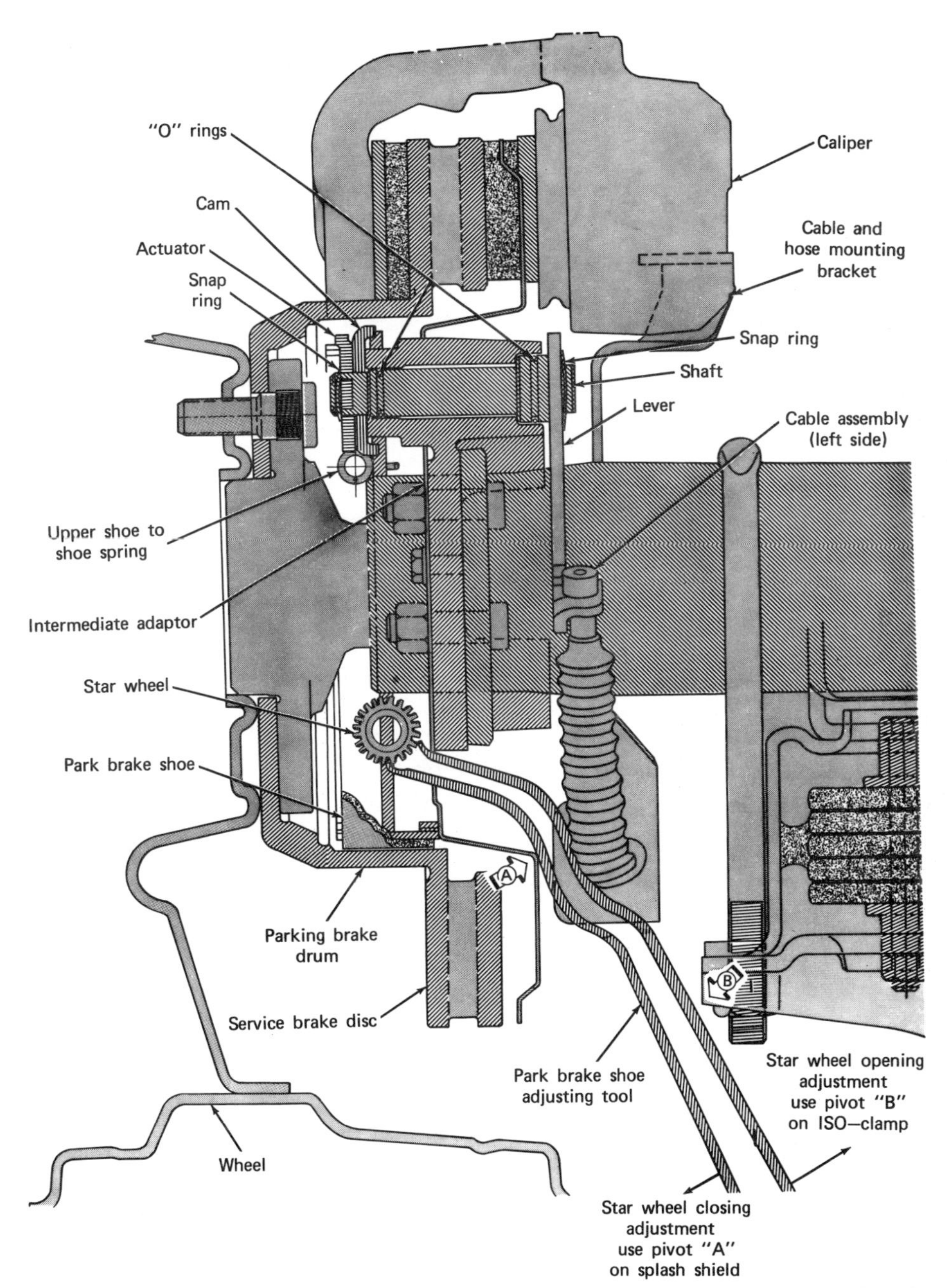

Figure 12.13 Sectional view of a parking brake that uses a drum in the center of a rotor. Note the special brake spoon and the pivot points that should be used in adjustment (Chrysler Corporation).

Figure 12.14 Adjusting the parking brake through an access hole in the drum (courtesy of the Chevrolet Service Manual, Chevrolet Motor Division).

ADJUSTING PARKING BRAKES THAT USE A DRUM IN A ROTOR

The procedures for adjusting parking brakes that use a drum in the center of a rotor may vary with different designs. For this reason, you should always consult the manufacturer's manual for the car on which you are working. However, here is a typical procedure:

1 Raise the car and support it by the suspension.

2 Set the parking brake operating lever in the fully released position.

3 Set the transmission shift lever in the neutral position.

4 Using a suitable tool, rotate a star wheel adjuster so that the shoes expand and seat against the drum.

5 Back off the star wheel adjuster about 12 notches.

6 Check to be sure the wheel is free of drag.

7 Check the operation of the parking brake operating lever.

8 Repeat operations 4 to 7 on the remaining wheel.

9 Lower the car.

10 Test the parking brakes.

Job 12f

ADJUST A PARKING BRAKE (DRUM IN ROTOR)

SATISFACTORY PERFORMANCE

A satisfactory performance on this job requires that you do the following:

1 Adjust the parking brake on the car assigned.
2 Following the steps in the "Performance Outline" and the manufacturer's procedure and specifications, complete the job within 150 percent of the manufacturer's suggested time.
3 Fill in the blanks under "Information."

PERFORMANCE OUTLINE

1 Raise and support the car.
2 Rotate the star wheel adjuster until the shoes seat against the drum.
3 Back off the star wheel adjuster.
4 Check the wheel for drag.
5 Repeat steps 2 to 4 on the remaining wheel.
6 Lower the car.
7 Check the operation of the brake.

INFORMATION

Vehicle identification ____________________

Reference used ____________ Page(s) ____________

Location of adjustment ____________________

Number of notches specified to be backed off ____________

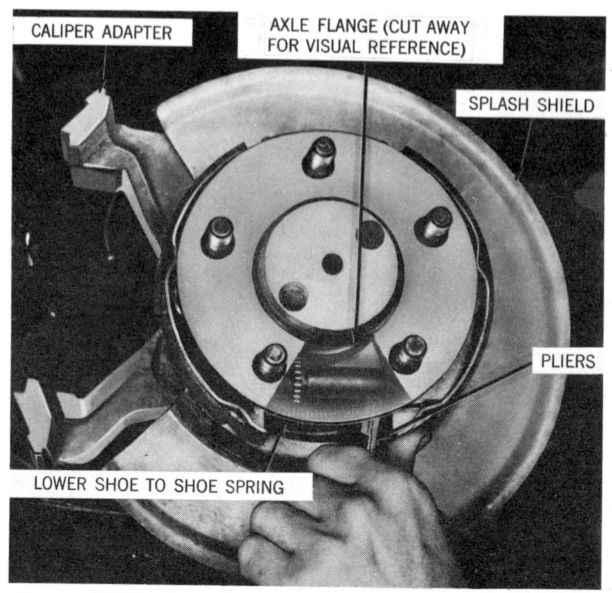

Figure 12.15 Removing the lower shoe-to-shoe spring (Chrysler Corporation).

PARKING BRAKE SHOE REPLACEMENT

The procedures for replacing parking brake shoes that use a drum in the center of a rotor vary according to the design. You should always check the manufacturer's manual for the procedure for specific cars. However, the following procedure is typical:

Removal

1 Raise the car and support it with stands.

2 Remove the rear wheel.

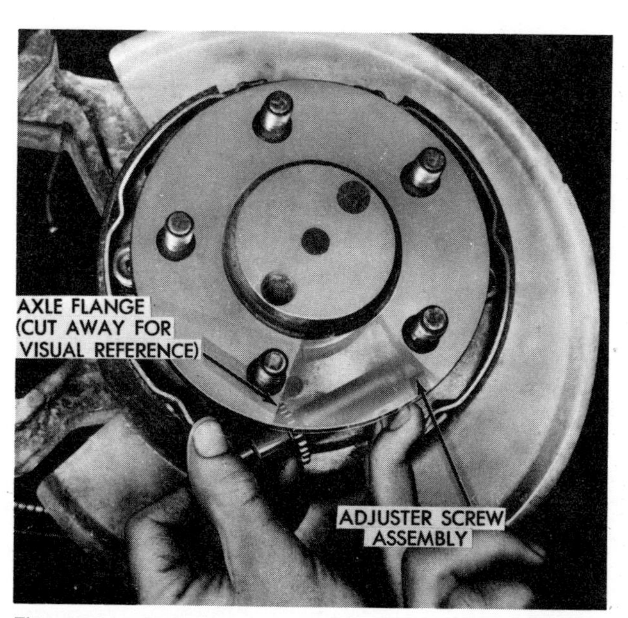

Figure 12.16 Removing the star wheel adjuster (Chrysler Corporation).

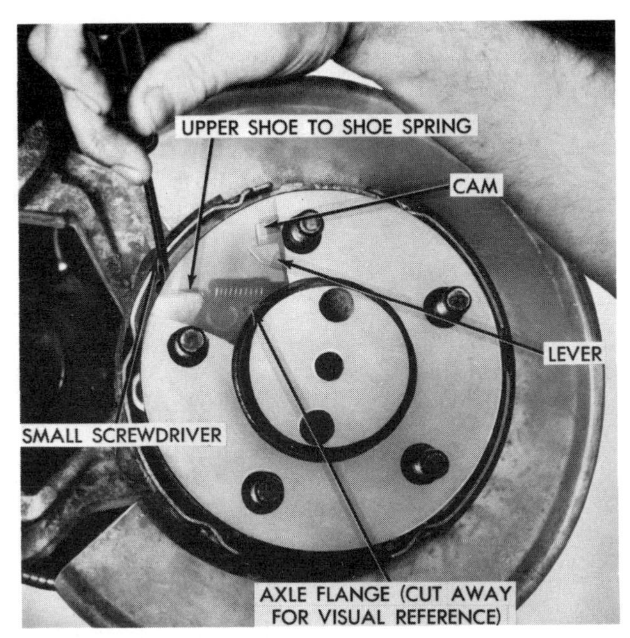

Figure 12.17 Removing the upper shoe-to-shoe spring (Chrysler Corporation).

3 Remove the caliper.

Note. Do not forget to support the caliper to keep it from hanging on the brake hose.

4 Remove the rotor.

5 Using a pair of pliers as shown in Figure 12.15, remove the spring conecting the lower ends of the shoes.

6 Spread the lower ends of the shoes apart, and remove the star wheel adjuster as shown in Figure 12.16.

7 Using a small screwdriver as shown in Figure 12.17, unhook one end of the upper shoe-to-shoe spring and remove it.

8 Spread the brake shoes apart as far as possible so they move out from under the axle flange.

9 Using a hold-down spring tool or a pair of pliers as shown in Figure 12.18, remove the hold-down assemblies.

10 Remove the shoes from the splash shield.

Note. Some manufacturers do not refer to these brake shoes as primary and secondary shoes. The upper ends of the shoes may be shaped differently, and the shoes may be called the left shoe and the right shoe. You should mark the location of the shoes when you remove them, and match them with the new shoes so you can be sure you are installing them in their proper places. Figure 12.19 illustrates the differences in left and right shoes.

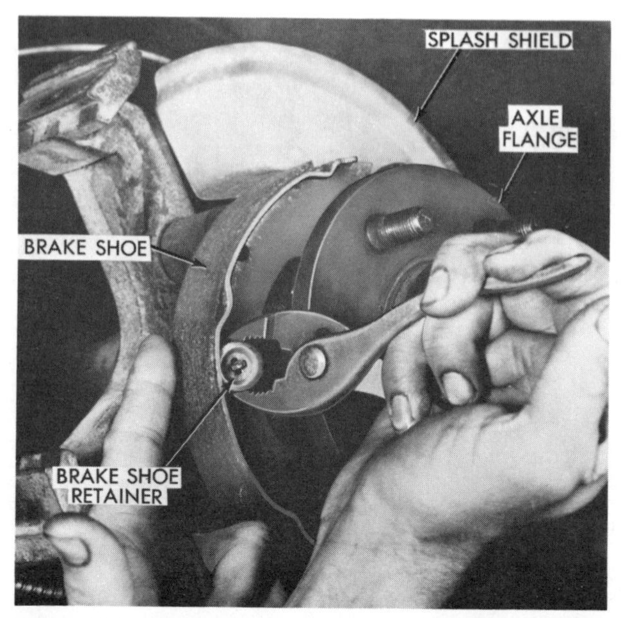

Figure 12.18 Removing the hold-down assemblies (Chrysler Corporation).

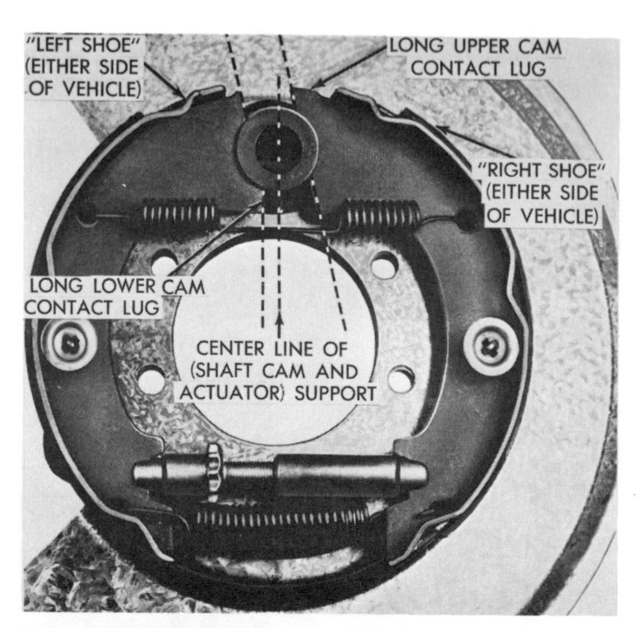

Figure 12.19 Parking brake shoes installed. The axle shaft has been removed so that the different shapes of the shoes can be seen (Chrysler Corporation).

Installation

1 Place a thin film of approved lubricant on the splash shield ledges or bosses.

2 Place the shoes on the splash shield, and install the hold-downs (Figure 12.18).

3 Push the shoes in under the axle flange as close as possible, and install the upper shoe-to-shoe spring (Figure 12.17).

4 Inspect and lubricate the star adjuster. Be sure the adjustment screw is turned all the way in.

5 Install the star adjuster by spreading the lower ends of the shoes (Figure 12.14).

6 Install the lower shoe-to-shoe spring (Figure 12.15).

7 Install the rotor.

8 Install the caliper.

9 Install the wheel.

10 Adjust the parking brake.

11 Check the operation of the brake.

12 Repeat operations 2 to 11 on the opposite wheel.

13 Lower the car.

14 Test the operation of the parking brake.

Job 12g

REPLACE PARKING BRAKE SHOES

SATISFACTORY PERFORMANCE

A satisfactory performance on this job requires that you do the following:

1 Replace the parking brake shoes on the car assigned.
2 Following the steps in the "Performance Outline" and the manufacturer's procedure and specifications, complete the job within 200 percent of the manufacturer's suggested time.
3 Fill in the blanks under "Information."

PERFORMANCE OUTLINE

1 Raise and support the car.
2 Remove a rear wheel.
3 Remove the caliper.
4 Remove the rotor.
5 Remove the shoes.
6 Lubricate the shoe ledges.
7 Install the replacement shoes.
8 Install the rotor.
9 Install the caliper.
10 Install the wheel.
11 Repeat steps 2 to 10 on the remaining wheel.
12 Adjust the parking brake.
13 Lower the car.
14 Check the operation of the brake.

INFORMATION

Vehicle identification ______________________________

Reference used ____________________ Page(s) ____________

Parking Brakes That Mechanically Actuate the Rear Calipers This type of parking brake pushes the caliper piston out by means of a *thrust screw.* The thrust screw is fitted with a self-adjusting nut that is pressed into the base of the piston. The self-adjusting nut allows the thrust screw to lengthen automatically as the piston moves out in its bore to compensate for lining wear. This action maintains the proper parking brake adjustment throughout the life of the lining. As shown in Figure 12.20, all the parts of this type of brake except the lever are enclosed in the caliper assembly.

When the parking brake is applied, the lever at the rear of the caliper moves and causes a thrust bearing to rotate. This rotation causes a set of steel balls to roll up the sides of ramped pockets. As the balls roll up their ramps they provide a wedging action that applies a force to the thrust screw. This force pushes the piston out, applying the brake.

Figure 12.20 A cutaway view of a rear caliper with a mechanically operated parking brake (Ford Motor Company).

ADJUSTMENT The brake assembly is self-adjusting, but cable stretch and wear in the cable-and-lever system may require that a cable adjustment be made. An adjustment nut is provided at the equalizer, but a different procedure from the one you followed previously is required to obtain the proper adjustment. The adjustment procedure that follows is typical. But you should always check the manual for the particular procedure for the car on which you are working.

1 Raise the car and support it by the suspension with car stands.

2 Set the parking brake operating lever in the full released position.

3 Set the transmission shift lever in neutral.

4 Tighten the adjusting nut at the equalizer until the levers on the calipers barely start to move.

5 Loosen the adjusting nut just enough to allow the levers to return to their stops.

6 Apply and release the parking brakes several times.

7 Check the levers at the calipers by pulling them to the rear.

Note. If the levers move to the rear even slightly, the adjustment is too tight. Loosen the adjustment nut. If the levers do not move, the adjustment is correct.

8 Lower the car.

9 Check the operation of the parking brake.

Job 12h

ADJUST THE PARKING BRAKE CABLES ON A CAR WITH MECHANICALLY ACTUATED REAR CALIPERS

SATISFACTORY PERFORMANCE

A satisfactory performance on this job requires that you do the following:

1 Adjust the parking brake cables on the car assigned.

2 Following the steps in the "Performance Outline" and the manufacturer's procedure and specifications, complete the job within 150 percent of the manufacturer's suggested time.

3 Fill in the blanks under "Information."

PERFORMANCE OUTLINE

1 Raise and support the car.

2 Set the parking brake lever and the gear shift lever in their proper positions.

3 Adjust the length of the cables.

4 Apply and release the brake several times.

5 Check the position of the levers at the calipers.

6 Readjust the cables if necessary.

7 Lower the car.

8 Check the operation of the brake.

INFORMATION

Vehicle identification ____________________

Reference used ________________ Page(s) __________

Location of adjustment ____________________

Wrench size of adjusting nut ____________________

SUMMARY

In this chapter you have learned about the operation of the various parking brake systems. You have developed skills in the adjustment and repair of the components in these systems. You can identify all the parts and their function. You can replace cables and shoes, and you can perform all the required adjustments.

SELF-TEST

Each incomplete statement or question in this test is followed by four suggested completions or answers. In each case select the *one* that best completes the statement or answers the question.

1 When the parking brakes are applied, the driver's effort is multiplied through
 A. hydraulics
 B. cables
 C. leverage
 D. servo action

2 When brake shoes are being installed on the rear wheels of a car fitted with drum brakes, the link spring is placed between the link and the
 A. primary shoe
 B. secondary shoe
 C. parking brake lever
 D. anchor pin

3 Two mechanics are discussing parking brake systems.
 Mechanic A says that the service brakes should be checked and adjusted before the parking brake is adjusted.
 Mechanic B says that when parking brakes are being adjusted, the car should be supported by car stands placed under the frame.
 Who is right?
 A. A only
 B. B only
 C. Both A and B
 D. Neither A nor B

4 Two mechanics are discussing parking brake systems.
 Mechanic A says that a battery cable should be disconnected before a front parking brake cable secured under the instrument panel is replaced.
 Mechanic B says that the application of both rear brakes is equalized by the use of a link and spring.
 Who is right?
 A. A only
 B. B only
 C. Both A and B
 D. Neither A nor B

5 Parking brake shoes are sometimes refered to as
  I. primary and secondary shoes
  II. left and right shoes
 A. I only
 B. II only
 C. Either I or II
 D. Neither I nor II

6 Parking brakes are properly adjusted when
  I. they are fully applied and holding after the operating lever has been moved through less than half its possible travel
  II. they are fully released when the operating lever is in the released position
 A. I only
 B. II only
 C. Both I and II
 D. Neither I nor II

7 Excess slack in the brake cables can be eliminated in most instances by an adjustment at the
 A. operating lever
 B. equalizer
 C. star wheel adjusters
 D. link

8 On some cars less effort is required to apply the parking brakes while the service brakes are held on than while the service brakes are off. This method of brake application will work on cars fitted with
  I. drum brake systems
  II. combination brake systems

A. I only
B. II only
C. Both I and II
D. Neither I nor II

9 On cars fitted with parking brakes that use drums in the center of the rear rotors, an access hold for the star wheel adjuster is provided in the
   I. drum
   II. splash shield
A. I only
B. II only
C. Either I or II
D. Neither I nor II

10 Two mechanics are discussing parking brakes.
Mechanic A says that the axle flange should be removed before the parking brake shoes are replaced in a system that incorporates brake drums in the center of the rear rotors.
Mechanic B says that in a brake of this type the shoes are expanded by a cam.
Who is right?
A. A only
B. B only
C. Both A and B
D. Neither A nor B

Chapter 13 Power Assist Units and Antiskid Devices

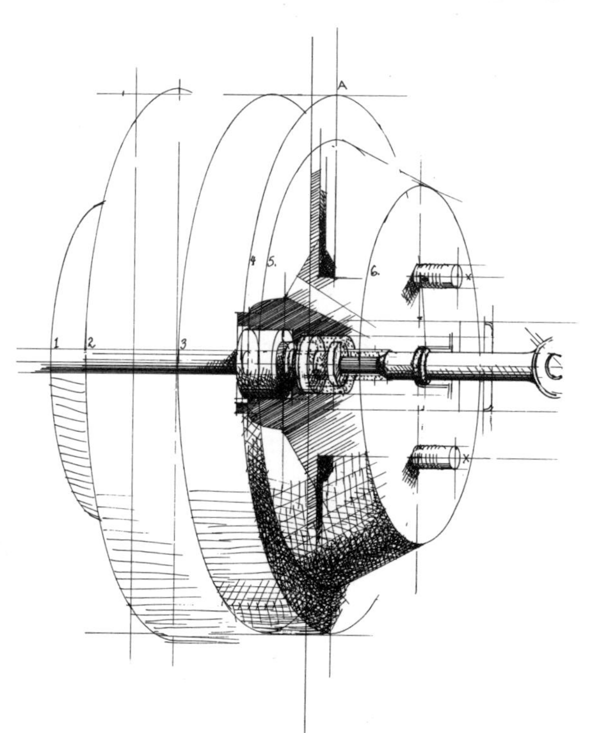

Power brakes have gained wide acceptance with car buyers because of the reduced driver effort that such brakes require. Modern power assist units have proved to be relatively troublefree, but certain brake problems can be solved only by knowing the way in which those units function.

The development of electronic devices that can instantly sense when a wheel is about to skid has added a new dimension to highway safety. Antiskid systems automatically regulate braking action so that the wheels never lock. This eliminates skidding and enables a driver to bring a car to a stop in a straight line, even on slippery road surfaces.

Your objectives in this chapter are to:

1
Identify the operating positions of a vacuum suspended power brake booster.

2
Check the operation of vacuum brake boosters.

3
Replace vacuum brake booster units.

4
Identify the major components of hydraulically operated brake boosters.

5
Identify the functions of the parts in an antiskid device.

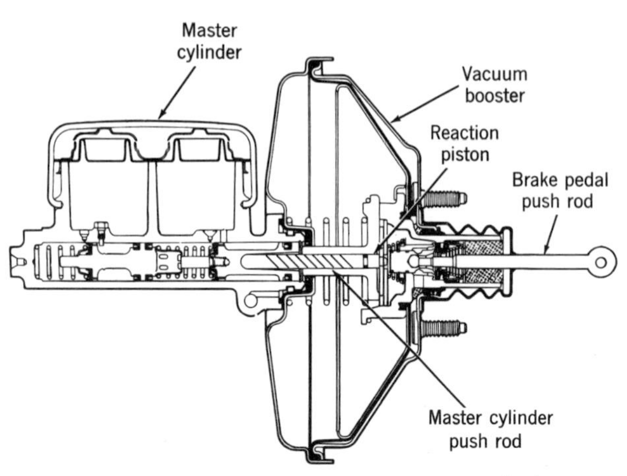

Figure 13.1 A vacuum suspended booster unit (courtesy American Motors Corporation).

POWER ASSIST UNITS Power assist units greatly multiply the force applied to the brake pedal. Most power assist units are *vacuum brake boosters.* They use engine intake manifold vacuum and atmospheric pressure to multiply the driver's effort. Some power assist units are *hydraulic boosters.* In those units the pressure for the brake booster is provided by the hydraulic pump that also provides pressure for the power steering.

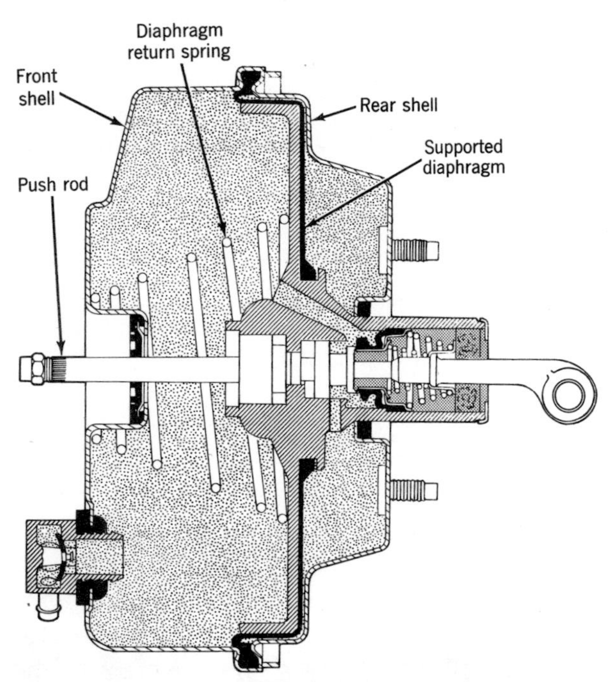

Figure 13.2 Major components of the power chamber (Ford Motor Company).

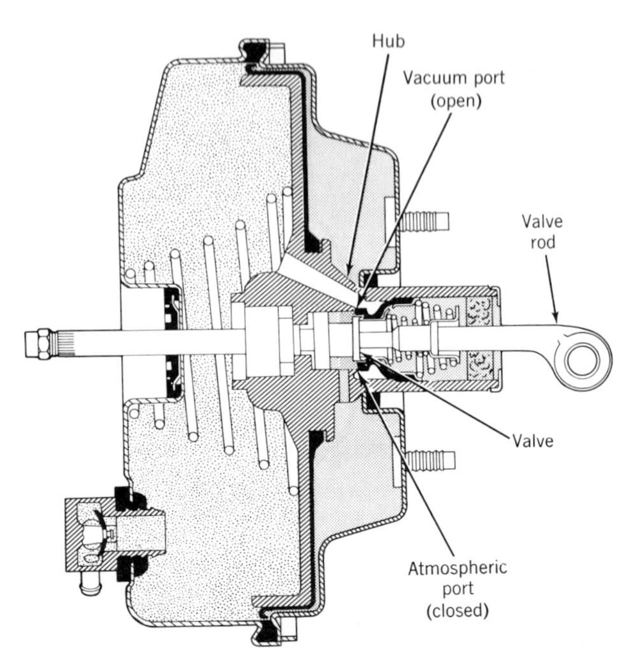

Figure 13.3 Major components of the control valve (Ford Motor Company).

Vacuum Brake Boosters Most vacuum brake boosters now in use are called *vacuum suspended* units. The term *vacuum suspended* describes the condition of the unit when the car engine is running and the brakes are released. When in that condition, vacuum is present on both sides of a diaphragm so that the diaphragm is suspended in a vacuum. Figure 13.1 shows a cross-sectional view of a typical vacuum suspended brake booster.

Though all the booster components are combined in a single unit, or assembly, a vacuum suspended booster consists of two subassemblies. Those subassemblies are the *power chamber* and the *control valve.*

The Power Chamber The power chamber applies the force to the master cylinder piston. The chamber consists of a front shell, a rear shell, a supported diaphragm, a diaphragm return spring, and a pushrod. These parts are shown in Figure 13.2.

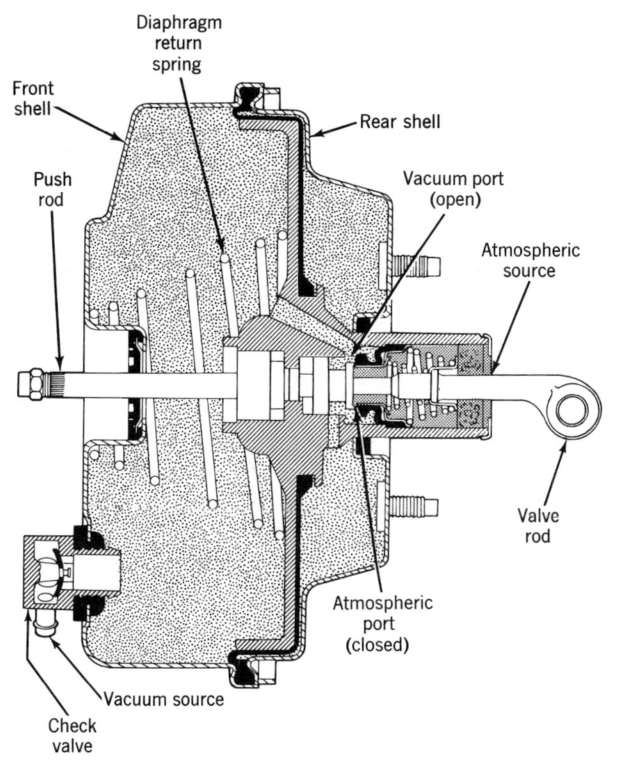

Figure 13.4 Booster in the released position (Ford Motor Company).

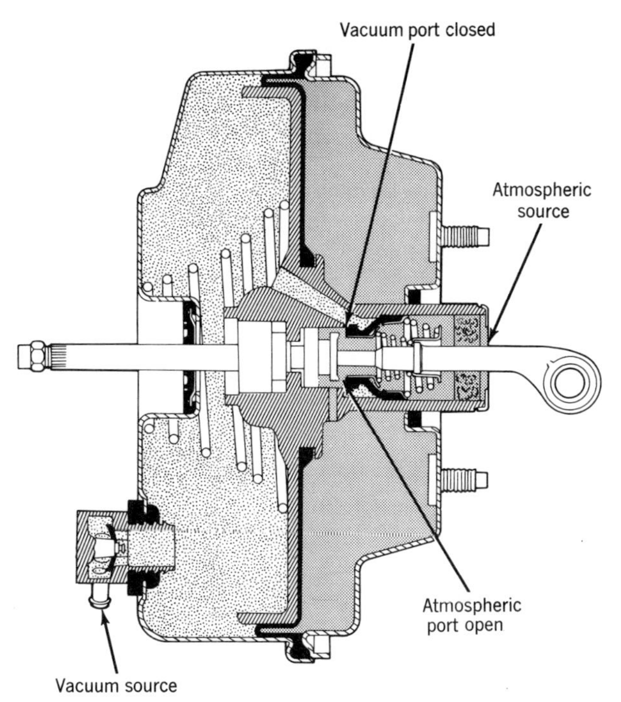

Figure 13.5 Booster in the applied position (Ford Motor Company).

The Control Valve The control valve determines the amount of force the power chamber applies to the master cylinder piston. It does this by opening and closing two ports: (1) the vacuum port and (2) the atmospheric port. The control valve is a spool-type valve built into the hub of the diaphragm. It is actuated by the driver through a rod attached to the brake pedal (Figure 13.3).

Operation of Vacuum Suspended Booster
All vacuum suspended boosters function in three modes, or positions. Those positions are (1) released, (2) applied, and (3) holding.

Released Position Figure 13.4 shows the position of the booster parts when the unit is in the released position. When the brake pedal is released, engine intake manifold vacuum evacuates the air from the front shell of the power chamber through the check valve. Since the valve rod is in the rear position, the valve plunger is also held to the rear. This holds the vacuum port open, allowing vacuum to evacuate the air from the rear shell. Because manifold vacuum is present on both sides of the diaphragm, the pressures on both sides of it are equal. This allows the diaphragm spring to hold the diaphragm to the rear. Then the pushrod exerts no force on the master cylinder piston.

Applied Position When the driver depresses the brake pedal, the valve rod pushes the valve plunger forward. This closes the vacuum port and opens the atmospheric port, as shown in Figure 13.5. Atmospheric pressure enters the rear shell of the power chamber through the air filter near the valve rod. Since manifold vacuum is maintained in front of the diaphragm, the air pressure behind the diaphragm pushes the diaphragm forward. The movement of the diaphragm is transmitted by the pushrod to the master cylinder piston.

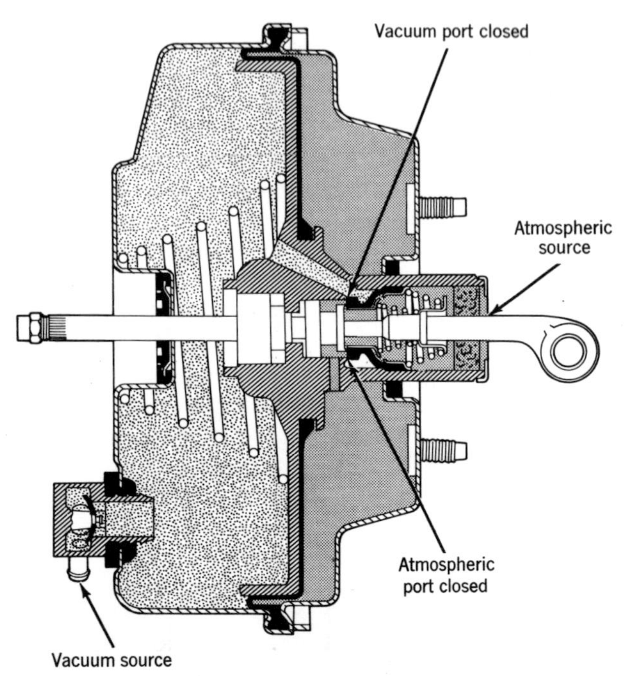

Figure 13.6 Booster in the holding position (Ford Motor Company).

Holding Position Most driving conditions require only gradual braking. To provide such braking, a position between the released and applied positions must be provided. This is why the control valve is mounted in the hub of the diaphragm. When the diaphragm moves forward, the valve body moves away from its control rod. The valve body closes the atmospheric port as shown in Figure 13.6. This valve action regulates the pressures on the diaphragm, giving the driver control over the degree of braking. Additional pedal movement reopens the atmospheric port and causes the pushrod to exert more force on the master cylinder piston.

Reducing the pressure on the brake pedal closes the atmospheric port. This causes a pressure drop on the rear of the diaphragm and decreases the force applied to the master cylinder piston. If the pedal is held in one position, the movement of the diaphragm centers the valve, closing both the vacuum port and the atmospheric port. This maintains a continuous holding force on the pushrod.

Since a vacuum suspended booster unit requires only that a valve be moved to apply or release the brakes, the driver would experience no "feel" when pressing on the brake pedal unless there were some special provision for it. The absence of feel would make it very difficult to bring a car to a smooth, controlled stop. To provide feel, a reaction device is built into the booster. This device may consist of a rubber disc or an assembly of spring steel levers. It is fitted between the diaphragm and the pushrod. When the brakes are applied, the reaction device transfers a small percentage of the pushrod output force back through the valve rod to the brake pedal. The transferred force is proportional to the force applied to the master cylinder piston. It enables the driver to judge how much force must be applied to the pedal.

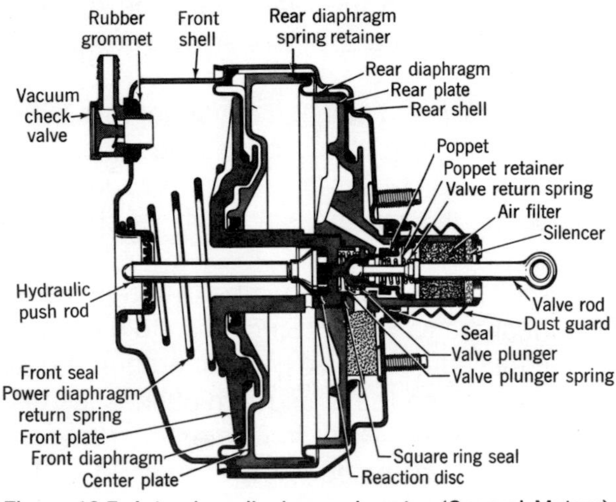

Figure 13.7 A tandem diaphragm booster (General Motors).

Tandem Diaphragm Boosters Tandem diaphragm boosters are vacuum suspended boosters that use two diaphragms. The diaphragms are mounted in tandem, or one behind the other, as shown in Figure 13.7.

Both diaphragms share a common shell, but they are separated by a partition, or center plate. The two diaphragms double the area against which atmospheric pressure can push. This provides a very high output force with little pedal effort on the part of the driver.

The operation of tandem diaphragm boosters is the same as that of single diaphragm boosters. The two types of boosters are tested in the same manner. All the differences between them are internal.

Because safety is the most important consideration in the operation of any part of a brake system, all power assist units provide a fail-safe feature. The valve rod is directly in line with the pushrod, and if sufficient effort is applied by the driver, it will force the diaphragm forward. This moves the pushrod and applies the brakes even if the booster unit fails. The driver becomes aware of any such failure as the pedal effort required to stop the car greatly increases.

Job 13a

IDENTIFY THE OPERATING POSITIONS OF A VACUUM SUSPENDED POWER BRAKE BOOSTER

SATISFACTORY PERFORMANCE

A satisfactory performance on this job requires that you do the following:

1 Identify the positions of a booster unit by placing the number of each position in front of the words that describe the condition of the control valve.

2 Correctly match the three positions with the control valve conditions within 10 minutes.

PERFORMANCE SITUATION

1 Released position
2 Applied position
3 Holding position

Booster Position	Valve Conditions	
	Vacuum Valve	*Atmospheric Valve*
______	Open	Open
______	Closed	Closed
______	Open	Closed
______	Closed	Open

Checking a Vacuum Booster You can quickly check the operation of a vacuum booster unit in the following manner:

1 With the car engine turned off, depress the brake pedal several times. This action "bleeds off" any vacuum that may be present in the booster system.

2 Maintain a light pressure on the brake pedal and start the engine. If the booster is operating properly, the pedal will drop slightly under your foot. If no drop is felt, the booster unit is not operating.

Before you replace what may appear to be a defective booster unit, you should inspect the vacuum lines and hoses and their connections. Since the difference in pressure that causes the diaphragm to move depends on intake manifold vacuum, any restriction or leak in the connecting vacuum lines and hoses affects booster operation. A kink often causes a restriction whereas a loose connection or a cracked hose leaks. A quick inspection of the vacuum hoses and their connections could save you a lot of time.

As the engine is the source of vacuum, you should connect a vacuum gauge to the intake manifold so that the vacuum can be measured. Check the shop manual for the procedure and specifications for the car on which you are working. A reading lower than that specified could indicate a need for new manifold gaskets or for other engine repairs. If a low reading is obtained, further diagnosis of the brake system should be halted until any needed engine work has been performed. Continue your checks when the manifold vacuum is at an acceptable level.

If the vacuum reading at the manifold is satisfactory, the vacuum hose should be disconnected from the booster check valve, and the gauge should be installed on the hose. Any drop in vacuum measured at this point usually indicates a leak in the connecting lines or hoses.

A continuous hissing noise at the valve rod behind the brake pedal may be caused by a leaking diaphragm or a defective valve. In either case the booster unit must be overhauled or replaced.

Job 13b

CHECK THE OPERATION OF A BRAKE BOOSTER

SATISFACTORY PERFORMANCE

A satisfactory performance on this job requires that you do the following:

1 Perform the checks listed below on the car assigned.

2 Finish the checks and list your findings in the "Performance Record" within 15 minutes.

PERFORMANCE RECORD

1 Check the operation of the booster. Did the pedal drop when the engine was started? ____________

____

2 Check the condition of the vacuum lines. Are the lines tight and in good condition? ____________

____

3 Check for leakage. Are there any air or fluid leaks? ____________

____

4 Do you consider the booster unit serviceable in its present condition? ____________

____

5 What service do you recommend, if any?

____

____

INFORMATION

Vehicle identification ____________________________

REMOVAL AND INSTALLATION OF VACUUM BOOSTER UNITS

For removing, adjusting, and installing a typical vacuum suspended booster unit, you should consult an appropriate manual. Here is the general procedure for removal.

Removal

1 Remove the bolts or nuts that secure the master cylinder to the front shell of the booster unit.

Note. On some cars you will not have to disconnect the brake lines. Usually you can pull the master cylinder away from the booster unit until the pushrod is clear. Use care so as not to kink or strain the tubing.

2 Disconnect the vacuum hose from the check valve.

3 Remove the bolt or pin that secures the valve rod to the brake pedal linkage.

4 Remove the bolts or nuts that hold the booster unit to the firewall.

5 Remove the booster unit by lifting it upward and forward.

Push Rod Adjustment. **Before installing any booster unit, either new or rebuilt, you should check the pushrod length. If the pushrod length does not match the specifications of the car maker, it must be adjusted. A rod that is too long holds the master cylinder piston forward. The piston blocks the compensating ports, causing the brakes to drag and eventually to lock. If the rod is too short, the driver has to depress the brake pedal too far before brake application begins.**

The general procedure for pushrod adjustment is as follows:

1 Consult an appropriate manual for the proper pushrod length.

2 Using a piece of stiff cardboard, fabricate a gauge as shown in Figure 13.8. Dimension "A" must be the pushrod length specified by the manufacturer.

3 Hold the gauge against the booster shell as shown in Figure 13.9. The gauge should make even contact with the shell and the booster screw.

4 If the adjustment is incorrect, turn the screw to gain the proper adjustment. Be sure to hold the pushrod as shown in Figure 13.10.

Installation The general procedure for installing a vacuum booster unit is:

1 Place the booster unit in position.

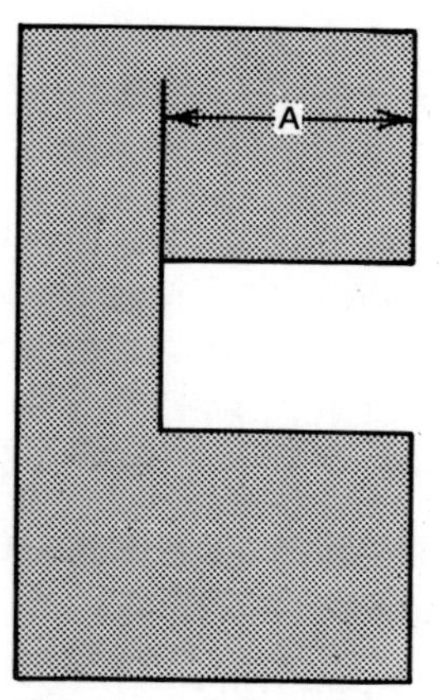

Figure 13.8 A pushrod gauge (Ford Motor Company).

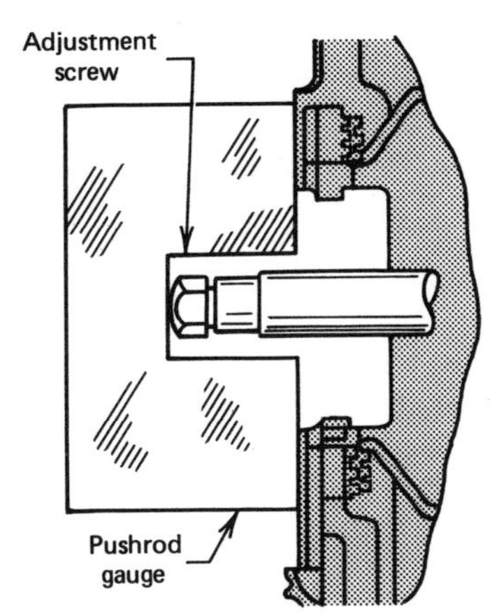

Figure 13.9 The use of a pushrod gauge (Ford Motor Company).

2 Start the bolts or nuts, taking care that the valve rod is aligned with the pedal linkage.

3 Tighten the bolts or nuts.

4 Install the pin or bolt that connects the valve rod to the pedal linkage.

5 Install the vacuum hose.

6 Position the master cylinder on the front shell of the booster unit, and install the nuts or bolts.

Note. **If you have disconnected any lines or hoses in the hydraulic system, you must bleed the system.**

7 Test the operation of the booster unit.

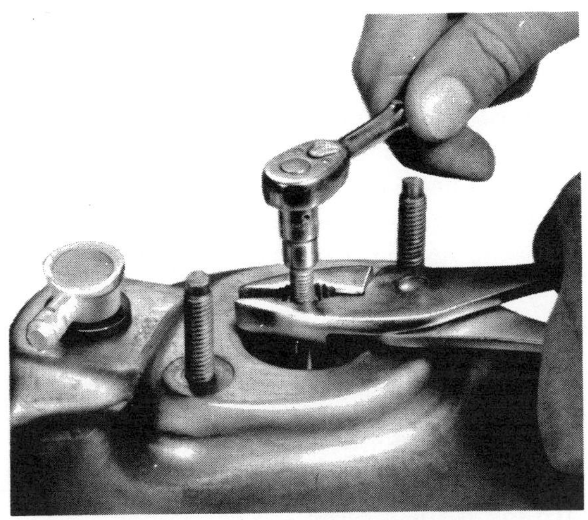

Figure 13.10 Holding the pushrod while turning the adjusting screw (courtesy American Motors Corporation).

Job 13c

REPLACE A VACUUM BOOSTER UNIT

SATISFACTORY PERFORMANCE

A satisfactory performance on this job requires that you do the following:

1 Replace a vacuum booster unit on the car assigned.
2 Following the steps in the "Performance Outline" and the manufacturer's procedure and specifications, complete the job within 150 percent of the manufacturer's suggested time.

PERFORMANCE OUTLINE

1 Remove the master cylinder from the booster.
2 Disconnect the vacuum line.
3 Disconnect the valve rod from the pedal linkage.
4 Remove the booster.
5 Check the push rod length and adjust it if needed.
6 Install the booster.
7 Connect the valve rod to the linkage.
8 Install the vacuum line.
9 Install the master cylinder.
10 Test the operation of the unit.

INFORMATION

Vehicle identification ______________________

Booster Unit Overhaul Most power assist units require the use of special tools and spring retaining devices for disassembly and assembly. Some manufacturers advise against any attempt to overhaul these units and do not list parts for them in their catalogs. For these reasons, most mechanics usually replace a faulty booster unit with a new unit or with one that has been overhauled by a shop that specializes in rebuilding booster units.

If you have to overhaul a power assist unit, you should first study the procedure outlined by the unit manufacturer. Proceed only if the proper tools and equipment are available. Any attempt to disassemble a booster unit without using the proper tools and holding devices may release spring-loaded parts and cause you personal injury.

Hydraulically Operated Brake Boosters Some cars, especially those equipped with four-wheel disc brakes, are fitted with a hydraulically operated booster unit. The pressure to operate the unit is provided by the power-steering pump. The unit fits in the place normally occupied by a conventional vacuum booster unit, and it serves the same function. It increases the force a driver exerts on the brake pedal, and it applies that force to the piston of a

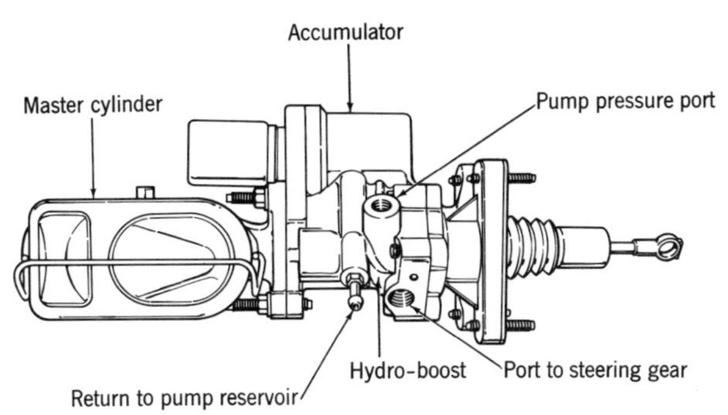

Figure 13.11 Top view of a hydraulically operated brake booster (Ford Motor Company).

master cylinder. A typical hydraulically operated booster is shown in Figure 13.11.

Construction A typical hydraulically operated booster contains the following assemblies:

1 A sliding valve that regulates the pressure during braking.
2 A system of levers that moves the valve when the driver depresses the brake pedal.
3 A boost piston that provides the output force to the master cylinder pushrod.
4 A reserve system.

Note. The reserve system consists of an accumulator that stores a small amount of fluid under pressure. The stored fluid makes possible a few powered brake applications if for some reason pressure from the power steering pump becomes unavailable.

The assemblies of a typical hydraulically operated booster are shown in Figure 13.12.

Operation When the brake pedal is in the released position, the spool-shaped valve is held open by a spring. The valve controls two passages. One passage allows the fluid from the power steering pump to flow directly to the power steering gear. The fluid pressure is not applied to the boost chamber. The other passage provides an opening from the boost chamber to a line that runs back to the power

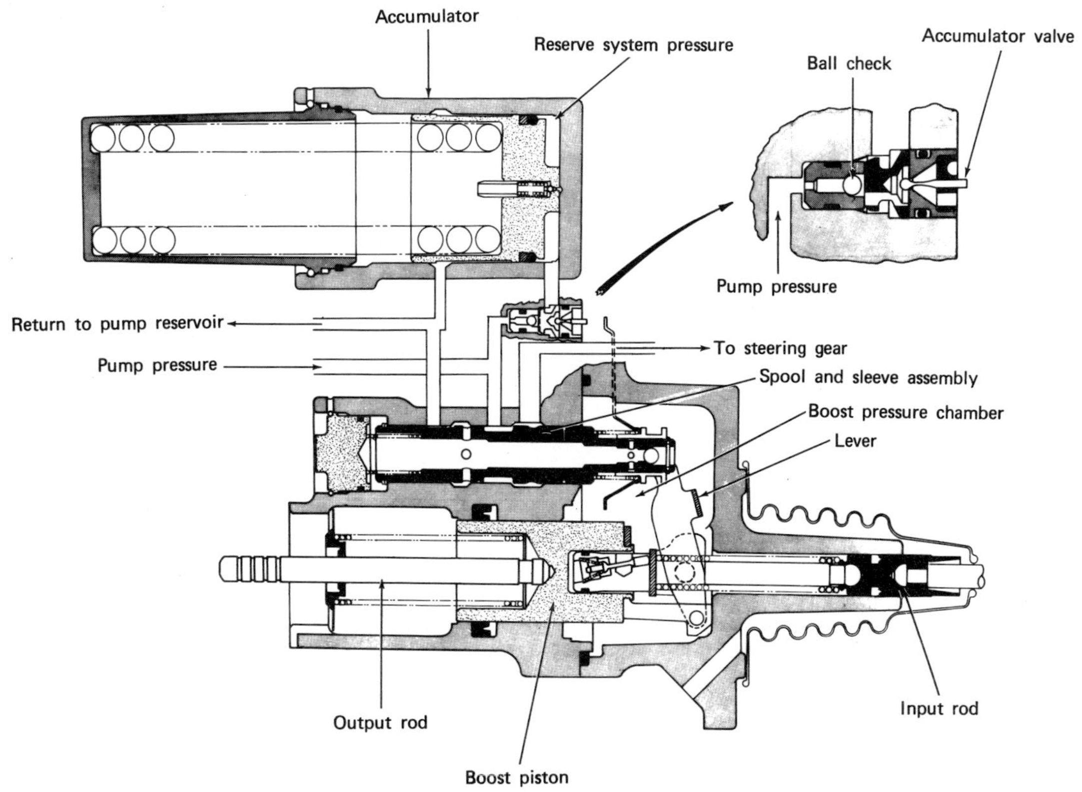

Figure 13.12 A sectional view of a hydraulically operated brake booster (Ford Motor Company).

steering pump reservoir so that the chamber cannot maintain any pressure.

When the brake pedal is depressed, the input rod starts operating the lever system. The lever system moves the valve, admitting fluid to the boost chamber and, at the same time, closing the passage to the pump reservoir. This allows pressure to build up in the boost chamber. As the pressure increases, it forces the boost piston forward. The movement of the boost piston is transmitted to the master cylinder piston through the output rod.

When the brake pedal is released, the valve returns to the open position. In that position the valve cuts off the fluid pressure to the boost chamber and allows the pressure in the chamber to escape through the line to the reservoir.

The reserve system accumulator is filled by pump pressure, which forces fluid into the accumulator cylinder. The fluid forces a piston down in the cylinder bore against a very strong coil spring. A system of check valves keeps the accumulator filled and pressurized. The accumulator stores enough fluid under pressure to provide for two or three stops without benefit of fluid from the power steering pump. After the fluid in the accumulator has been depleted, the brakes can still be applied. But the pedal effort required is greatly increased.

Service

You can test a hydraulic booster in the same way that you test a vacuum booster. With the car engine off, bleed off the pressure in the accumulator by depressing the brake pedal several times. Hold your foot on the brake pedal, and start the engine. If the unit is functioning properly, the pedal will drop slightly under your foot. This is usually followed by a slight surge of the pedal. If a unit does not pass this simple test, you should check the manufacturer's manual for the detailed diagnostic procedure. Since the unit depends on the power steering pump for pressure, a check of that system will have to be made.

Do not attempt to remove or disassemble any part of a hydraulic booster without first studying the procedure outlined by the manufacturer. The accumulator can store fluid under pressures exceeding 1500 psi. The sudden release of this pressure can cause you serious injury.

Job 13d

IDENTIFY THE MAJOR COMPONENTS OF A HYDRAULICALLY OPERATED BRAKE BOOSTER

SATISFACTORY PERFORMANCE

A satisfactory performance on this job requires that you do the following:

1 Identify the major components of a hydraulically operated brake booster unit by matching the numbered parts on the drawing with the list of part names.

2 Correctly identify 5 of the 6 numbered parts within 10 minutes.

PERFORMANCE SITUATION

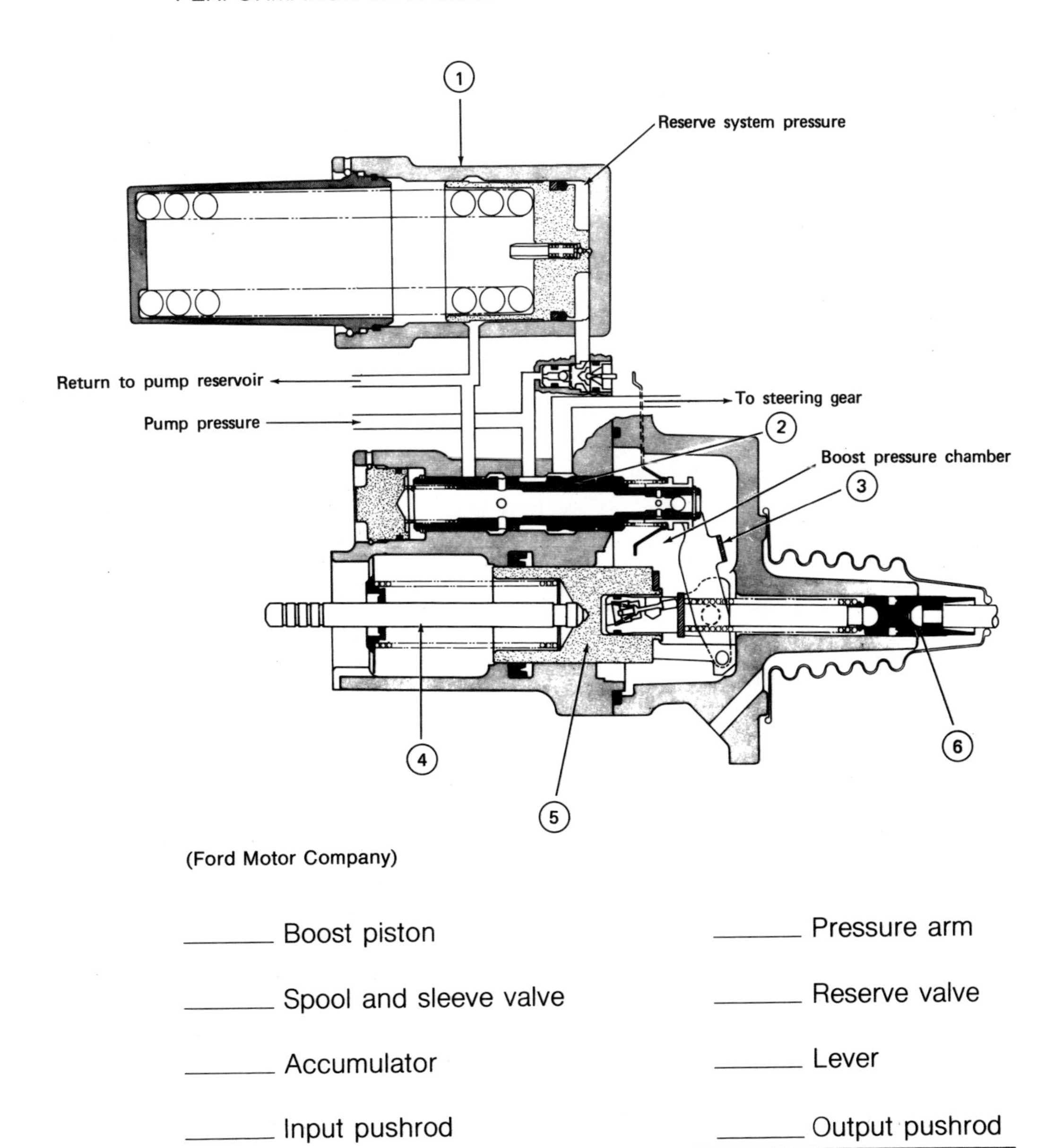

(Ford Motor Company)

_______ Boost piston

_______ Spool and sleeve valve

_______ Accumulator

_______ Input pushrod

_______ Pressure arm

_______ Reserve valve

_______ Lever

_______ Output pushrod

ANTISKID DEVICES You are already aware of the importance of the friction between the brake linings and the drums or rotors. Equally important is the friction between the tires and the road. If the friction between the brake linings and the drums or rotors is greater than the friction between the tires and the road, the tires skid.

Many factors contribute to a skid. Among them are tire condition, road condition, brake condition, the way in which the brakes are applied, and weight shift during braking. Some of those factors can be controlled, but some cannot. The car owner can replace worn tires and maintain the proper tire pressure but cannot control the condition of the road. A mechanic can provide and maintain an efficient braking system but cannot prevent situations that require panic stops.

Neither the car owner nor the mechanic can control weight shift. As stated earlier, the center of gravity of a car shifts forward during braking. This shift places more of the car weight on the front tires than on the rear tires. The extra weight increases the coefficient of friction between the front tires and the road. At the same time, the weight removed from the rear tires decreases the coefficient of friction between them and the road. Because of this, the rear tires are more apt to skid than the front tires.

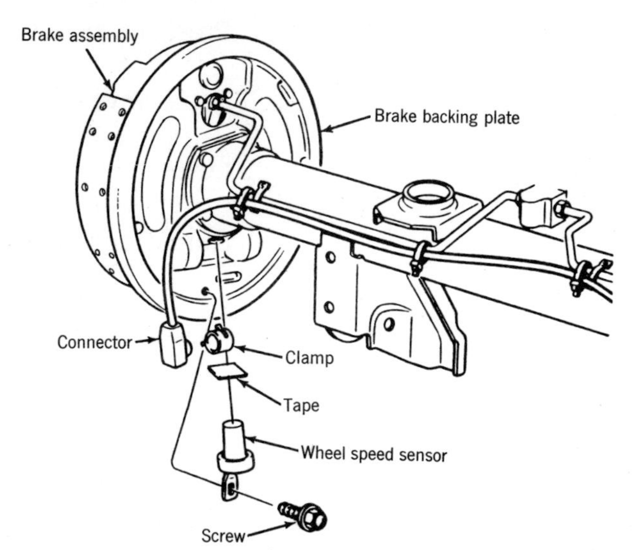

Figure 13.13 A wheel speed sensor and attaching parts (General Motors).

A tire that is skidding can slide in any direction. A slight bump or dip in the road surface can easily cause the rear end of a skidding car to swerve to one side. Since the rear wheels cannot be steered, the direction in which the car moves may be uncontrollable. Even if the direction of movement can be maintained, the stopping distance is usually increased. It is increased because the friction between the tire and the road is usually less when the tire is sliding than when it is rolling. This decrease in friction is especially noticeable on wet or icy roads.

Of course, skidding can be very dangerous. For this reason, antiskid devices are added to many brake systems. Such devices (1) maintain the directional stability of a car, and (2) decrease the required stopping distance.

Most antiskid devices use a small computer to determine the proper amount of braking for each situation. The computer measures the difference between the rate of deceleration of a rotating wheel and the rate of deceleration of the car. This difference is usually referred to as *slip.* If the car is traveling 60 mph and the wheel is going 50 mph, there is a slip of about 15 percent. The most efficient braking is obtained within the slip range of 10 to 20 percent. When the slip is less than 10 percent, the stopping distances increases. When the slip exceeds 20 percent, the wheel usually locks and skids.

Optimal braking is obtained when the slip range of 10 to 20 percent is maintained throughout the entire period of braking. During normal braking on dry, well-paved road surfaces, a skilled driver can maintain this slip range by applying just the right amount of force to the brake pedal. However, it is impossible for any driver to maintain this slip range during a panic stop, especially if the road surface is wet or icy.

Antiskid devices maintain the proper slip range automatically. They do so by pumping the brakes from 3 to 5 times per second whenever a rear wheel begins to lock under heavy braking.

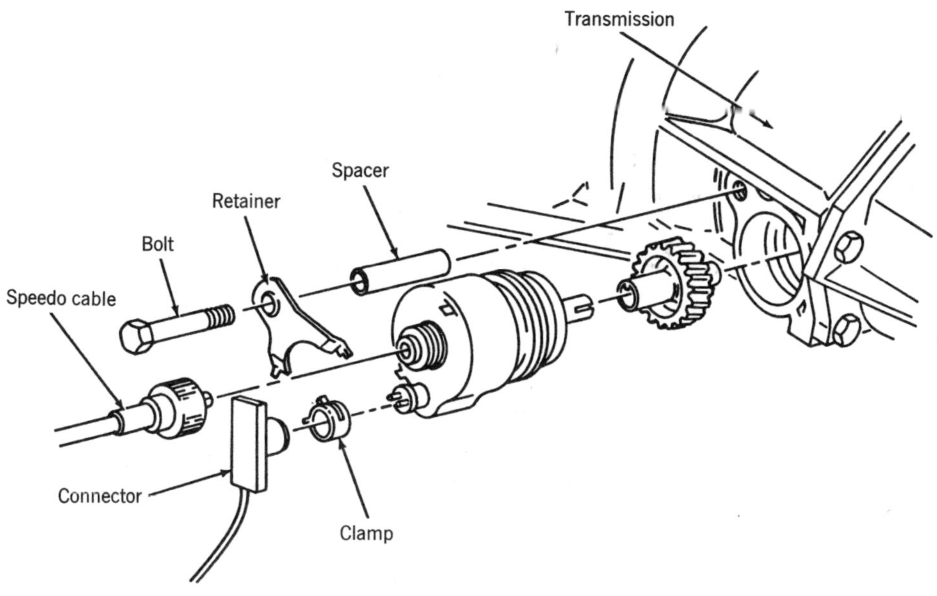

Figure 13.14 A transmission-mounted speed sensor and attaching parts (General Motors).

This way the locking point is never reached, or if it is reached, the brakes are released so fast that the wheel does not actually stop turning. This action allows the wheels to keep turning even though the brakes are applied as tight as possible.

Components A typical antiskid device contains three major components: (1) sensors, (2) a computer, and (3) an actuator.

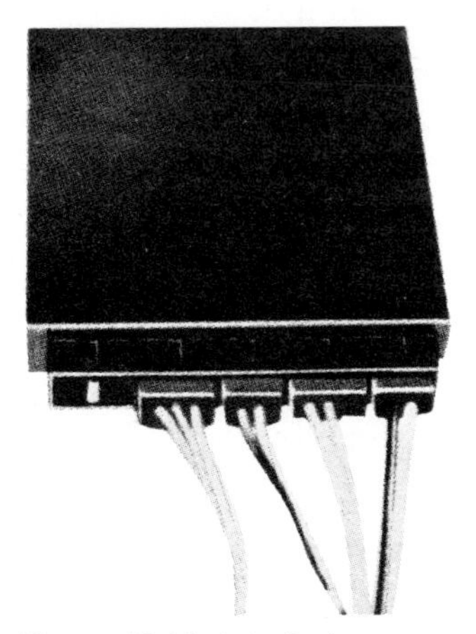

Figure 13.15 A typical computer, or controller (Ford Motor Company).

The Sensors The sensors detect the speed of the wheels and communicate it to the computer by means of a signal. Some systems use a sensor mounted at each rear wheel. This type of sensor is shown in an exploded view in Figure 13.13. Other systems use a single sensor mounted at the rear of the transmission. An exploded view of such a sensor is shown in Figure 13.14. Still other systems use a similar sensor mounted at the differential pinion. This type sensor detects the average speed of both rear wheels.

The Computer The computer, sometimes called the controller, receives the signal from the sensor or sensors. From that signal, it determines when a wheel is about to lock. It then determines the correct braking cycle and transmits this information to the actuator. A typical computer is shown in Figure 13.15.

The Actuator The actuator is an electrically triggered device that pumps the brakes. As shown in Figure 13.16, it is made up of several subassemblies. They function to allow the alternate use of engine manifold vacuum and air pressure to produce the pumping action.

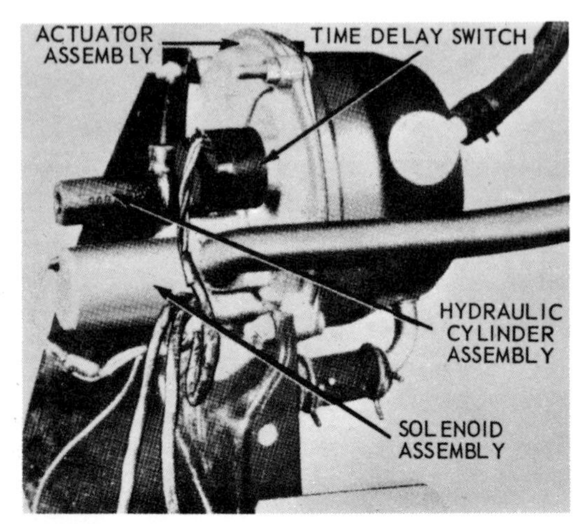

Figure 13.16 An actuator assembly (Ford Motor Company).

Figure 13.17 shows all the components of a typical antiskid device connected in their circuits.

Operation Antiskid devices operate only when a wheel starts to lock. When braking is gradual, the wheels decelerate relatively slowly without locking. The slow change of the signal from the sensor or sensors to the computer indicates that the wheels are coming to a stop within a normal range of deceleration. This range is acceptable to the computer, which sends no signal to the actuator. As long as braking is such that the slip rate does not exceed about 15 percent, the computer does not send a signal. Therefore, the actuator does not pump the brakes.

Whenever a rear wheel starts to lock, the wheel decelerates abruptly. The rapid change in the signal to the computer indicates that a skid is about to occur. The computer then sends a signal to the actuator, and the actuator does two things.

1 It closes a valve between the master cylinder and the rear brake system. This traps the pressure in the rear brake system, preventing the pumping action of the actuator from causing the brake pedal to pulsate.
2 It moves a piston outward from a chamber, or cylinder. Since the chamber is part of the hydraulic system for the rear brakes, the volume of the system is increased. The increased volume decreases the pressure in the rear brake system.

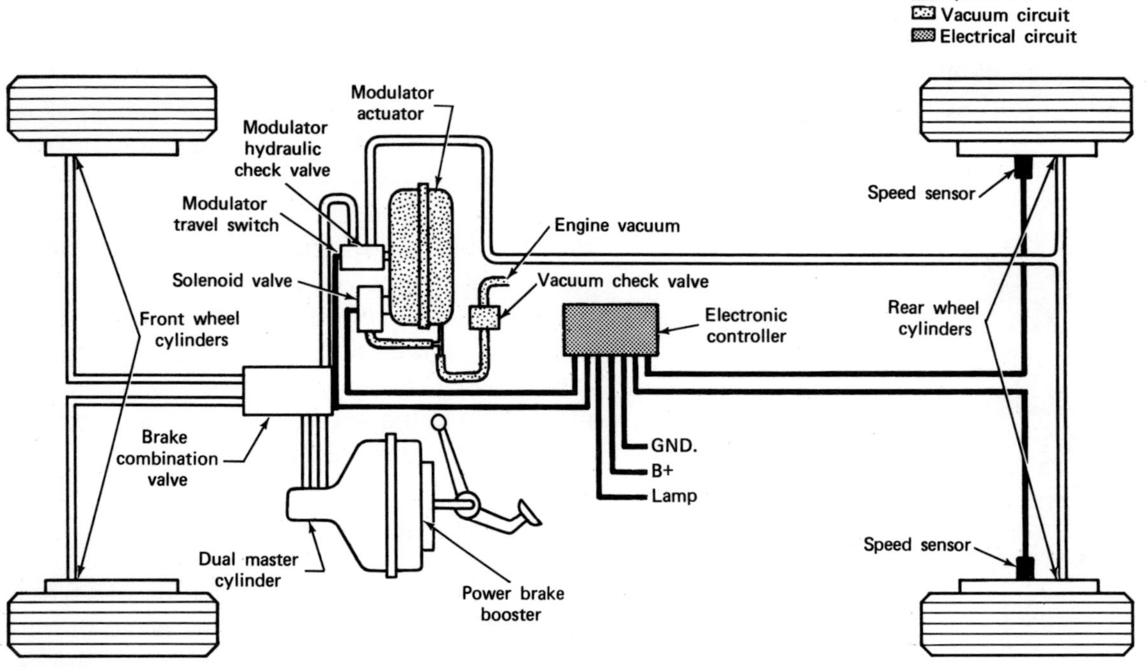

Figure 13.17 Control circuits of a typical antiskid device (General Motors).

When the pressure in the rear brake system is decreased, the brakes release slightly. This allows the wheel to accelerate. As the speed of the wheel approaches the speed of the car, the computer receives a "normal" signal from the sensor, and it stops signalling the actuator. The actuator moves the piston back into the cylinder. The pressure is restored, and the wheel again begins decelerating. This action occurs very fast. The entire cycle, from the partial pressure release to full pressure and back to partial release, happens in one-fifth to one-third of a second. The cycle continues as long as the wheel tends to lock. If the driver holds full force on the brake pedal throughout a stop, the antiskid device functions until the car speed drops to about five mph. Then the device disengages. If the driver eases off on the brake pedal at any time during the stop, the device instantly disengages.

Service

Service on antiskid devices is based entirely on the diagnosis of a particular problem. The diagnosis consists of operational checks and electrical-resistance checks in a certain sequence and to specifications set by the manufacturer of the particular device being checked. Before attempting to service any antiskid device, you should study the manual covering that particular device.

Job 13e

IDENTIFY THE FUNCTION OF THE PARTS IN AN ANTISKID DEVICE

SATISFACTORY PERFORMANCE

A satisfactory performance on this job requires that you do the following:

1 Identify the functions of the components of an antiskid device by placing the number of each component in front of the component function.

2 Correctly match the three components with their functions within 10 minutes.

PERFORMANCE SITUATION

1 Sensor

2 Computer

3 Actuator

_______ determines when a wheel is decelerating too rapidly.

_______ sends a signal to the actuator.

_______ sends a signal to the computer.

_______ shuts off the rear brake system from the master cylinder.

_______ pumps the brakes at the rear of the car.

_______ indicates wheel speed.

_______ controls intake manifold vacuum and atmospheric pressure.

SUMMARY

By completing this chapter, you have learned how vacuum suspended booster units function. You can diagnose their faults and adjust and install replacement units. You understand the operation of hydraulically operated booster units and are aware of their critical servicing requirements. You also understand the operation of a typical antiskid device, and you have an understanding of the function of its principal components.

SELF-TEST

Each incomplete statement or question in this test is followed by four suggested completions or answers. In each case select the *one* that best completes the statement or answers the question.

1 Two mechanics are discussing vacuum brake boosters.
Mechanic A says that when the engine is running and the driver's foot is off the brake pedal, a vacuum is present on both sides of the booster diaphragm.
Mechanic B says that when the brakes are in the applied position, vacuum is present on the front side of the diaphragm.
Who is right?
A. A only
B. B only
C. Both A and B
D. Neither A nor B

2 In a vacuum suspended booster unit, the control valve is built into the
A. power chamber
B. hub of the diaphragm
C. front shell
D. rear shell

3 Air at atmospheric pressure enters the booster unit during the
A. applied position
B. advanced position
C. holding position
D. retarded position

4 To provide the driver with a sense of "feel" during braking, vacuum boosters use a
A. pressure restrictor
B. diaphragm spacer
C. reaction device
D. vacuum diverter

5 A continuous hissing noise heard at the valve rod behind the brake pedal can be caused by a leaking
A. check valve
B. diaphragm
C. front shell
D. vacuum line

6 Two mechanics are discussing vacuum booster brakes.
Mechanic A says that the booster pushrod length must be checked whenever the master cylinder is replaced.
Mechanic B says the booster pushrod length must be checked whenever the booster is replaced.
Who is right?
A. A only
B. B only
C. Both A and B
D. Neither A nor B

7 A hydraulically operated booster unit stores fluid under pressure by means of
A. a boost piston
B. a pressure chamber
C. an actuator
D. an accumulater

8 Maximum braking is obtained when the wheels of a car are decelerating in a slip range of
A. 0 to 10 percent
B. 10 to 20 percent
C. 20 to 30 percent
D. 30 to 40 percent

9 When the actuator in an antiskid device receives a signal from the computer, it
I. shuts off the rear brake system from the master cylinder
II. pumps the rear brakes to control the deceleration of the rear wheels
A. I only
B. II only

C. Either I or II
D. Both I and II

10 Rear wheels tend to lock and skid when a fast stop is attempted from high speed. This is because
A. the metering valve cuts down the pressure to the front brakes
B. the center of gravity of the car shifts to the front
C. the rear brakes share a common line from the master cylinder
D. the proportioning valve provides more pressure for the rear wheel cylinders or calipers

Appendix

BRAKE INSPECTION CHECKLIST A complete inspection combines all the inspection procedures you studied and performed throughout your work in the various systems. The inspection regulations of the department of motor vehicles in some states require the removal of all brake drums, and this procedure is recommended wherever possible. The inspection points and things to look for are listed here. A brake inspection covering these points should be made not only as a part of preventative maintenance, but as a part of diagnostic procedure.

BRAKE SHOES
1 The linings should not exhibit excessive wear. Bonded lining should be at least $\frac{1}{32}$ in. (0.8 mm) thick. Riveted lining should be thick enough so that the rivet heads cannot contact the drum surface.
2 The linings should not be glazed or cracked, and they should not show signs of separation from the shoe platform.
3 The linings should not be contaminated by grease or brake fluid.
4 The holddown springs and the retracting springs should not be discolored or distorted.
5 The self-adjusting mechanism should be operative.

WHEEL CYLINDERS
1 No signs of leakage should be evident. The boots should be peeled back to disclose any leaked fluid that may be trapped behind them.
2 The pistons should move freely in their bores.

DRUMS
1 The drums should not be cracked, scored, or out-of-round.
2 Grease retainers should show no signs of leakage or damage.
3 Wheel bearings should exhibit no wear or roughness, and they should be properly lubricated and adjusted.

BRAKE HOSES AND LINES
1 There should be no leaks or stains indicating seepage.
2 Hoses should show no signs of abrasion or cracking.
3 Lines should be free of rust and corrosion.

BRAKE CABLES
1 The cables should move freely in their housings.
2 They should not be kinked or frayed.

VALVES
1 Valves should show no signs of leakage.
2 They should be securely mounted.

MASTER CYLINDER
1 Fluid levels should be at the proper height.
2 No leakage should be evident.

DIAGNOSIS OF BRAKING PROBLEMS

Usually, the cause of a problem must be found by diagnosis. In most instances, a thorough inspection will disclose the problem. A proper diagnosis, however, should always include road testing.

Because of the close relationship of all the systems in an automobile, you should always remember that some braking problems are not caused by defects in the brake system. Many braking problems can be traced to the tires, the wheel bearings, and the steering and suspension systems.

TIRE FACTORS

Unequal Tire Pressures Unequal tire pressures can cause a car to drift to one side when the brakes are applied. Any diagnosis of braking problems should include a check of tire pressures and their adjustment to the manufacturer's recommendations.

Worn Tires A tire with excessive tread wear may slide on the pavement when the brakes are applied. This will result in uneven braking. Worn tires should be replaced before you attempt further diagnosis.

Tires With Different Tread Patterns As the coefficients of friction of tires with differing tread patterns may vary, a drift or pull to one side may result from mismatching tires on the same axle. Rotate or change the tires to match the tread patterns before attempting further diagnosis.

WHEEL BEARINGS

Another common cause of braking complaints is loose wheel bearings. If a check indicates that a front wheel is too loose on its spindle, remove the bearings for inspection. Wheel bearings do not become loose except through wear. Adjusting a loose bearing without first inspecting it for possible damage is an unsafe practice.

SUSPENSION AND STEERING SYSTEMS

Braking applies a considerable load to the steering and suspension systems. In many instances, braking complaints can be traced to loose or worn steering and suspension parts. Improper wheel alignment can also cause problems that are often blamed on the brake system. Always include a check of the steering and suspension systems in your diagnosis of braking problems.

DIAGNOSTIC CAUSE AND REPAIR LIST

The following list covers typical braking complaints and gives the most probable causes for those complaints along with the required corrections, or repairs.

DRUM BRAKES

Excessive Pedal Travel, Pedal Goes to the Floor, Pedal Too Low

PROBABLE CAUSE	CORRECTION
1 Excessive lining-to-drum clearance	1 Adjust brakes.
2 Self-adjusters not working	2 Repair or replace adjusters.
3 Leaks in hoses or lines	3 Replace defective parts.

4 Leaking wheel cylinder	4 Overhaul or replace.
5 Leaking master cylinder	5 Overhaul or replace.
6 Air in system	6 Bleed system.
7 Low fluid level	7 Fill reservoirs and bleed if necessary.
8 Improper fluid in system; rubber parts have deteriorated	8 Flush system. Replace all rubber parts in system.

Soft or Spongy Pedal

PROBABLE CAUSE	CORRECTION
1 Air in system	1 Bleed system.
2 Improper fluid in system; rubber parts have deteriorated	2 Flush system. Replace all rubber parts in system.
3 Lining not fitted to drum	3 Arc grind lining.
4 Drums worn or machined thin	4 Replace drums.

High Pedal Effort Needed, Excessive Effort Needed to Stop Car

PROBABLE CAUSE	CORRECTION
1 Improper adjustment	1 Adjust brakes.
2 Glazed lining	2 Replace lining.
3 Poor lining and drum surfaces	3 Machine drums and lining.
4 Lining contaminated with grease or brake fluid	4 Repair leakage, replace linings.
5 Improper fluid in system	5 Flush system. Replace all rubber parts.
6 Pistons frozen in master cylinder or in wheel cylinder	6 Overhaul or replace.
7 Brake pedal linkage binding	7 Free and lubricate linkage.

Sensitive Brakes, Grabby Brakes, Braking Too Severe

PROBABLE CAUSE	CORRECTION
1 Improper brake adjustment	1 Adjust brakes.
2 Linings contaminated with grease or brake fluid	2 Replace linings.
3 Lining loose on shoe	3 Replace linings.
4 Loose wheel bearings	4 Inspect and adjust bearings.
5 Loose backing plate	5 Tighten mounting bolts.

Bouncing Brake Pedal, Pulsating Pedal, Pedal Vibration

PROBABLE CAUSE	CORRECTION
1 Drum out of round	1 Machine drum.

2 Drum loose on hub or axle flange	2 Replace drum. Check hub or flange for damage. Replace hub if necessary.
3 Loose wheel bearings	3 Inspect and adjust.
4 Bent rear axle shaft	4 Replace axle shaft.

Brakes Drag, Car Does Not Roll Free, All Brakes Drag

PROBABLE CAUSE	CORRECTION
1 Compensating port in master cylinder blocked	1 Blow open with compressed air.
2 Master cylinder pistons not fully returning	2 Free pedal linkage. Adjust push rod, pedal stop, or both to provide free play.
3 Improper fluid in system; rubber parts deteriorated	3 Flush system. Replace all rubber parts in system.

One Brake Drags, One Wheel Drags

PROBABLE CAUSE	CORRECTION
1 Weak or broken shoe retracting springs	1 Replace springs.
2 Improper adjustment	2 Adjust brakes.
3 Loose wheel bearings	3 Inspect and adjust.
4 Frozen wheel cylinder piston	4 Overhaul or replace.
5 Drum out of round	5 Machine drum.
6 Plugged line or hose	6 Replace line or hose.

Rear Brakes Drag, Rear Wheels Drag

PROBABLE CAUSE	CORRECTION
1 Improper adjustment	1 Adjust service brakes and parking brakes.
2 Parking brake cable frozen	2 Free and lubricate cables. Replace cables if necessary.

Car Pulls to One side

PROBABLE CAUSE	CORRECTION
1 Unequal tire pressures	1 Correct tire pressures.
2 Different tread designs	2 Rotate tires to obtain similar tread on the same axle.
3 Linings contaminated with grease or brake fluid	3 Repair leak. Replace linings.
4 Improper adjustment	4 Adjust brakes.
5 Loose wheel bearings	5 Inspect and adjust.
6 Loose backing plates	6 Tighten mounting bolts.
7 Foreign matter in drums	7 Clean brake assemblies.

8 Frozen wheel cylinder piston	8 Overhaul or replace.
9 Weak or broken shoe retracting springs	9 Replace springs.
10 Dissimilar drum surfaces	10 Machine drums.
11 Defective suspension parts	11 Inspect and replace as needed.
12 Loose or defective steering parts	12 Inspect and adjust or replace as needed.
13 Unequal camber	13 Align front end.
14 Primary and secondary linings reversed	14 Install correctly.

One Wheel Locks, One Tire Skids

PROBABLE CAUSE	CORRECTION
1 Excessive tire tread wear	1 Replace tire.
2 Linings contaminated with grease or brake fluid	2 Repair leak. Replace linings.

Brakes Squeak When Applied While Car is at Rest

PROBABLE CAUSE	CORRECTION
1 Lack of lubrication at ledges on backing plates	1 Lubricate ledges on backing plates.
2 Excessive lining-to-drum clearance	2 Adjust brakes.

Brake Squeal or Squeak When Applied While Car is in Motion

PROBABLE CAUSE	CORRECTION
1 Lining excessively worn	1 Replace lining.
2 Improper fit between lining and drum	2 Machine drums and lining.
3 Lining loose on shoes	3 Replace lining or shoes.
4 Shoes bent or distorted	4 Replace shoes.
5 Backing plates bent	5 Replace backing plate.
6 Weak or broken hold-down springs	6 Replace hold-down springs.

Shudder or Chatter on Application

PROBABLE CAUSE	CORRECTION
1 Linings contaminated with grease, brake fluid, or dirt	1 Clean linings if dirty. Replace if contaminated with grease or brake fluid.
2 Drums out of round	2 Machine drums.
3 Drum surface imperfections	3 Machine drums.
4 Loose wheel bearings	4 Inspect and adjust.

Click or Snap Heard When Applying Brakes

PROBABLE CAUSE	CORRECTION
1 Grooves worn in ledges in backing plates	1 File or grind ledges smooth, or replace backing plate. Lubricate ledges.
2 Bent shoes	2 Replace shoes.

Repeated Clicking Heard When Braking

PROBABLE CAUSE	CORRECTION
1 Shoes being pulled off backing plate by threaded effect of a poorly turned brake drum	1 Machine drums. Use very fine finish cut, or finish by grinding.
2 Weak or broken hold-down springs	2 Replace hold-down springs.

Scraping or Grinding Noise Heard When Braking

PROBABLE CAUSE	CORRECTION
1 Lining worn or broken from shoe	1 Replace lining. Machine drums.
2 Foreign matter trapped between lining and drum	2 Clean brake assembly.
3 Poor drum surfaces	3 Machine drums.
4 Weak or broken hold-down springs allowing edge of shoe to hit drum	4 Replace hold-down springs.

Thump When Brakes are Applied

PROBABLE CAUSE	CORRECTION
1 Excessive lining-to-drum clearance allowing shoe to travel too far to touch anchor pin	1 Adjust brakes.
2 Loose backing plate	2 Tighten mounting bolts.

DISC BRAKES Car Pulls to One Side

PROBABLE CAUSE	CORRECTION
1 Unequal tire pressures	1 Correct tire pressures.
2 Different tread designs	2 Rotate tires to obtain similar tread on the same axle.
3 Linings contaminated with grease or brake fluid	3 Repair leak, and replace brake shoes.
4 Frozen or sticky caliper piston	4 Overhaul or replace caliper.

5 Loose caliper mounting	5 Tighten mounting bolts.
6 Unequal camber	6 Align front end.
7 Loose suspension or steering system parts	7 Tighten or replace as needed.
8 A malfunction in the rear brake system	8 Check rear system. Repair as required.

Pedal Bounce, Pedal Pulsation, Chatter, Roughness

PROBABLE CAUSE	CORRECTION
1 Excessive lateral runout	1 Machine rotor.
2 Loose wheel bearings	2 Inspect and adjust.
3 Rotor surfaces not parallel	3 Machine rotor.
4 Rear drums out of round	4 Machine drums.

High Pedal Effort Needed, Excessive Effort Needed to Stop Car

PROBABLE CAUSE	CORRECTION
1 Power booster not functioning	1 Repair or replace booster unit.
2 Partial system failure	2 Check front and rear systems for failure. Repair as needed.
3 Excessive lining wear	3 Replace shoes.
4 Frozen or sticky caliper pistons	4 Overhaul or replace calipers.

Excessive Pedal Travel, Pedal Goes to Floor, Pedal Too Low

PROBABLE CAUSE	CORRECTION
1 Partial system failure	1 Check front and rear systems for failure. Repair as needed.
2 Air in system	2 Bleed system.
3 Rear brakes not properly adjusted	3 Adjust rear brakes.
4 Low fluid level	4 Check for cause of fluid loss, and repair. Fill reservoir. Bleed if necessary.

Dragging Brakes, Heavy Drag on Front Wheels

PROBABLE CAUSE	CORRECTION
1 Master cylinder pistons not fully returning	1 Free pedal linkage. Adjust push rod, pedal stop, or both to provide free play.
2 Compensating port in master cylinder blocked	2 Blow open with compressed air.
3 Brake line to front wheels restricted	3 Replace line.
4 Frozen or sticky caliper pistons	4 Overhaul or replace calipers.

PROBABLE CAUSE	CORRECTION
5 Replacement master cylinder has a check valve in the front wheel brake outlet.	5 Remove check valve.

Erratic, Grabbing, or Uneven Braking Action

PROBABLE CAUSE	CORRECTION
1 Refer to all causes listed for "Car Pulls to One Side"	1 Refer to corrections listed for "Car Pulls to One Side."
2 Defective metering valve or proportioning valve	2 Replace defective valves.
3 Defective power assist unit	3 Repair or replace booster unit.

REFERENCE MATERIAL Throughout this text you have been constantly advised to consult various manuals. As a brake mechanic, you should have these manuals available so you can refer to them for information you may need on a particular job. You should start now to build your reference library so that you will be well equipped when you start to work.

MANUFACTURER'S SERVICE MANUALS Most car manufacturers publish complete shop manuals for each model year of the cars they build. Information on the availability and price of these manuals can be obtained by contacting the manufacturers at the following addresses:

American Motors	American Motors Corporation 14250 Plymouth Road Detroit, Michigan 48232
Buick	Buick Motor Division General Motors Corporation Flint, Michigan 48550
Cadillac	Cadillac Motor Car Division General Motors Corporation 2860 Clark Avenue Detroit, Michigan 48232
Chevrolet	Chevrolet Motor Division General Motors Corporation General Motors Building Detroit, Michigan 48202
Chrysler	Chrysler Corporation P. O. Box 1919 Detroit, Michigan 48231
Dodge	Dodge Division Chrysler Motors Corporation P. O. Box 857 Detroit, Michigan 48231

Ford	Ford Motor Company The American Road Dearborn, Michigan 48121
Jeep	Jeep Corporation American Motors Corporation Toledo, Ohio 43600
Lincoln and Mercury	Lincoln-Mercury Division Ford Motor Company 3000 Schaefer Road Dearborn, Michigan 48216
Oldsmobile	Oldsmobile Division General Motors Corporation 920 Townsend Street Lansing, Michigan 48921
Plymouth	Plymouth Division Chrysler Motors Corporation P. O. Box 857 Detroit, Michigan 48231
Pontiac	Pontiac Motor Division General Motors Corporation 1 Pontiac Plaza Pontiac, Michigan 48053

COMPREHENSIVE SHOP MANUALS

The most commonly used specifications and the procedures for the most often performed service operations for recent-model cars are compiled in these manuals. Published yearly, the manuals provide the best single source of reference material you will need in your daily work. The two most widely used manuals are as follows:

Motor Auto Repair Manual	Motor 250 West 55th Street New York, New York, 10019
Chilton's Auto Repair Manual	Chilton Book Company 401 Walnut Street Philadelphia, Pennsylvania 19106

MANUALS AND CATALOGS PUBLISHED BY MANUFACTURERS OF BRAKE SYSTEM PARTS, TOOLS, AND EQUIPMENT

The makers of the parts, tools, and equipment you use in servicing brake systems offer a variety of reference materials. Many of those materials are available free of charge upon request from the following companies:

American Brakeblok Division Abex Corporation 1650 West Big Beaver Road Troy, Michigan 48084	Brake lining

AC-Delco Division General Motors Corporation 3044 West Grand Blvd. Detroit, Michigan, 48202	Brake parts and lining
Ammco Tools, Inc. 2100 Commonwealth Avenue North Chicago, Illinois 60064	Brake tools, drum and rotor lathes and grinders
Barrett Brake Service Equipment John Bean Division FMC Corporation Lansing, Michigan 48900	Brake tools, drum and rotor lathes and grinders
Bear Manufacturing Corporation 2830 Fifth Street Rock Island, Illinois 61201	Brake tools, drum and rotor lathes and grinders
The Bendix Corporation 1217 South Walnut Street South Bend, Indiana 46620	Brake parts, power assist units
Blackhawk Manufacturing Company Applied Power Industries, Inc. Box 8720 Milwaukee, Wisconsin 53227	Jacks, car stands
Clayton Manufacturing Company P. O. Box 550 El Monte, California 91734	Brake testing equipment
EIS Corporation Middlefield, Connecticut 06455	Brake parts
The Gates Rubber Company 999 South Broadway Denver, Colorado 80217	Hydraulic hoses
Gatke Corporation 228 North LaSalle Street Chicago, Illinois 60601	Brake lining
Gibson Products Corporation Division of Rolero, Inc. P. O. Box 7187 Cleveland, Ohio 44128	Brake parts
Grey-Rock Division Raybestos Manhattan, Inc. Box 9140 Bridgeport, Connecticut 06603	Brake lining
Ingersol Rand/Proto Tool Company 2309 Santa Fe Avenue Los Angeles, California 90058	Hand and power tools

K-D Manufacturing Company Hempland Road Lancaster, Pennsylvania 17604	Hand tools
Kelsey Products Division Kelsey-Hayes Company 38481 Huron River Drive Romulus, Michigan 48174	Brake parts
Lockheed Products Wagner Electric Corporation 6400 Plymouth Avenue St. Louis, Missouri 63100	Brake parts
Morak Brakes, Inc. 9902 Avenue D Brooklyn, New York 11236	Brake parts and lining
Owatonna Tool Company P. O. Box 268 Owatonna, Minnesota 55060	Hand tools and pullers
Raybestos Division Raybestos Manhattan, Inc. Bridgeport, Connecticut 06603	Brake lining
The Russell Manufacturing Company Middletown, Connecticut 06457	Brake lining
Snap-On Tools Corporation 8028 28th Avenue Kenosha, Wisconsin 53140	Hand and power tools
Star Machine and Tool Company 201 Sixth Street, S.E. Minneapolis, Minnesota 55414	Drum and rotor lathes and grinders
Van Norman Machine Company 3640 Main Street Springfield, Massachusetts 01107	Drum and rotor lathes and grinders
Walker Manufacturing Company 1201 Michigan Boulevard Racine, Wisconsin 53402	Jacks and car stands
Weaver Division Dura Corporation 2171 South 9th Street Springfield, Illinois 62705	Jacks and car stands
S. K. Wellman Corporation 200 Egbert Road Bedford, Ohio 44146	Brake lining

Glossary

Access slots Openings in the backing plates or drums that allow access to the star wheel adjusters.

Accumulator A chamber in a hydraulically operated brake booster unit that stores a small quantity of fluid under pressure.

Aligning cups Devices used in mounting floating drums on the arbor of a brake drum lathe.

Anchor pin A steel pin, or stud, mounted on the backing plate. The anchor pin keeps the brake shoes from turning with the drum.

Arbor The rotating shaft of a lathe on which a drum or rotor is mounted for machining.

Arc grinder See *brake shoe grinder.*

Arc grinding The machining operation by which the lining on a pair of brake shoes is ground to fit the drum in which they will be used.

Asbestos A nonflammable, heat-resistant mineral used in making brake lining.

Axle (rear) A shaft that transmits the driving force from the differential to the rear wheels.

Backing plate A pressed-steel plate upon which the brake shoes, wheel cylinder, and anchor pin are mounted.

Bail The spring-wire loop used to secure the cover on most master cylinder reservoirs.

Ball bearing An antifriction bearing that uses a series of steel balls held between inner and outer bearing races.

Barrel shape A drum defect caused by excessive wear at the center of the friction surface.

Bearing cone The inner race for a ball or roller bearing.

Bearing cup The outer race for a ball or roller bearing.

Bearing race The inner or outer ring that provides the smooth, hard contact surface for the balls or rollers in a bearing. See also *cup.*

Bell mouth A drum defect caused by excessive wear, expansion, or both at the open end of a brake drum.

Bellows seal An expanding diaphragm used as a seal between the master cylinder reservoir and the reservoir cover. It prevents air from contacting the fluid, yet it allows the fluid to change in volume.

Bench bleeding A method of purging air from a master cylinder prior to the installation of the cylinder.

Bleeder hose A length of rubber tubing used in bleeding brakes.

Bleeder jar A glass or transparent plastic container used to detect the escape of air while bleeding brakes.

Bleeder screw See *bleeder valve.*

Bleeder tank See *pressure bleeder.*

Bleeder tubes Short, curved pieces of tubing used to facilitate the bleeding of master cylinders.

Bleeder valve A valve placed in a hydraulic system where it can be opened to allow the release of air.

Bleeder wrench A tool used to open bleeder valves.

Bleeding The procedure by which air is purged from a hydraulic system.

Boiling point The exact temperature at which a liquid begins to turn to a vapor.

Bonded lining Brake lining that is attached to a brake shoe by an adhesive.

Bonding agent The cement used to secure bonded lining to a brake shoe.

Boot A flexible rubber or plastic cover used over the open ends of master cylinders and wheel cylinders to keep out water and other foreign matter.

Bore The walls of a cylinder. Also used to refer to the diameter of a cylinder.

Bore diameter The diameter of a cylinder.

Bosses Raised ledges, or platforms, on a backing plate, which support the brake shoes.

Brake drum A ring-shaped housing that rotates around fixed brake shoes and is slowed or stopped when the shoes are expanded.

Brake drum lathe A machine that is used to refinish the inner surface of a brake drum.

Brake fade The loss of braking friction caused by excessive heat.

Brake feel A feeling transmitted from the brake system back to the driver during braking.

Brake fluid A special liquid used in hydraulic brake systems.

Brake hose Flexible tubing used to transmit pressure in the hydraulic part of a brake system.

Brake line Special rigid steel tubing used to transmit pressure in the hydraulic part of a brake system.

Brake lining A friction material, usually asbestos, that is fastened to the brake shoes.

Brake shoe The metal form to which the brake lining is attached.

Brake shoe grinder A machine used to grind the lining on a brake shoe so that it will fit a particular drum.

Brake spoon A tool used to turn star wheel adjusters and thus to adjust the brake lining-to-drum clearance.

Caliper The actuating device of a disc brake. A hydraulic clamp that forces brake shoes into contact with the disc, or rotor.

Car stands Pedestal-type supports for holding up a car once the car has been raised.

Castellated nut A nut having slots through which a cotton pin may be passed to secure the nut to its bolt or stud.

Center of gravity The point about which the weight of a car is evenly distributed. The point of balance.

Centering cones Devices used to mount floating drums on the arbor of a brake drum lathe.

Coefficient of friction A relative measurement of the friction developed between two objects in contact with each other.

Coil spring A length of spring-steel wire wound in the shape of a spiral.

Combination brake system A dual brake system that uses disc brakes at the front wheels and drum brakes at the rear wheels.

Combination valve A valve used in combination brake systems that combines two or more valves in a common housing. A combination valve may contain a pressure differential valve, a proportioning valve, and a metering valve.

Compensating port A small hole connecting the master cylinder reservoir with the master cylinder bore. The compensating port is open while the brakes are released, and it provides a means of allowing for expansion and contraction of the brake fluid.

Cotter pin A round locking pin formed by a folded semi-circular steel wire. The pin is locked by spreading the paired ends of the wire.

Cup expanders Metal discs formed to fit inside piston cups and to keep the lips of the cups in tight contact with the cylinder walls while the hydraulic system is not pressurized.

Cylinder hone A rotating tool that uses abrasive stones to remove minor imperfections and to polish the bores of wheel cylinders and master cylinders.

Dampening belt A rubber belt that is wound around the outside of a brake drum or rotor before the drum or rotor is machined. The belt dampens out vibrations that may affect the quality of the finished surface.

Dial indicator A precision instrument that indicates linear measurement on a dial face.

Diaphragm A flexible membrane that separates two chambers and yet allows the volume of each chamber to change.

Disc See *rotor.*

Disc brake A brake system that utilizes a disc or rotor that is slowed or stopped by the clamping action of calipers on the rotor.

DOT Department of Transportation.

Double flare The expanded end of tubing that is folded back to provide a double thickness.

Drum brake A brake system that utilizes a drum or ring and curved shoes that expand to contact the inside wall of the drum or ring.

Dual brake system A brake system that utilizes two separate hydraulic systems.

Dual master cylinder A master cylinder that has two reservoirs and two pistons, usually in tandem. Dual master cylinders are used with dual brake systems.

Duo-servo A brake design that provides servo action regardless of the direction of drum rotation.

Dust covers Small plugs made of rubber or metal, used to cover the access holes in backing plates and drums.

Eccentric Off center. A drum defect caused by unequal wear, drum distortion, or both.

Energy The ability to do work.

Equalizer A device used in parking brake systems to equalize the pull of both rear brake cables.

Feeler gauge A thin strip of metal of known thickness used to measure the clearance between two parts.

Filler port A large hole connecting the master cylinder reservoir with the master cylinder bore. The filler port permits fluid to flow from the reservoir into the hydraulic system.

Fixed anchor A nonadjustable anchor pin. It may be riveted or welded to the backing plate, or it may pass through the backing plate and attach to a part of the suspension system.

Flaring tool A tool used to give the ends of tubing a flared shape.

Floating caliper A single-piston caliper positioned by pins, bolts, or ways.

Floating drum A brake drum that is not secured to a hub.

Flushing A method of cleaning a hydraulic system by pumping alcohol or brake fluid through the system.

Foot pound A unit of measurement for torque. In tightening, one foot pound is the torque obtained by a pulling force of one pound applied to a wrench handle 12 inches long.

Force A push or pushing effort measured in pounds.

Friction The resistance to motion between two objects in contact with each other.

Gravity bleeding A method of purging air from a hydraulic system by allowing the fluid to force air out of an opened bleeder valve by its own weight.

Hard spots Scattered bumps on the friction surface of a brake drum that become apparent after machining Hard spots are caused by excessive heat and pressure, which change the molecular structure of the cast iron.

Heat dissipation The transfer of heat. In brake systems the heat produced by braking is transferred to the air.

Heel The end of the brake shoe nearest the anchor pin.

Herringbone pattern The characteristic pattern cut by a tool bit when a brake drum is machined without the proper use of a dampening belt.

High pedal The condition in which the brakes are applied when the brake pedal is depressed only a slight amount.

Hold-down A device that uses spring tension to hold a brake shoe against a backing plate.

Hold-down spring A spring used in a hold-down.

Hone To remove metal with a fine abrasive stone.

Horses See *car stands.*

Hub The central part of a wheel. The housing for the bearings upon which the wheel rotates around the spindle.

Hubbed drum A brake drum that is mounted on a hub.

Hydraulic Using fluids to transmit force and motion.

Hydraulics The science of the use of fluids to transmit force and motion.

Hydraulic brakes Brakes that are actuated by hydraulic pressure.

Hydroscopic The tendency to absorb water.

Integral Made in one piece.

Jack A device for raising a car.

Kinetic energy The energy of motion.

Low pedal The condition in which the brake pedal must travel very far or very close to the floor before the brakes are applied.

Lubricant Any material, usually liquid or semiliquid, that reduces friction when placed between two moving parts.

Manual bleeding A method of purging air from a hydraulic system by manually operating the brake pedal.

Master cylinder The part of the hydraulic system that converts the force of the driver to hydraulic pressure.

Mechanical brakes A brake system that is actuated mechanically, usually by rods or cables.

Metering valve A valve used in combination brake systems that shuts off the flow of fluid to the front calipers during light pedal applications. It acts to delay the operation of the front brakes until the rear brakes have started to apply.

Millimeter A metric unit of measurement equal to 0.039370 in. Usually abbreviated as mm, as in 1 mm.

NIASE National Institute for Automotive Service Excellence.

Out-of round A drum defect in which the friction surface is not round, but has worn or warped into an oval or eliptical shape.

Pad A common term for a brake shoe used in disc brakes.

Parallelism The parallel alignment of the two surfaces of a disc brake rotor.

Parking brake A mechanical brake system used to keep a parked car from moving.

Pascal's Law A basic law of hydraulics. "When pressure is exerted on a confined fluid, the pressure is transmitted equally and in all directions."

Pedal reserve See *high pedal.*

Penetrating oil A very thin oil that is used to penetrate rust and corrosion and to free rusted parts.

Piston A movable plug that fits in a cylinder.

Piston cup A rubber cup-shaped part that seals a cylinder and eliminates leakage between the piston and the cylinder walls.

Piston stops Tabs, or protrusions, on a backing plate positioned to prevent the wheel cylinder pistons from leaving the wheel cylinder.

Pits The holes or roughness left on a surface as a result of rust or corrosion.

Play Movement between two parts.

Power brakes A hydraulic brake system that utilizes engine intake manifold vacuum or an external hydraulic power source to boost the braking effort of the driver.

Pressure The amount of force applied to a definite area. It is measured in pounds per square inch (psi).

Pressure bleeder A tank that stores brake fluid under pressure. When connected to the master cylinder, the fluid is forced through the system and facilitates bleeding.

Pressure bleeding A method of purging air from a hydraulic system by forcing fluid through the system by means of a pressure bleeder.

Pressure differential valve A spool-type valve used in dual brake systems to detect any difference in pressure between the systems. Its motion usually operates a switch that sends current to a warning lamp on the instrument panel.

Primary brake shoe In self-energizing brakes, the brake shoe that is pulled away from the anchor by the rotation of the drum. Usually it is the forward shoe.

Proportioning valve A valve used in dual brake systems that decreases the pressure at the rear brakes in proportion to pedal force. It operates at relatively high pressure and minimizes the possibility of rear wheel lock during panic stops.

Pull The tendency of a car to pull or lead to one side when the brakes are applied.

Pushrod (master cylinder) The rod that transmits the movement and force of the driver from the brake pedal lever to the master cylinder piston.

Pushrod (wheel cylinder) The rod that transmits the movement and force of the wheel cylinder piston to the brake shoe.

Race See *bearing race.*

Radial load A load that is applied at 90° to an axis of rotation.

Radii adapter A mounting device that is used to center a drum or rotor on the arbor of a lathe. A radii adapter centers the drum or rotor through contact with the bearing races.

Reservoir A storage area for extra brake fluid. Usually integral with the master cylinder.

Residual pressure The slight pressure that remains in a hydraulic system after the brake pedal has been released.

Residual pressure (check) valve A valve that is in the master cylinder and that acts to maintain a slight pressure in the system at all times.

Retaining spring A spring that is used to connect the lower ends of a pair of brake shoes and hold them in contact with the star wheel adjuster.

Retracting spring A spring that is used to pull the brake shoes away from the drum when the brake pedal is released. The retracting spring also pushes the wheel cylinder piston back into its bore and thus returns the brake fluid to the master cylinder.

Reverse bleeding A method of purging air from a hydraulic system by forcing fluid into the system at a bleeder valve and by allowing the air to escape at the master cylinder.

Rivet A fastening device used to secure brake lining to a brake shoe. A headed pin that is placed through holes in two objects. The end opposite the head is expanded to secure the pin.

Roller bearing An antifriction bearing that uses a series of steel rollers held between inner and outer bearing races.

Rotor A disc that is attached to a wheel or hub to provide a friction surface for a brake system.

Rotor lathe A machine that is used to refinish the surfaces of a brake rotor.

Runout Any variation in the movement of the surface of a rotating object.

SAE Society of Automotive Engineers.

Score A scratch or groove. Commonly found on the friction surfaces of drums and rotors.

Sealed bearing A bearing that has been lubricated and sealed at the time of manufacture.

Secondary brake shoe In self-energizing brakes, the brake shoe that is pushed into contact with an anchor by the rotation of the drum. Usually the rear shoe.

Self-adjusting brake A brake that automatically maintains the proper lining-to-drum clearance.

Self-energizing brake A brake design in which the brake shoes, through leverage and wedging action, are applied with a greater force than that furnished by the wheel cylinder.

Servo action A braking action in which one shoe serves to add to the application force of another. This action provides a high brake application force without requiring high pedal effort on the part of the driver.

Sliding caliper A single piston caliper that is positioned by machined surfaces on its anchor plate.

Snap ring A split ring that is held in a groove by its own tension. Internal split rings are used in grooves cut around the bore of a hole. External snap rings are used in grooves cut around a shaft.

Soft pedal The soft, springy feeling detected when the brake pedal is depressed and there is air present in the hydraulic system.

Spindle That part of the front suspension system about which a front wheel rotates. A shaft or pin about which another part rotates.

Splash shield A metal deflector used to protect a disc brake rotor from road splash. It may be formed to improve the flow of air over the rotor.

Spongy pedal See *soft pedal.*

Spoon See *brake spoon.*

Star wheel adjuster A threaded device that is used to expand the brake shoes so that the lining is held close to the drum.

Static pressure See *residual pressure.*

Taper A lack of parallelism. A drum or rotor defect in which the thickness of the drum or rotor at the outer edge differs from its thickness at the inner edge.

Tapered roller bearing An antifriction bearing that uses a series of tapered steel rollers held between tapered inner and outer races. Tapered roller bearings are used where both radial and thrust loads must be handled.

Thrust load A load that is applied in line with an axis of rotation.

Tire rotation Moving the wheel and tire assemblies to different locations to equalize wear patterns.

Tire tread The portion of a tire that contacts the road surface.

Toe The end of a brake shoe that is not adjacent to its anchor.

Tolerance A permissible variation, usually stated as extremes of a specification.

Tool bit The hardened steel or carbide blade that cuts away metal during machining.

Torque A force that tends to produce a twisting or turning motion.

Torque sequence The order in which a series of bolts or nuts should be tightened.

Torque wrench A wrench or handle that indicates the amount of torque applied to a bolt or nut. A tool used to tighten bolts and nuts to a specific torque.

Tubing bender A tool used to bend tubing without kinking or deforming its walls.

Tubing cutter A tool used to cut tubing. In operation, the tubing is held between a pair of rollers and a sharp wheel. The tool is moved around the tubing, and the wheel cuts the tubing cleanly and without distortion.

Tubing wrench A wrench used to turn fittings on tubing. A tubing wrench distributes the turning forces evenly around the fitting and minimizes the possibility of damage.

Ventilated rotor A disc brake rotor that is formed with cooling fins cast between its friction surfaces.

Wheel cylinder The output cylinder in a hydraulic brake system. Hydraulic pressure forces pistons to move outward from the cylinder, pushing the brake shoes into contact with the brake drum.

Answer Key with Text References

Chapter 1
1 C page 1-2
2 C page 1-2
3 C page 1-2
4 C page 1-2
5 B page 1-3

Chapter 2
1 A page 2-2
2 A page 2-2
3 C page 2-2
4 B page 2-3
5 C page 2-4
6 C page 2-5
7 B page 2-5

Chapter 3
1 D page 3-5
2 C page 3-2
3 D page 3-5
4 C page 3-2
5 A page 3-7
6 B page 3-5
7 A page 3-2

Chapter 4
1 C page 4-2
2 D page 4-6, 4-7
3 C page 4-12
4 D page 4-20
5 C page 4-7
6 C page 4-20
7 B page 4-9
8 B page 4-15
9 C page 4-17, 4-21
10 D page 4-17

Chapter 5
1 C page 5-7
2 B page 5-2
3 D page 5-2
4 A page 5-3
5 C page 5-9, 5-10, 5-16, 5-17
6 A page 5-3
7 C page 5-12, 5-19, 5-20
8 B page 5-12
9 C page 5-13, 5-19
10 D page 5-16

Chapter 6
1 D page 6-2
2 D page 6-18
3 C page 6-4
4 A page 6-5
5 C page 6-3, 6-6
6 B page 6-7
7 D page 6-9
8 B page 6-11
9 C page 6-16
10 C page 6-16

Chapter 7
1 B page 7-4
2 B page 7-4
3 C page 7-7
4 D page 7-7
5 C page 7-8, 7-15
6 A page 7-8
7 C page 7-12
8 B page 7-14
9 C page 7-14, 7-15
10 C page 7-14

Chapter 8
1 D page 8-7
2 A page 8-2
3 B page 8-2
4 C page 8-2
5 B page 8-3
6 B page 8-14
7 D page 8-14
8 B page 8-13
9 C page 8-15
10 C page 8-15, 8-16

Chapter 9
1 B page 9-3
2 B page 9-2
3 C page 9-4
4 C page 9-4
5 A page 8-15
6 A page 9-9
7 C page 9-10
8 D page 9-11
9 B page 9-16
10 A page 9-4

Chapter 10
1 D page 10-2
2 D page 10-2
3 D page 10-3, 10-4
4 B page 10-8
5 C pages 10-10
6 A page 10-11
7 B pages 10-14, 10-13
8 D pages 10-13
9 A page 10-10
10 B page 10-15, 10-16

Chapter 11
1 A page 11-5, 11-2
2 B page 11-2
3 C page 11-4
4 D page 11-14
5 A page 11-5
6 B page 11-10
7 C page 11-15
8 A page 11-16
9 A page 11-16
10 B page 11-15

Chapter 12
1 C page 12-2
2 A page 12-4
3 A page 12-6
4 A page 12-10, 12-4
5 C page 12-4, 12-19
6 C page 12-5
7 B page 12-6
8 C page 12-2
9 C page 12-14
10 B page 12-19, 12-14

Chapter 13
1 C page 13-3
2 B page 13-3
3 A page 13-3
4 C page 13-4
5 B page 13-6
6 B page 13-8
7 D page 13-12
8 B page 13-14
9 D page 13-16
10 B page 13-14

Index